Lecture Notes in Computer Science 9802

FoLLI Publications on Logic, Language and Information

Subline of Lectures Notes in Computer Science

More information about this series at http://www.springer.com/series/7407

Michał Skrzypczak

Descriptive Set Theoretic Methods in Automata Theory

Decidability and Topological Complexity

Springer

Author

Michał Skrzypczak
University of Warsaw
Warsaw
Poland

This work was carried out at:
Faculty of Mathematics, Informatics, and Mechanics,
University of Warsaw,
Warsaw, Poland
and accepted there as a PhD thesis.

ISSN 0302-9743 ISSN 1611-3349 (electronic)
Lecture Notes in Computer Science
ISBN 978-3-662-52946-1 ISBN 978-3-662-52947-8 (eBook)
DOI 10.1007/978-3-662-52947-8

Library of Congress Control Number: 2016944816

LNCS Sublibrary: SL1 – Theoretical Computer Science and General Issues

Printed on acid-free paper

This Springer imprint is published by Springer Nature
The registered company is Springer-Verlag GmbH Berlin Heidelberg

Preface

This thesis is devoted to the study of problems of automata theory from the point of view of descriptive set theory. The analyzed structures are ω-words and infinite trees. Most of the results presented here have the form of an effective decision procedure that operates on representations of regular languages.

Special effort is put into providing effective characterizations of regular languages of infinite trees that are definable in weak monadic second-order logic (WMSO). Although no such characterization is known for all regular languages of infinite trees, the thesis provides characterizations in some special cases: for game automata, for languages of thin trees (i.e., trees with countably many branches), and for Büchi automata. Additionally, certain relations between WMSO-definable languages and Borel sets are proved.

Another problem studied in the thesis is the alternating index problem (also called the Rabin–Mostowski index problem). Again, the problem in its full generality seems to be out of the reach of currently known methods. However, a decision procedure for the class of game automata is proposed in the thesis. These automata form the widest class of automata for which the problem is currently known to be decidable.

The thesis also addresses the problem of providing an algebraic framework for regular languages of infinite trees. For this purpose, the notion of prophetic thin algebras is introduced. It is proved that finite prophetic thin algebras recognize exactly the bi-unambiguous languages — languages L such that both L and the complement L^{c} can be recognized by unambiguous automata. Additionally, a new conjecture about the definability of choice functions is stated. It is proved that this conjecture is strongly related to the class of prophetic thin algebras. In particular, the conjecture implies an effective characterization of the class of bi-unambiguous languages.

Finally, the thesis studies contemporary quantitative extensions of the class of regular languages. First, lower bounds (that match upper bounds) on the topological complexity of MSO+U-definable languages of ω-words are given. These lower bounds can be used to prove that MSO+U logic is undecidable on infinite trees in a specific sense. It is also shown that languages of ω-words recognizable by certain counter automata have a separation property with respect to ω-regular languages. The proof relies on topological methods in the profinite monoid.

June 2016 Michał Skrzypczak

Acknowledgements

I would like to emphasize my gratitude to my supervisors Mikołaj Bojańczyk and Igor Walukiewicz for their assistance, without which this thesis would not have been possible. Both of them were always helpful and ready to discuss scientific problems, share their experience, and exchange new ideas. Additionally, I would like to thank them for encouraging me to attend a number of conferences, workshops, and scientific visits that allowed me to establish international cooperation. Moreover, I would like to thank Henryk Michalewski for posing a number of inquiring and inspiring questions that provided good motivation in developing some of the results presented in the thesis.

I owe the formation of my scientific background to the Automata Group in Warsaw, especially to: Damian Niwiński, Mikołaj Bojańczyk, Henryk Michalewski, and Filip Murlak. Additionally, I wish to thank Szczepan Hummel, Tomasz Idziaszek, Denis Kuperberg, and Nathanaël Fijalkow for the time they devoted to our discussions.

During my PhD studies I had the opportunity to make several scientific visits. First and foremost I would like to thank my supervisor Igor Walukiewicz for a number of invitations to LaBRI. Moreover, I would like to thank Jacques Duparc and Alex Rabinovich for inviting me to work together (the results obtained during these visits are not included in the thesis).

I wish to express my gratitude to the reviewers of the thesis. Their suggestions helped me to increase the readability of the final version of the document.

Last but not least, I list my co-authors, whom I would like to thank for the cooperation: M. Bilkowski, M. Bojańczyk, U. Boker, A. Facchini, O. Finkel, T. Gogacz, S. Hummel, T. Idziaszek, D. Kuperberg, O. Kupferman, H. Michalewski, M. Mio, F. Murlak, D. Niwiński, A. Rabinovich, A. Radziwończyk-Syta, and S. Toruńczyk.

My work during my PhD studies was supported by Poland's National Science Centre (decision DEC-2012/05/N/ST6/03254) and by ERC Starting Grant SOSNA.

Contents

Subclasses of Regular Languages

Chapter 1
Basic Notions

In this chapter we introduce basic notions that are used across this thesis. Section 1.1 introduces formally ω-words, infinite trees, and operations that transform them. In Sect. 1.2 we introduce the syntax and the semantics of logics that will be used in the rest of the thesis. Section 1.3 contains a brief introduction of perfect information two-player games. In Sect. 1.4 we define automata models that will be used later. Section 1.5 presents the framework of recognition from the algebraic point of view. In Sect. 1.6 basic topological concepts are introduced. Finally, Sect. 1.7 lists known properties of regular languages of ω-words and infinite trees.

Most of the material presented in this chapter is standard. Therefore, a reader familiar with automata theory and topology may skip most of the formal definitions. The following sections contain some less standard concepts: ranks of well-founded ω-trees are introduced in Sect. 1.6.3, the boundedness theorem is stated in Sect. 1.6.4, simple co-inductive definitions are defined in Sect. 1.6.5, various classes of regular languages (e.g. unambiguous, Büchi, ...) are defined in Sect. 1.7.1, and Sect. 1.7.4 introduces the languages $W_{i,j}$ that are complete for respective classes of the alternating index hierarchy.

The following choices are taken in the thesis:

- min-parity condition is used (i.e. a sequence of priorities $(p_n)_{n\in\mathbb{N}}$ is accepting if the least priority appearing infinitely often is even), see page 7,
- the classes of the Rabin-Mostowski alternating index hierarchy are denoted using symbols $\mathbf{\Pi}^{\mathrm{alt}}_j$ and $\mathbf{\Sigma}^{\mathrm{alt}}_j$ (indices $(0, j)$ and $(1, j+1)$ respectively), see page 22,
- transitions of alternating automata are defined as positive Boolean combinations of atomic transitions (e.g. $(q_1, \mathrm{L}) \vee ((q_2, \mathrm{R}) \wedge (q_3, \mathrm{L}))$), see page 7,
- the players in the games are denoted $\exists$ and $\forall$, usually $\exists$ takes the role of the *prover* and $\forall$ is the *refuter*.

1.1 Structures

In this section we introduce the objects that will be studied in the thesis — mainly ω-words and infinite trees.

M. Skrzypczak, *Set Theoretic Methods in Automata Theory*, LNCS 9802
DOI: 10.1007/978-3-662-52947-8_1

We use the axiom of choice whenever needed, without explicitly noting this fact. Therefore, the proofs of the thesis are done in Zermelo–Fraenkel set theory with the axiom of choice (shortly ZFC).

A set is countable if its cardinality is at most $\aleph_0$. ω is the first infinite ordinal. $\varnothing$ stands for the empty set. By $\mathbb{N}$ we denote the natural numbers, we use the symbols $\mathbb{N}$ and ω interchangeably, depending on the context. $|X|$ stands for the cardinality of a set X, if X is finite then $|X| \in \mathbb{N}$. By $\mathsf{P}(X)$ we denote the powerset of X — the set of all subsets of X. If X and Y are two disjoint sets then we write $X \sqcup Y$ for the union of the two, emphasising the fact that the union is disjoint. We use the notation $\exists!_x\ \varphi$ to express that there exists a unique x satisfying φ.

By ω_1 we denote the first uncountable ordinal. An ordinal η is countable if and only if $\eta < \omega_1$. The addition of ordinals is defined in such a way that $\omega+1 > 1+\omega = \omega$. The multiplication of ordinals is defined in such a way that $\omega+\omega = \omega\cdot 2 > 2\cdot\omega = \omega$. An ordinal of the form $\eta + 1$ is called *successor ordinal*. Ordinals $\eta > 0$ that are not successor ordinals are called *limit ordinals*. Sometimes we identify an ordinal with the set of smaller ordinals, e.g. $\omega = \{0, 1, \ldots\}$, $n = \{0, 1, \ldots, n-1\}$, and $2 = \{0, 1\}$.

Letter A is used to denote an *alphabet* — a non-empty finite set of *letters* $a \in A$.

Let $f: X \to Y$ be a function. By $\mathrm{dom}(f) \stackrel{\text{def}}{=} X$ we denote the domain X of f and by $\mathrm{rg}(f) \subseteq Y$ we denote the set of values of f. If $X' \subseteq \mathrm{dom}(f)$ then $f{\restriction}_{X'}$ stands for the restriction of f to the set X' (i.e. $\mathrm{dom}(t{\restriction}_{X'}) = X'$). By $f: X \rightharpoonup Y$ we denote a partial function from X to Y, i.e. a function $f: \mathrm{dom}(f) \to Y$ with $\mathrm{dom}(f) \subseteq X$.

If a space X is known from the context and $L \subseteq X$ then L^{c} stands for the complement of L, i.e. $L^{\mathrm{c}} \stackrel{\text{def}}{=} X \setminus L$.

1.1.1 *Finite Words and ω-words*

Let X be a non-empty countable set. The family of all finite words over X is denoted by X^*. The empty word is denoted by ϵ. The length of a finite word u is denoted as $|u|$. The set of all non-empty finite words over X is denoted X^+. The successive letters of a word $u \in X^*$ are $u_0, u_1, \ldots, u_{|u|-1}$. The n'th letter of a word u is $u(n)$ or u_n. By X^n we represent the set of words of length precisely n. Similarly, $X^{\leqslant n}$ contains words of length at most n. For an element $x \in X$, $\sharp_x(u)$ stands for the number of occurrences of x in a finite word $u \in X^*$.

An ω-word over X is a mapping $\alpha: \omega \to X$, the set of all such ω-words is X^ω. By $X^{\leqslant\omega}$ we denote the set of all finite and ω-words over X.

The prefix order on $X^{\leqslant\omega}$ is denoted $\preceq$. If X is linearly ordered then the lexicographic order on $X^{\leqslant\omega}$ is denoted $\leqslant_{\mathrm{lex}}$. We implicitly assume that every alphabet A is linearly ordered.

Concatenations. If u is a finite word of length at least n or an ω-word, then $u{\restriction}_n \in X^n$ is the finite word obtained by taking the first n letters of u, i.e. $u{\restriction}_n \stackrel{\text{def}}{=} u_0u_1\ldots u_{n-1}$. The concatenation of two words u, α (where u is finite and α may be infinite) is denoted by $u \cdot \alpha$ or simply $u\alpha$. Similarly, if L is a language of finite or ω-words then

$$u \cdot L = \{u \cdot \alpha : \alpha \in L\}.$$

For a non-empty finite word w by w^∞ we denote the ω-word $w \cdot w \cdot \cdots$ An ω-word of the from $u \cdot w^\infty$ for non-empty finite words u, w is called *regular*.

If $\alpha \in A^\omega$, $\beta \in B^\omega$ are two ω-words then $\alpha \otimes \beta \in (A \times B)^\omega$ is the ω-word obtained as the product of two: for $n \in \omega$ we define $(\alpha \otimes \beta)(n) = (\alpha(n), \beta(n)) \in A \times B$.

One of the crucial features of ω-sequences is expressed by Ramsey's theorem — it is possible to decompose such a sequence in a *monochromatic* way. This technique was used by Büchi in his complementation lemma [Büc62]. In the following, by $[\mathbb{N}]^2$ we denote the set of all unordered pairs of natural numbers.

Theorem 1.1 (Ramsey). *Let C be a finite set of* colours *and $\alpha\colon [\mathbb{N}]^2 \to C$ be a function assigning to every pair of numbers $\{n, m\} \in [\mathbb{N}]^2$ a colour $\alpha(\{n, m\}) \in C$. Then there exists an infinite* monochromatic *set: a set $S \subseteq \mathbb{N}$ such that*

$$\alpha\big(\{n, m\}\big) = \alpha\big(\{n', m'\}\big) \quad \textit{for all } \{n, m, n', m'\} \subset S.$$

1.1.2 Infinite Trees

In this thesis we are mainly interested in infinite trees: both binary and ω-branching, partial and complete. Therefore, in this section we will introduce the following four notions (the brackets denote optional parts of the name):

- complete ω-trees $\omega\mathrm{Tr}_X$,
- (partial) ω-trees $\omega\mathrm{PTr}_X$,
- (complete binary) trees Tr_X,
- partial (binary) trees PTr_X.

ω-trees. A partial ω-tree (shortly *ω-tree*) $\tau \in \omega\mathrm{PTr}_X$ is a partial function $\tau\colon \mathrm{dom}(\tau) \to X$ with a prefix-closed domain $\mathrm{dom}(\tau) \subseteq \omega^*$. Elements of $\mathrm{dom}(\tau)$ are called *nodes* of the ω-tree. For a pair of ω-trees $\tau \in \omega\mathrm{PTr}_X$, $\tau' \in \omega\mathrm{PTr}_{X'}$ of the same domain $\mathrm{dom}(\tau) = \mathrm{dom}(\tau')$ let $\tau \otimes \tau' \in \omega\mathrm{PTr}_{X \times X'}$ be given by $(\tau \otimes \tau')(u) = (\tau(u), \tau'(u))$; in that case we call τ' a *labelling* of τ (by X'). A set $Y \subseteq \mathrm{dom}(\tau)$ can treated as a labelling of τ by $\{0, 1\}$, i.e. an element of $\omega\mathrm{PTr}_{\{0,1\}}$. If the set $X = \{x\}$ is singleton then we can identify an ω-tree $\tau \in \omega\mathrm{PTr}_X$ with its domain $\tau \subseteq \omega^*$; in such a case we also skip the set X and write $\tau \in \omega\mathrm{PTr}$.

A node of the form $(u \cdot i) \in \mathrm{dom}(\tau)$ is called a *child* of u in τ. A node $u \in \mathrm{dom}(\tau)$ is a *leaf of an ω-tree* $\tau \in \omega\mathrm{PTr}_X$ if it has no children in τ. A node $u \in \mathrm{dom}(\tau)$ is *branching* if it has at least two distinct children in τ. If u, u' are distinct children of the same node then they are *siblings*. If $\tau \in \omega\mathrm{PTr}_X$ is an ω-tree and $u \notin \mathrm{dom}(\tau)$ but all the prefixes of u are nodes of τ then we say that *u is off τ*. In particular, $\epsilon \in \omega^*$ is off $\varnothing \in \omega\mathrm{PTr}$.

If the domain of an ω-tree $\tau \in \omega\mathrm{PTr}_X$ is ω^* then t is called a *complete ω-tree*; the set of all such ω-trees is denoted $\omega\mathrm{Tr}_X$.

For a pair of ω-trees $\tau, \tau' \in \omega\mathrm{PTr}_X$ we write $\tau \subseteq \tau'$ if $\mathrm{dom}(\tau) \subseteq \mathrm{dom}(\tau')$ and for every $u \in \mathrm{dom}(\tau)$ we have $\tau(u) = \tau'(u)$.

Binary Trees. A particular case of an ω-tree is a binary tree. We use special symbols to denote the alphabet of the *directions* in the domain of a binary tree: we write $\mathtt{L}$ for 0 and $\mathtt{R}$ for 1. Hence, a *direction* is an element $d \in \{\mathtt{L}, \mathtt{R}\}$, the opposite direction is denoted $\bar{d}$.

A labelled complete binary tree (shortly *tree*) over X is an ω-tree $t \in \omega\mathrm{PTr}_X$ with $\mathrm{dom}(t) = \{\mathtt{L}, \mathtt{R}\}^*$. The space of all such trees is denoted by Tr_X. If $t \in \omega\mathrm{PTr}_X$ and $\mathrm{dom}(t) \subseteq \{\mathtt{L}, \mathtt{R}\}^*$ then t is called a *partial tree*; the set of all partial trees over X is denoted PTr_X. Again, if X is a singleton then we skip it.

Decompositions. If $\tau \in \omega\mathrm{PTr}_X$ is an ω-tree and $u \in \omega^*$ then by $\tau{\restriction}_u \in \omega\mathrm{PTr}_X$ we denote the subtree of τ rooted in u, formally:

$$\mathrm{dom}\big(\tau{\restriction}_u\big) \stackrel{\mathrm{def}}{=} \{w : uw \in \mathrm{dom}(\tau)\}, \quad \tau{\restriction}_u(w) \stackrel{\mathrm{def}}{=} \tau(uw).$$

By the definition, if $u \notin \mathrm{dom}(\tau)$ then $\tau{\restriction}_u = \emptyset$. An ω-tree is *regular* if it has only finitely many different subtrees.

If $u \in \mathrm{dom}(\tau)$ or u is off τ then by $\tau[u \leftarrow \tau']$ we denote the ω-tree obtained by plugging an ω-tree $\tau' \in \omega\mathrm{PTr}_A$ into τ with the root of τ' put in u:

$$\begin{aligned}
\mathrm{dom}\big(\tau[u \leftarrow \tau']\big) &\stackrel{\mathrm{def}}{=} \{w \in \mathrm{dom}(\tau) : u \not\preceq w\} \sqcup \{uw : w \in \mathrm{dom}(\tau')\},\\
\tau[u \leftarrow \tau'](w) &\stackrel{\mathrm{def}}{=} \tau(w) \text{ for } u \not\preceq w,\\
\tau[u \leftarrow \tau'](uw) &\stackrel{\mathrm{def}}{=} \tau'(w).
\end{aligned}$$

In particular, we have $\tau[u \leftarrow \tau']{\restriction}_u = \tau'$. Observe that if $\tau, \tau' \in \mathrm{Tr}_X$ are binary trees and $u \in \{\mathtt{L}, \mathtt{R}\}^*$ then $\tau{\restriction}_u$ and $\tau[u \leftarrow \tau']$ are binary trees (elements of Tr_X).

For $a \in A$ by $a(t_{\mathtt{L}}, t_{\mathtt{R}}) \in \mathrm{Tr}_A$ we denote the tree consisting of the root ϵ labelled by the letter a and two subtrees $t_{\mathtt{L}}, t_{\mathtt{R}} \in \mathrm{Tr}_A$ respectively.

Branches. Let τ be an ω-tree. A finite sequence $u \in \mathrm{dom}(\tau)$ such that u is a leaf of τ is called a *finite branch of* τ. An infinite sequence α such that for every i the prefix $\alpha{\restriction}_i$ is a node of τ is called an (infinite) *branch of* τ. If τ is a tree in PTr_X then the branches of τ are over the alphabet $\{\mathtt{L}, \mathtt{R}\}$. Sometimes we identify a branch α with the set of nodes $\{\alpha{\restriction}_i\}_{i\in\mathbb{N}}$ that form a path.

We now recall a simple yet powerful lemma about ω-trees.

Lemma 1.1 (König's lemma). *Let $\tau \subseteq \omega^*$ be an ω-tree. Assume that every node $u \in \tau$ has only finitely many children in τ (i.e. τ is finitely-branching). Then τ contains an infinite branch if and only if τ is infinite (as a set).*

1.2 Logic

In this section we introduce the logics studied in the thesis. The logics are introduced in the usual way.

The thesis is devoted mostly to *Monadic Second-Order* (MSO) logic. This logic is an extension of First-Order (FO) logic with monadic quantifiers ranging over subsets of the domain. Formally, assume a structure with a domain Θ over a signature Σ. The syntax of MSO allows:

- the equality $x = y$, the predicates from Σ, and the predicate $x \in X$,
- Boolean operators $\vee, \wedge, \neg$,
- first-order quantifiers $\exists_x, \forall_x$ over elements of Θ,
- monadic second-order quantifiers $\exists_X, \forall_X$ over subsets $X \subseteq \Theta$.

WMSO logic has the same syntax as MSO. The difference is the semantics: the monadic second-order quantifiers of WMSO range over finite subsets of the domain. Since finiteness is definable in MSO on ω-words and infinite trees, the expressive power of WMSO is contained in the expressive power of MSO. First-Order logic (FO) can be defined as a restriction of MSO by disallowing the monadic second-order quantifiers.

Relational Structures. Fix an alphabet A. An ω-word $\alpha \in A^\omega$ can be seen as a relational structure with:

- the domain ω,
- the binary relation $\leqslant$,
- the successor function $s(i) = i + 1$, and
- predicates $P_a(x)$ for $a \in A$ — $P_a(x)$ holds for $x \in \omega$ if $\alpha(x) = a$.

A tree $t \in \mathrm{Tr}_A$ can be seen as a relational structure with:

- the domain $\{\mathrm{L}, \mathrm{R}\}^*$,
- the binary relations $\preceq$ and $\leqslant_{\mathrm{lex}}$,
- two successor functions $s_{\mathrm{L}}(u) = u\mathrm{L}$, $s_{\mathrm{R}}(u) = u\mathrm{R}$, and
- predicates $P_a(x)$ for $a \in A$ — $P_a(x)$ holds for $x \in \{\mathrm{L}, \mathrm{R}\}^*$ if $t(x) = a$.

Since the successor functions can be defined using the orders, sometimes we assume that the signature contains only the orders and the predicates $P_a(x)$.

Languages. We write $\Theta \models \varphi$ if a sentence φ is satisfied by a structure Θ. For a sentence φ on ω-words over an alphabet A we define

$$\mathrm{L}(\varphi) \stackrel{\text{def}}{=} \{\alpha \in A^* : \alpha \models \varphi\}.$$

Similarly, if φ is a formula on infinite trees over an alphabet A then

$$\mathrm{L}(\varphi) \stackrel{\text{def}}{=} \{t \in \mathrm{Tr}_A : t \models \varphi\}.$$

In both cases we say that $\mathrm{L}(\varphi)$ is the *language of* φ. A language L is MSO-*definable* (resp. WMSO-*definable*, FO-*definable*) if there exists a sentence of MSO (resp. of WMSO, of FO) φ such that $\mathrm{L}(\varphi) = L$.

1.3 Games

One of the most important tools in studying regular languages are games of infinite duration. A generic infinite duration game is defined by a tuple $\mathcal{G} = \langle V_\exists, V_\forall, v_\mathrm{I}, E, W\rangle$ where:

- $V_\exists$ and $V_\forall$ are disjoint sets. We put $V \stackrel{\text{def}}{=} V_\exists \sqcup V_\forall$. Elements of V are called *positions* of $\mathcal{G}$. Elements of V_P are called *positions belonging to* P, for a player $P \in \{\exists, \forall\}$.
- $v_\mathrm{I} \in V$ is an *initial position*.
- $E \subseteq V \times V$ is an *edge relation*. We assume that for every $v \in V$ the set $vE \stackrel{\text{def}}{=} \{v' : (v, v') \in E\}$ is finite and non-empty.
- $W \subseteq V^\omega$ is a *winning condition*.

Strategies. For simplicity, by $V^* \cdot V_P$ we denote the set of finite sequences of vertices such that the last vertex belongs to V_P for a player $P \in \{\exists, \forall\}$.

A *strategy* of a player $P \in \{\exists, \forall\}$ is a function $\sigma : V^* \cdot V_P \to V$ such that for every $u \in V^* \cdot V_P$ we have $(u, u\sigma(u)) \in E$. An infinite sequence $\pi \in V^\omega$ such that for every i we have $(\pi(i), \pi(i+1)) \in E$ is called a *play*. A play π is *consistent* with a strategy σ if whenever $\pi(i) \in V_P$ then $\pi(i+1) = \sigma(\pi\restriction_{i+1})$. A play π is *winning* for $\exists$ if $\pi \in W$, otherwise π is *winning* for $\forall$. A strategy σ of a player P is *winning* if every play π consistent with σ is winning for P.

A game is *determined* if one of the players has a winning strategy. In general not every infinite duration game is determined. The following theorem shows that all *topologically simple* games are determined (see Sect. 1.6.1 for an introduction to Borel sets).

Theorem 1.2 (Martin [Mar75]). *If W is a Borel subset of V^ω then the game $\mathcal{G}$ is determined.*

1.3.1 Positional Strategies

A strategy σ of a player P is *positional* if the value $\sigma(uv)$ for $u \in V^*$ and $v \in V_P$ depends only on v. A strategy σ of a player P is *finite memory* if there exist:

- a finite set M called the *memory structure*,
- an element $m_\mathrm{I} \in M$,
- a function $\delta : M \times V \to M$, such that

$\sigma(uv)$ for $u \in V^*$ and $v \in V_P$ depends only on v and $\delta(m_{\mathrm{I}}, uv)$ defined inductively:

$$\delta(m, \epsilon) \stackrel{\text{def}}{=} m$$
$$\delta(m, uv) \stackrel{\text{def}}{=} \delta\big(\delta(m, u), v\big).$$

We will be particularly interested in games that are *tree-shaped* — for every $v \in V$ there is at most one path from v_{I} to v in the graph (V, E). In such a game every strategy is positional and such a strategy can be identified with its domain $\mathrm{dom}(\sigma) \subseteq V$ — the set of positions accessible via σ from v_{I}.

Sometimes we will be interested in finite approximations of strategies. A *finite strategy σ for a player P* in a tree-shaped game is a finite subset of the arena such that for every $v \in \sigma$ either no element of vE is in σ (v is a *leaf* of σ) or:

- if $v \in V_P$ then exactly one of the elements of vE is in σ,
- otherwise all the elements of vE are in σ.

1.3.2 Parity Games

A min-parity game (shortly parity game) is an infinite duration game with the winning condition W of a special form. Assume that $\Omega: V \to \{i, i+1, \dots, j\}$ is a function that assigns to every position of a game its *priority*. A play π *satisfies the parity condition* if

$$\liminf_{n\to\infty} \Omega(\pi(n)) \equiv 0 \pmod 2,$$

i.e. if the smallest priority that occurs infinitely often during π is even. We define the winning condition W_Ω of a parity game $\langle V_\exists, V_\forall, v_{\mathrm{I}}, E, \Omega\rangle$ as the set of plays satisfying the parity condition. We define the *index* of a game $\mathcal{G}$ as the pair (i, j) — the range of priorities used in this game.

The crucial property of parity games is that they are positionally determined, as expressed by the following theorem.

Theorem 1.3 ([EJ91, Mos91, JPZ08]). *If $\mathcal{G}$ is a parity game (not necessarily finite) then one of the players has a positional winning strategy in $\mathcal{G}$.*

If $\mathcal{G}$ is finite then a winning strategy can be effectively constructed.

1.4 Automata

The fundamental results of Büchi and Rabin say that both satisfiability problems of MSO formulae on ω-words and infinite trees[1] are decidable. In both cases the proof

[1] A simple interpretation argument shows that MSO is decidable also on ω-trees.

goes through a construction of appropriate automata with expressive power equal to MSO logic. In this section we define various models of automata for infinite objects that will be used in this thesis.

Alternating Automata. We start with a definition of the most general variant of automata, namely the alternating ones. For the sake of simplicity we focus on the min-parity acceptance condition. We introduce the ω-word and infinite tree automata uniformly. An *alternating automaton* is a tuple $\langle A^{\mathcal{A}}, Q^{\mathcal{A}}, q_{\mathrm{I}}^{\mathcal{A}}, \delta^{\mathcal{A}}, \Omega^{\mathcal{A}} \rangle$ where:

- $A^{\mathcal{A}}$ is an alphabet.
- $Q^{\mathcal{A}}$ is a finite set of *states*.
- $q_{\mathrm{I}}^{\mathcal{A}} \in Q^{\mathcal{A}}$ is an *initial state*.
- $\delta^{\mathcal{A}}$ is a *transition function* assigning to a pair $(q, a) \in Q^{\mathcal{A}} \times A^{\mathcal{A}}$ the transition $b = \delta^{\mathcal{A}}(q, a)$ built using the following grammar

$$b ::= \top \mid \bot \mid b \vee b \mid b \wedge b \mid b_0$$

  where b_0 is an *atomic transition* defined below.
- $\Omega^{\mathcal{A}} \colon Q^{\mathcal{A}} \to \mathbb{N}$ is a *priority function*.

An *atomic transition* b_0 of an ω-word automaton is a pair $(q, 1)$ for $q \in Q^{\mathcal{A}}$. An atomic transition of a tree automaton is a pair (q, d) where $q \in Q^{\mathcal{A}}$ and $d \in \{\mathtt{L}, \mathtt{R}\}$.

If an automaton $\mathcal{A}$ is known from the context, we omit the superscript $\mathcal{A}$.

Acceptance Game. Fix an alternating automaton $\mathcal{A}$, a state $q_0 \in Q$, and a tree $t \in \mathrm{Tr}_A$. We define the game $\mathcal{G}(\mathcal{A}, t, q_0)$ as follows:

- $V = \{\mathtt{L}, \mathtt{R}\}^* \times (S_\delta \cup Q)$, where S_δ is the set of all subformulae of formulae in $\mathrm{rg}(\delta)$ (all the formulae that appear in the transitions of $\mathcal{A}$);
- all the positions of the form $(u, b_1 \vee b_2)$ belong to $\exists$ and the remaining ones to $\forall$;
- $v_{\mathrm{I}} = (\epsilon, q_0)$;
- E contains the following pairs (for all $u \in \{\mathtt{L}, \mathtt{R}\}^*$):
  - $\big((u, b), (u, b)\big)$ for $b \in \{\top, \bot\}$,
  - $\big((u, b), (u, b_i)\big)$ for $b = b_1 \wedge b_2$ or $b = b_1 \vee b_2$ and $i = 1, 2$,
  - $\big((u, q), (u, \delta(q, t(u)))\big)$ for $q \in Q$,
  - $\big((u, b_0), (ud, q)\big)$ for an atomic transition $b_0 = (q, d)$;
- $\Omega(u, \top) = 0$, $\Omega(u, \bot) = 1$, $\Omega(u, q) = \Omega^{\mathcal{A}}(q)$ for $q \in Q$, $u \in \mathrm{dom}(t)$, and for other positions Ω is $\max(\mathrm{rg}(\Omega^{\mathcal{A}}))$.

In the case of an ω-word α the game $\mathcal{G}(\mathcal{A}, \alpha, q_0)$ is almost the same, the differences are:

- $V = \omega \times (S_\delta \cup Q)$,
- the initial position v_{I} is $(0, q_0)$,
- for an atomic transition $b_0 = (q, 1)$ we put into E the edge $\big((i, b_0), (i + 1, q)\big)$.

An automaton $\mathcal{A}$ *accepts* an ω-word α (resp. tree t) from $q_0 \in Q$ if $\exists$ has a winning strategy in $\mathcal{G}(\mathcal{A}, \alpha, q_0)$ (resp. $\mathcal{G}(\mathcal{A}, t, q_0)$). By $\mathrm{L}(\mathcal{A}, q_0)$ we denote the set of structures accepted by the automaton $\mathcal{A}$ from a state q_0. We write $\mathrm{L}(\mathcal{A})$ for $\mathrm{L}(\mathcal{A}, q_{\mathrm{I}}^{\mathcal{A}})$ and $\mathcal{G}(\mathcal{A}, t)$ (resp. $\mathcal{G}(\mathcal{A}, \alpha)$) for $\mathcal{G}(\mathcal{A}, t, q_{\mathrm{I}}^{\mathcal{A}})$ (resp. $\mathcal{G}(\mathcal{A}, \alpha, q_{\mathrm{I}}^{\mathcal{A}})$). An automaton $\mathcal{A}$ *recognises* a language L if $\mathrm{L}(\mathcal{A}) = L$.

A state $q \in Q^{\mathcal{A}}$ is *non-trivial* if it recognises a *non-trivial language* i.e. if $\mathrm{L}(\mathcal{A}, q) \neq \emptyset$ and $\mathrm{L}(\mathcal{A}, q)^{\mathrm{c}} \neq \emptyset$. Without loss of generality we implicitly assume that all our alternating automata have only non-trivial states (possibly except the initial state), as expressed by the following fact.

Fact 1.4. *Every alternating automaton recognising a non-trivial language can be effectively transformed into an equivalent alternating automaton without trivial states.*

Additionally, each transition of an alternating automaton can be simplified so that it does contain neither $\top$ nor $\bot$ under $\vee$ or $\wedge$.

Proof. Let $\mathcal{A}$ be an alternating automaton. We just remove trivial states of $\mathcal{A}$. If q is trivial then in each transition we replace each subterm of the form (q, d) by $\bot$ or $\top$ (depending on whether $\mathrm{L}(\mathcal{A}, q) = \emptyset$ or $\mathrm{L}(\mathcal{A}, q)^{\mathrm{c}} = \emptyset$).

Finally, we can simplify the transition expression using the standard laws: $(\top \wedge b) = b$, $(\bot \wedge b) = \bot$, $(\top \vee b) = \top$, $(\bot \vee b) = b$. After this step the automaton is still an alternating automaton recognising the same language but it does not contain any trivial states. ∎

Deterministic and Non-deterministic Automata. An ω-word automaton is *deterministic* if all its transitions are *ω-word deterministic*, i.e. of the form $(q, 1)$. A tree-automaton is *deterministic* if all its transitions are *tree deterministic*, i.e. of the form $(q_{\mathrm{L}}, \mathrm{L}) \wedge (q_{\mathrm{R}}, \mathrm{R})$. An automaton is *non-deterministic* if its transitions are disjunctions of deterministic transitions.

Note that if $\mathcal{A}$ is a non-deterministic automaton then the transition function can be written as a relation:

- $\delta \subseteq Q \times A \times Q$ in the case of ω-words — an element (q, a, q') of δ represents that $\delta(q, a) = \ldots \vee (q', 1) \vee \ldots$
- $\delta \subseteq Q \times A \times Q \times Q$ in the case of trees — an element $(q, a, q_{\mathrm{L}}, q_{\mathrm{R}})$ of δ represents that $\delta(q, a) = \ldots \vee \big((q_{\mathrm{L}}, \mathrm{L}) \wedge (q_{\mathrm{R}}, \mathrm{R})\big) \vee \ldots$

For simplicity, we sometimes assume that a non-deterministic automaton $\mathcal{A}$ has a set of initial states $I^{\mathcal{A}} \subseteq Q^{\mathcal{A}}$. Clearly, such an automaton can be equipped with an additional initial state $q_{\mathrm{I}}^{\mathcal{A}}$ and the transition relation can take care of guessing from which state $q \in I^{\mathcal{A}}$ to start.

A *run* of a non-deterministic ω-word automaton over an ω-word $\alpha \in \left(A^{\mathcal{A}}\right)^{\omega}$ is an ω-word $\rho \in \left(Q^{\mathcal{A}}\right)^{\omega}$ such that for every $i \in \omega$ the triple $(\rho(i), \alpha(i), \rho(i+1))$ is a transition of $\mathcal{A}$.

A *run* of a non-deterministic tree automaton over a tree $t \in \mathrm{Tr}_{A^{\mathcal{A}}}$ is a tree $\rho \in \mathrm{Tr}_{Q^{\mathcal{A}}}$ such that for every $u \in \{\mathrm{L}, \mathrm{R}\}^*$ the quadruple $(\rho(u), t(u), \rho(u\mathrm{L}), \rho(u\mathrm{R}))$ is a transition of $\mathcal{A}$.

A non-deterministic automaton $\mathcal{A}$ accepts an ω-word α (resp. an infinite tree t) if there exists a run ρ of $\mathcal{A}$ on α (resp. t) such that:

- ρ *is parity-accepting*: the sequence $\Omega(\rho(0)), \Omega(\rho(1)), \ldots$ satisfies the min-parity condition (resp. the min-parity condition is satisfied on all infinite branches of ρ).
- The *value of* ρ defined as $\rho(0)$ (resp. $\rho(\epsilon)$ in the case of infinite trees) equals $q_{\mathrm{I}}^{\mathcal{A}}$. If there is a set of initial states $I^{\mathcal{A}}$ then the value of ρ is required to belong to $I^{\mathcal{A}}$.

A run that satisfies both the above conditions is called *accepting*. Clearly the above definition is equivalent to the one given for alternating automata — a run can be seen as a strategy of $\exists$ in the respective game.

A non-deterministic automaton is *unambiguous* if it has at most one accepting run on every input. In particular, every deterministic automaton is unambiguous. For more intermediate classes of automata in-between deterministic and non-deterministic ones see e.g. [CPP07, HP06, BKKS13].

1.4.1 Parity Index of an Automaton

In this section we define the index of an automaton. These definitions are used in Sect. 1.7.2 to introduce the Rabin-Mostowski index hierarchy.

Let $\mathcal{A}$ be an alternating tree automaton. Let $\mathrm{Graph}(\mathcal{A})$ be the directed edge-labelled graph over the set of vertices $Q^{\mathcal{A}}$ such that there is an edge $p \xrightarrow{(a,d)} q$ whenever (q, d) occurs in $\delta^{\mathcal{A}}(p, a)$. Additionally, vertices of $\mathrm{Graph}(\mathcal{A})$ are labelled by values of $\Omega^{\mathcal{A}}$. We write $p \xrightarrow{u} q$ if there is a path in $\mathrm{Graph}(\mathcal{A})$ whose edge-labels yield the word u.

The *(Rabin-Mostowski) index* of a parity automaton $\mathcal{A}$ is the pair (i, j) where i is the minimal and j is the maximal priority of the states of $\mathcal{A}$. In that case $\mathcal{A}$ is called an *(i, j)-automaton*. Since shifting all priorities by an even number does not influence the language recognised by an automaton, we can always assume that i is either 0 or 1. An automaton is a *Büchi* automaton if $(i, j) = (0, 1)$; it is a *co-Büchi* automaton if $(i, j) = (1, 2)$.

An alternating automaton $\mathcal{A}$ is a $\mathrm{Comp}(i, j)$-automaton (see [AS05]) if each strongly-connected component in $\mathrm{Graph}(\mathcal{A})$ has priorities between i and j or between $i+1$ and $j+1$. It follows from the definition that each $\mathrm{Comp}(i, j)$-automaton is an $(i, j{+}1)$ automaton, and can be transformed into an equivalent $\mathrm{Comp}(i{+}1, j{+}2)$-automaton by shifting the priorities. The $\mathrm{Comp}(0, 0)$-automata are more widely known as *weak alternating automata*.

1.5 Algebra

This thesis is based mainly on automata. However, in some contexts it is convenient to use the algebraic approach to recognition. Therefore, we introduce the basic concepts, namely semigroups, monoids, Wilke algebras, and ω-semigroups. We assume the reader to be familiar with basic notions of universal algebra. Also, we use multi-sorted algebras, a thorough introduction to these algebras with respect to recognition is given in [Idz12].

This section is used only in certain chapters of the thesis (namely in Part II and Chap. 11) and may be skipped during the first reading.

Assume that M, N are two algebraic structures with the same operations. A function $f\colon M \to N$ is a *homomorphism* if it preserves all the operations: for every operation P of arity n and every choice of arguments $(x_1, \ldots, x_n) \in M^n$ we have

$$f\big(P(x_1, \ldots, x_n)\big) = P\Big(f(x_1), \ldots, f(x_n)\Big).$$

1.5.1 Semigroups and Monoids

A semigroup is an algebraic structure M equipped with an operation $\cdot\colon M^2 \to M$ that is associative $(a\cdot(b\cdot c) = (a\cdot b)\cdot c)$. A monoid is a semigroup with a distinguished element $1 \in M$ that satisfies $1 \cdot a = a \cdot 1 = a$. The operation $\cdot$ is called *product* and 1 is called the *neutral element*.

An element $e \in M$ of a semigroup is called *idempotent* if $e \cdot e = e$. If M is finite then for every element $s \in M$ there is a unique idempotent in the set $\{s^n : n \in \mathbb{N}\}$, this idempotent is called the *idempotent power of* s and denoted[2] $s^\sharp$.

Observe that the set of all finite words A^* over an alphabet A has a natural structure of an infinite monoid with the operation of concatenation and 1 defined as the empty word. Similarly, A^+ is a semigroup.

If a function $f\colon M \to N$ between two monoids is a homomorphism of semigroups then it is also a homomorphism between monoids (i.e. f must preserve 1).

Additional structural properties of monoids (namely Green's relations) are introduced in Sect. 8.4.1 on page 151. They are only used in one construction in Chap. 8.

1.5.2 Wilke Algebras

Now we introduce Wilke algebras that form one of the equivalent formalisms for recognition of ω-regular languages, see [Wil93] and [PP04].

A Wilke algebra is a pair (H, V) with the following operations (for $h \in H$ and $s, s' \in V$):

- $s \cdot s' \in V$,
- $s \cdot h \in H$,
- $s^\infty \in H$.

[2] Often the notion s^ω is used, to avoid confusion with ω-words we use $s^\sharp$.

such that $(V, \cdot)$ is a semigroup and the following axioms are satisfied:

$$\begin{aligned} s \cdot (s' \cdot s'') &= (s \cdot s') \cdot s'' \\ s \cdot (s' \cdot h) &= (s \cdot s') \cdot h \\ (s \cdot s')^{\infty} &= s \cdot (s' \cdot s)^{\infty} \\ \forall_{n \geqslant 1} \left(s^n\right)^{\infty} &= s^{\infty} \end{aligned}$$

An ω-semigroup is a Wilke algebra with an additional operation $\prod \colon V^{\omega} \to H$ such that

$$\begin{aligned} \prod(s, s, \ldots) &= s^{\infty} \\ s \cdot \prod(s_0, s_1, \ldots) &= \prod(s, s_0, s_1, \ldots) \\ \prod(s_0 \cdot \ldots \cdot s_{k_1},\ s_{k_1+1} \cdot \ldots \cdot s_{k_2},\ \ldots) &= \prod(s_0, s_1, s_2, \ldots) \end{aligned}$$

For every alphabet A the pair (A^{ω}, A^{+}) has a natural structure of an ω-semigroup. Additionally, (A^{ω}, A^{+}) is a *free ω-semigroup on A*, as expressed by the following fact.

Fact 1.5 ([PP04, Proposition 4.5]). *Let A be an alphabet and (H, V) be an ω-semigroup. For every function $f \colon A \to V$ there is a unique extension $\bar{f} \colon (A^{\omega}, A^{+}) \to (H, V)$ of f that is a homomorphism of ω-semigroups.*

The following theorem shows that finite Wilke algebras can be seen as representations of arbitrary finite ω-semigroups.

Theorem 1.6 (Wilke [PP04, Theorem 5.1]). *Every Wilke algebra has a unique extension by an operation $\prod$ into an ω-semigroup.*

However, the following example shows that there are functions $f \colon (A^{\omega}, A^{+}) \to (H, V)$ that are homomorphisms of Wilke algebras but not homomorphisms of ω-semigroups. Therefore, it is important in Fact 1.5 to require $\bar{f}$ to be a homomorphism of ω-semigroups.

Example 1.1. Let $A = \{a, b\}$, $H = \{h_a, h_b\}$, and $V = \{s_a, s_b\}$. Let $f \colon (A^{\omega}, A^{+}) \to (H, V)$ be defined as follows:

- for $u \in A^{+}$ let $f(u) = s_a$ if and only if u contains letter a,
- for a regular $\alpha \in A^{\omega}$ let $f(\alpha) = h_a$ if and only if α contains infinitely many letters a,
- for a non-regular $\alpha \in A^{\omega}$ let $f(\alpha) = h_b$.

The function f induces uniquely a structure of Wilke algebra on (H, V) in such a way that f becomes a homomorphism of Wilke algebras. By Theorem 1.6, the Wilke algebra (H, V) can be uniquely extended by an operation $\prod$ into an ω-semigroup. However, f is not a homomorphism of ω-semigroups, otherwise we would have

$$h_b = f(a\,ba\,bba\,bbba\,bbbba\,\cdots) = \prod(s_a, s_a, \ldots) = f(a\,a\,a\cdots) = h_a.$$

1.5.3 Recognition

Let $f\colon M \to N$ be a homomorphism between two algebraic structures (M and N may be multi-sorted here). Let F be a subset of one of the sorts of N and L be a subset of the respective sort of M. We say that f *recognises L using F* if

$$f^{-1}(F) = L.$$

Similarly, f *recognises L* if it recognises L using some F contained in the respective sort of N.

1.5.4 Ramsey's Theorem for Semigroups

In this section we present an application of Ramsey's theorem (see Theorem 1.1) to the ω-word case.

Theorem 1.7. *Let M be a finite semigroup and $f\colon A^* \to M$ be a homomorphism. Then for every ω-word $\alpha \in A^\omega$ there exists a sequence of finite words $u_0, u_1, u_2, \ldots$ and two elements s, e of the semigroup M such that:*

(i) $\alpha = u_0u_1u_2\ldots$,
(ii) $f(u_0) = s$,
(iii) $f(u_n) = e$ *for every* $n > 0$,
(iv) $s \cdot e = s$ *and* $e \cdot e = e$.

A pair (s, e) satisfying Condition (iv) above is often called a *linked pair*, see [PP04]. To simplify the properties in the above theorem we introduce the following definition.

Definition 1.1. *For a given homomorphism $f\colon A^* \to M$ we say that the* type *(or f-*type*) of a decomposition $\alpha = u_0u_1\ldots$ is $t = (s, e)$ if (s, e) is a linked pair, $f(u_0) = s$, and $f(u_n) = e$ for all $n > 0$.*

Of course not every decomposition has some type. However, Theorem 1.7 implies that for every ω-word α and homomorphism f there exists some decomposition of α of some type $t = (s, e)$. A priori there may be two decompositions of one ω-word of two distinct types.

1.6 Topology

In this section we introduce topological notions that will be used later. Most of the presented definitions and facts are basic and standard. Some more involved concepts are presented in Sects. 1.6.4 and 1.6.5.

A topological space $(X, \mathcal{U})$ is called Polish if it is separable (i.e. it contains a countable dense set) and the topology $\mathcal{U} \subseteq \mathsf{P}(X)$ comes from a complete metric on X. Elements $U \in \mathcal{U}$ are called *open* sets, the complement of an open set is *closed*. If a set is both closed and open then it is a *clopen*. A family $\mathcal{B} \subseteq \mathcal{U}$ is called a *basis of the topology* if for every $x \in X$ and $U \in \mathcal{U}$ such that $x \in U$ there exists $B \in \mathcal{B}$ such that $x \in B \subseteq U$. In that case $\mathcal{U}$ coincides with the family of unions of elements of $\mathcal{B}$. A space is *zero-dimensional* if the family of clopen sets is a basis of the topology.

If $D \subseteq X$ is a subset of a topological space X then by $\overline{D}$ we denote the closure of D — the intersection of all closed subsets of X that contain D.

A subset $K \subseteq X$ of a topological space is called *compact* if for every family $\mathcal{F}$ of open sets such that $K \subseteq \bigcup \mathcal{F}$ there exists a finite subfamily $\mathcal{F}' \subseteq \mathcal{F}$ such that $K \subseteq \bigcup \mathcal{F}'$.

Product Spaces. Let Z be a non-empty countable set. Z^ω with the product topology is a zero-dimensional Polish space. The family of sets of the form $u \cdot Z^\omega$ for $u \in Z^*$ is a basis for the topology of Z^ω. If Z is a singleton then Z^ω is also a singleton; if Z is finite then Z^ω is homeomorphic (i.e. topologically isomorphic) to the Cantor set 2^ω; if Z is countably infinite then Z^ω is homeomorphic to the Baire space ω^ω.

The set $(Z \sqcup \{\bot\})^{\omega^*} = \prod_{u \in \omega^*} (Z \sqcup \{\bot\})$ equipped with the natural product topology is a zero-dimensional Polish space. Observe that $\omega\mathrm{PTr}_Z$ is a subset of $(Z \sqcup \{\bot\})^{\omega^*}$. By a standard argument (see [Kec95, Theorem 3.8 in Chap. 3.B]) the spaces of partial ω-trees $\omega\mathrm{PTr}_Z$ and partial binary trees PTr_Z as well as their complete variants $\omega\mathrm{Tr}_Z$ and Tr_Z are zero-dimensional Polish spaces. The families of clopen sets of these spaces coincide with the finite Boolean combinations of sets of the form

$$\{\tau : u \in \mathrm{dom}(\tau) \wedge \tau(u) \in Z'\} \quad \text{for } u \in \omega^* \text{ and } Z' \subseteq Z.$$

1.6.1 Borel and Projective Hierarchy

Let us fix an uncountable Polish space X. The Borel hierarchy is defined inductively:

- $\mathbf{\Sigma}^0_1(X)$ denotes the family of open subsets of X,
- $\mathbf{\Pi}^0_1(X)$ denotes the family of closed subsets of X (the complements of open sets),

for a countable ordinal η:

- $\mathbf{\Sigma}^0_\eta(X)$ is the family of countable unions of sets from $\bigcup_{\beta<\eta} \mathbf{\Pi}^0_\beta(X)$,
- $\mathbf{\Pi}^0_\eta(X)$ is the family of countable intersections of sets from $\bigcup_{\beta<\eta} \mathbf{\Sigma}^0_\beta(X)$.

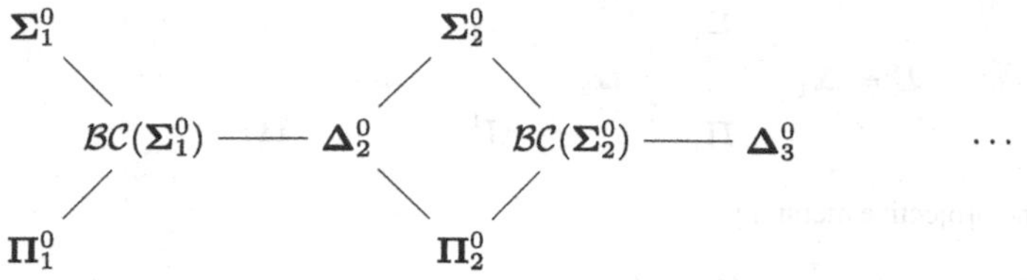

Fig. 1.1 The Borel hierarchy.

Note that for each η the family $\mathbf{\Sigma}^0_\eta(X)$ consists exactly of the complements of the sets from $\mathbf{\Pi}^0_\eta(X)$. $\mathbf{\Delta}^0_\eta(X)$ is defined as the intersection $\mathbf{\Sigma}^0_\eta(X) \cap \mathbf{\Pi}^0_\eta(X)$. Similarly, $\mathcal{BC}(\mathbf{\Sigma}^0_\eta)(X)$ is the family of finite Boolean combinations of sets from $\mathbf{\Sigma}^0_\eta(X)$. The families constitute a hierarchy — each family is included in all the families with greater subindex (see Fig. 1.1). An important fact about the hierarchy is that all the inclusions presented in Fig. 1.1 are strict.

The family of *Borel sets*, defined as

$$\mathcal{B}(X) = \bigcup_{\eta<\omega_1} \mathbf{\Sigma}^0_\eta(X)$$

is the least family closed under countable Boolean operations that contains all open sets. Proofs and details about the Borel hierarchy can be found e.g. in [Sri98, Chap. 3.6].

Projective Hierarchy. The class of Borel sets is not closed under projection. Each set that is a projection of a Borel set is called *analytic*, the family of analytic sets is denoted by $\mathbf{\Sigma}^1_1(X)$. Formally:

$$\mathbf{\Sigma}^1_1(X) \stackrel{\text{def}}{=} \left\{P \subseteq X : \exists_{B\in\mathcal{B}(X\times\omega^\omega)}\ P = \pi_1(B)\right\},$$

where π_1 is the projection on the first coordinate. The superscript 1 means that the class is a part of the projective hierarchy. The rest of the projective hierarchy is defined as follows (see Fig. 1.2):

$$\begin{aligned}
\mathbf{\Pi}^1_i(X) &\quad \text{consists of the complements of the sets from } \mathbf{\Sigma}^1_i(X),\\
\mathbf{\Sigma}^1_{i+1}(X) &\quad \text{consists of the projections of the sets from } \mathbf{\Pi}^1_i(X),\\
&\quad \text{i.e. } \mathbf{\Sigma}^1_{i+1}(X) \stackrel{\text{def}}{=} \left\{\pi_1(B) :\ B \in \mathbf{\Pi}^1_i(X\times\omega^\omega)\right\},\\
\mathbf{\Delta}^1_i(X) &\quad \text{is the intersection of } \mathbf{\Sigma}^1_i \text{ and } \mathbf{\Pi}^1_i.
\end{aligned}$$

The sets from the family $\mathbf{\Pi}^1_1(X)$ are called *co-analytic*. An important result in the theory, the theorem of Souslin (see e.g. [Kec95, Chap. 14.C]), states that if a set is analytic and co-analytic then it is in fact Borel. The Borel hierarchy together with

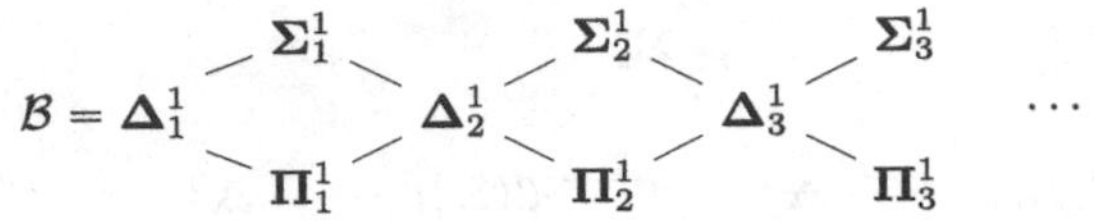

Fig. 1.2 The projective hierarchy.

$$\Sigma_1^0 \quad \Sigma_2^0 \quad \Sigma_3^0 \qquad \Sigma_\omega^0 \quad \Sigma_{\omega+1}^0 \qquad \Sigma_{\omega\cdot 2}^0 \qquad \Sigma_1^1 \quad \Sigma_2^1 \quad \Sigma_3^1$$
$$\times \quad \times \quad \cdots \quad \times \quad \cdots \quad \cdots \quad \times \quad \times \quad \cdots$$
$$\Pi_1^0 \quad \Pi_2^0 \quad \Pi_3^0 \qquad \Pi_\omega^0 \quad \Pi_{\omega+1}^0 \qquad \Pi_{\omega\cdot 2}^0 \qquad \Pi_1^1 \quad \Pi_2^1 \quad \Pi_3^1$$

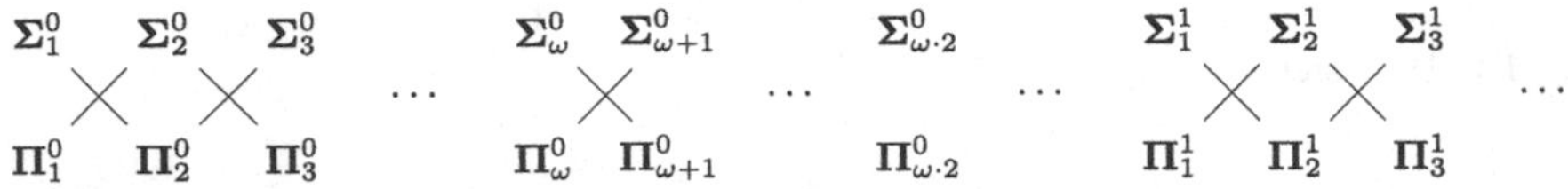

Fig. 1.3 The boldface hierarchy.

the projective hierarchy constitute the so-called *boldface hierarchy*, see the diagram on Fig. 1.3.

If the space is clear from the context we will omit it and write $\mathcal{B}$, $\boldsymbol{\Sigma}_\eta^0$, $\boldsymbol{\Pi}_\eta^0$ $\boldsymbol{\Sigma}_i^1$, $\boldsymbol{\Pi}_i^1$, etc.

1.6.2 Topological Complexity

For the needs of this thesis, a *topological complexity class* **C** is any of the classes of the boldface hierarchy, see Figs. 1.1 and 1.2.

Analogously to the complexity theory, we have the notions of *reductions* and *completeness*. Let X, Y be two topological spaces and let $K \subseteq X$ and $L \subseteq Y$. We say that a continuous mapping $f : X \to Y$ is a *reduction* of K to L if $K = f^{-1}(L)$. The fact that K can be continuously reduced to L is denoted by $K \leqslant_W L$. On Borel sets, the pre-order $\leqslant_W$ induces the so-called Wadge hierarchy (see [Wad83]) which greatly refines the Borel hierarchy and has the familiar ladder shape with pairs of mutually dual classes alternating with single self-dual classes.

It is a simple property of continuous mappings that if L belongs to a topological complexity class **C** then so does K for every $K \leqslant_W L$. A language L is called **C**-*hard* if every set $K \in \mathbf{C}$ can be reduced to L. We say that L is **C**-*complete* if additionally $L \in \mathbf{C}$ (i.e. L is the $\leqslant_W$-greatest element of **C**).

The following fact presents a standard way of using the above notions.

Fact 1.8. *If $\mathcal{C} \subsetneq \mathcal{D}$ are two (non-equal) topological complexity classes and L is $\mathcal{D}$-hard then $L \notin \mathcal{C}$.*

Proof. Assume to the contrary that $L \in \mathcal{C}$. Take any language $K \in \mathcal{D} \setminus \mathcal{C}$. Since L is $\mathcal{D}$-hard, we can write $K = f^{-1}(L)$ for some continuous mapping f. By the above observation, it implies that $K \in \mathcal{C}$, which gives a contradiction.   ∎

1.6.3 Ranks

Ranks form a powerful tool in analysis of descriptive properties of sets. In this section we introduce the most classical of the ranks — the rank on well-founded ω-trees. For an introduction to the theory of ranks see [Kec95, Chap. 2.E].

An ω-tree $\tau \in \omega\mathrm{PTr}$ is *well-founded* if it doesn't have an infinite branch. Otherwise τ is *ill-founded*. The set of all well-founded ω-trees is denoted $\mathrm{WF} \subseteq \omega\mathrm{PTr}$. The complement of WF is denoted IF.

It is possible to assign to each well-founded ω-tree $\tau \in \mathrm{WF}$ its *rank* — a measure of complexity of τ. If $\tau = \emptyset$ then $\mathrm{rank}(\tau) = 0$. Assume otherwise and let $(\tau_i)_{i\in\omega}$ be the sequence of subtrees of τ under the root: $\tau_i = \tau\restriction_{(i)}$ (if $i \notin \mathrm{dom}(\tau)$ then $\tau\restriction_{(i)} = \emptyset$). Put

$$\mathrm{rank}(\tau) = \sup_{i\in\omega} \Big(\mathrm{rank}(\tau_i) + 1\Big).$$

Since the domain of τ is countable, $\mathrm{rank}(\tau)$ is an ordinal number smaller than ω_1. By the definition, the rank is monotone: for $u \neq \epsilon$ we have $\mathrm{rank}(\tau) > \mathrm{rank}(\tau\restriction_u)$. Sometimes we call $\mathrm{rank}(\tau\restriction_u)$ the *rank of u (in τ)*.

Fact 1.9. *If* $\mathrm{rank}(\tau)$ *is a limit ordinal then the root* ϵ is infinitely branching in τ*: for infinitely many* $i \in \omega$ *we have* $i \in \mathrm{dom}(\tau)$.

Fact 1.10. *For every well-founded ω-tree* τ *and* $\eta \leqslant \mathrm{rank}(\tau)$ *there exists a node* $u \in \mathrm{dom}(\tau)$ *such that* $\mathrm{rank}(\tau\restriction_u) = \eta$.

Proof. For $\eta = \mathrm{rank}(\tau)$ we can take $u = \epsilon$. For $\eta < \mathrm{rank}(\tau)$ we proceed by induction on $\mathrm{rank}(\tau)$. ∎

1.6.4 The Boundedness Theorem

In this section we present the most fundamental result relating descriptive properties of a set and ranks — the boundedness theorem. First we recall that the ill-founded ω-trees is one of the crucial examples of a non-Borel set.

Theorem 1.11 ([Kec95, Theorem 27.1]). *The set* IF *of ill-founded ω-trees is* $\mathbf{\Sigma}^1_1$*-complete. Dually, the set* WF *of well-founded ω-trees is* $\mathbf{\Pi}^1_1$*-complete.*

The following theorem expresses the correspondence between the ranks of well-founded ω-trees and the topological complexity of sets.

Theorem 1.12 (The boundedness theorem (see [Kec95, Theorem 35.23])). *If* $X \subseteq \omega\mathrm{PTr}$ *is an analytic set and* $X \subseteq \mathrm{WF}$ *then there exists* $\eta < \omega_1$ *such that*

$$\forall_{\tau\in X}\ \mathrm{rank}(\tau) \leqslant \eta.$$

On the other hand, for every $\eta < \omega_1$ the set

$$\{\tau \in \omega\mathrm{PTr} : \mathrm{rank}(\tau) \leqslant \eta\} \quad \textit{is Borel.}$$

Sketch of the Proof. The second part of the statement can be proved by induction on η (it also follows from more general considerations of ranked sets).

Let us sketch a proof of the first part. First assume the contrary. The heart of the proof is to show that the following relation is analytic:

$$R_E \stackrel{\mathrm{def}}{=} \{(\tau, \tau') : \tau' \text{ is ill-founded or both are well-founded and } \mathrm{rank}(\tau) \leqslant \mathrm{rank}(\tau')\}.$$

Then WF has the following analytic definition

$$\mathrm{WF} = \{\tau : \exists_{\tau' \in X}\ (\tau, \tau') \in R_E\},$$

what contradicts the fact that WF is co-analytic complete. ■

A technique motivated by this proof is used in Sect. 6.3 (see page 105) to prove upper bounds on topological complexity of regular languages of thin trees.

1.6.5 Co-inductive Definitions

In some cases it is convenient to define a function using a *co-inductive definition.* In this section we formalise this notion for functions of the type $\mathrm{Tr}_A \to \mathrm{Tr}_B$. The crucial property is that every function defined in such a way is continuous. Whenever such a co-inductive definition is used, an explicit reference to this section is given. Therefore, one can skip this section when reading the thesis for the first time.

We state the properties of a co-inductive definition for binary trees for the sake of simplicity. The same construction works for ω-trees as well as partial trees. Although it is possible to formalize this notion in an abstract way using the language of category theory, we focus only on these concrete applications of co-induction.

Proposition 1.1. *Let A, B be two alphabets. Assume that for every $a \in A$ we have a triple $(t^{(a)}, u^{(a)}, w^{(a)})$, where $t^{(a)} \in \mathrm{Tr}_B$ is a tree and $u^{(a)}, w^{(a)}$ are two nodes of $t^{(a)}$ that are incomparable with respect to the prefix order $\preceq$ (in particular none of them is ϵ).*

There exists a unique function $f : \mathrm{Tr}_A \to \mathrm{Tr}_B$ such that for every $t \in \mathrm{Tr}_A$ such that $t = a(t_{\mathrm{L}}, t_{\mathrm{R}})$ we have:

$$f(t) = t^{(a)}\big[u^{(a)} \leftarrow f(t_{\mathrm{L}}), w^{(a)} \leftarrow f(t_{\mathrm{R}})\big]. \tag{1.1}$$

Moreover, the function f is continuous.

Proof. We will show how to uniquely define $f(t)(u)$ for a node $u \in \{\mathrm{L}, \mathrm{R}\}^*$ using Condition 1.1. It will imply that the function $f(t)$ is defined uniquely. Additionally, since $f(t)(u)$ will depend only on a finite part of t, it will imply that the function f is continuous.

We proceed by induction on the length of u (for all trees $t \in \mathrm{Tr}_A$ at once). Assume that for all $u' \prec u$ and all $t \in \mathrm{Tr}_A$ the value $f(t)(u')$ is already uniquely defined (and depends only on a finite part of t). Assume that $t = a(t_\mathrm{L}, t_\mathrm{R})$ for a letter $a \in A$ and two trees $t_\mathrm{L}, t_\mathrm{R} \in \mathrm{Tr}_A$.

If $u^{(a)} \preceq u$ (the case $w^{(a)} \preceq u$ is entirely dual) then let $u = u^{(a)} \cdot z$ for $z \in \{\mathrm{L}, \mathrm{R}\}^*$. Therefore, Condition (1.1) implies that $f(t)(u) = f(t_\mathrm{L})(z)$. Since z is shorter than u so this value is uniquely determined.

Now assume contrary, that u does not contain $u^{(a)}$ nor $w^{(a)}$ as a prefix. In that case Condition (1.1) implies that $f(t)(u) = t^{(a)}(u)$ and again this value is uniquely determined and depends only on the letter $t(\epsilon) = a$. ■

1.7 Regular Languages

In this section we collect standard properties of regular languages. Assuming some basic knowledge in automata theory the section can be skipped during the first reading. The presented facts are explicitly referenced whenever used. For a broad introduction to regular languages see [Tho96].

The following results summarize equivalent ways of defining various classes of regular languages.

Regular Languages. We start with a theorem about regular languages of finite words.

Theorem 1.13 (Trakhtenbrot [Tra62], Rabin Scott [RS59], cf. e.g. [PP04]). *The following conditions are effectively equivalent for a language $L \subseteq A^*$ of finite words:*

- *L is definable in* MSO,
- *L is definable in* WMSO,
- *L is recognised by a deterministic finite automaton*[3],
- *L is recognised by an alternating finite automaton,*
- *L is recognised by a homomorphism $f\colon A^* \to M$ into a finite monoid M.*

A language satisfying the above conditions is called a *regular language*.

ω-regular Languages. Now we give a characterization of regular languages of ω-words.

Theorem 1.14 (Büchi [Büc62], McNoughton [McN66], Mostowski [Mos84], Emmerson Jutla [EJ91], Wilke [Wil93]). *The following conditions are effectively equivalent for a language $L \subseteq A^\omega$ of ω-words:*

[3] Automata for finite words are not used in this thesis, therefore we skip a formal definition of them.

- *L is definable in* MSO,
- *L is definable in* WMSO,
- *L is recognised by a deterministic parity ω-word automaton,*
- *L is recognised by an alternating parity ω-word automaton,*
- *L is recognised by a homomorphism $f\colon (A^\omega, A^+) \to (H, V)$ into a finite ω-semigroup (H, V).*

A language satisfying the above conditions is called an *ω-regular language*.
As a consequence one obtains the decidability result of Büchi.

Theorem 1.15 (Büchi [Büc62]). *The* MSO *theory of the ω-chain $(\omega, \leqslant)$ is decidable. If an ω-regular language is non-empty then it contains a regular ω-word.*

Regular Tree Languages. The following theorem characterizes regular languages of infinite trees.

Theorem 1.16 (Rabin [Rab69], Muller Schupp [MS95]). *The following conditions are effectively equivalent for a language $L \subseteq \mathrm{Tr}_A$ of infinite trees:*

- *L is definable in* MSO,
- *L is recognised by a non-deterministic parity tree automaton,*
- *L is recognised by an alternating parity tree automaton.*

A language satisfying the above conditions is called a *regular tree language* (we avoid ambiguity here because regular languages of finite trees do not appear in this thesis).

As a consequence one obtains the celebrated result of Rabin.

Theorem 1.17 (Rabin [Rab69]). *The* MSO *theory of the complete binary tree is decidable. If a regular tree language is non-empty then it contains a regular tree.*

WMSO*-definable Languages.* The following theorem is a characterization of the WMSO-definable languages of infinite trees. The characterization is not effective in the sense that given any representation of a language L it is not known how to check whether L is WMSO-definable.

Theorem 1.18 (Rabin [Rab70], also Kupferman Vardi [KV99]). *The following conditions are effectively equivalent for a language $L \subseteq \mathrm{Tr}_A$ of infinite trees:*

- *L is definable in* WMSO,
- *L is recognised by a* Comp(0, 0)*-alternating tree automaton,*
- *both L and the complement L^c are recognised by alternating Büchi tree automata.*

Büchi Languages. The following theorem states the de-alternation result for Büchi automata. It also proves that Büchi automata correspond to the *existential* fragment of MSO with respect to WMSO.

Theorem 1.19 (Muller Schupp [MS95]). *The following conditions are effectively equivalent for a language $L \subseteq \mathrm{Tr}_A$ of infinite trees:*

- *L is recognised by a non-deterministic Büchi tree automaton,*
- *L is recognised by an alternating Büchi tree automaton,*
- *L is definable by a sentence of the form*

$$\exists_{X_1} \ldots \exists_{X_n} \varphi(X_1, \ldots, X_n)$$

where φ is a formula of WMSO.

Deterministic Languages. An easy construction of an appropriate automaton proves the following fact.

Fact 1.20. *If $\mathcal{A}$ is a deterministic tree automaton then $\mathrm{L}(\mathcal{A})$ can be recognised by an alternating $(1, 2)$-automaton.*

Games with ω-regular Winning Conditions. The following theorem expresses an important feature of games with ω-regular winning conditions.

Theorem 1.21 (Büchi Landweber [BL69], Gurevich Harrington [GH82], Emmerson Jutla [EJ91], Mostowski [Mos91]). *For a finite game $\mathcal{G}$ if the winning condition $W \subseteq V^\omega$ is ω-regular (over the alphabet V) then one of the players has a finite memory winning strategy in $\mathcal{G}$. Such a winning strategy can be effectively constructed.*

1.7.1 Classes of Regular Tree Languages

Now we define classes of regular tree languages that correspond to certain classes automata. A language L is:

- *deterministic* if L is recognised by a deterministic parity tree automaton,
- *unambiguous* if L is recognised by an unambiguous parity tree automaton,
- *bi-unambiguous* if both L and the complement L^c are unambiguous,
- *Büchi* if L is recognised by an alternating[4] Büchi tree automaton,
- *co-Büchi* if L is recognised by an alternating co-Büchi tree automaton.

An easy pumping argument shows that non-deterministic co-Büchi tree automata have very limited expressive power (e.g. they are weaker than alternating co-Büchi automata).

[4]Theorem 1.19 implies that equivalently one can take non-deterministic automata here.

1.7.2 Index Hierarchies

Now we introduce the classes of languages recognisable by automata of certain indices. We start with the alternating index hierarchy. For $i < j \in \mathbb{N}$, let[5]

- $\mathbf{RM}^{\mathrm{alt}}(i, j)$ be the class of regular tree languages recognised by alternating (i, j)-automata,
- $\mathbf{\Pi}_j^{\mathrm{alt}} \stackrel{\mathrm{def}}{=} \mathbf{RM}^{\mathrm{alt}}(0, j)$ (for $j = 1$ these are Büchi languages),
- $\mathbf{\Sigma}_j^{\mathrm{alt}} \stackrel{\mathrm{def}}{=} \mathbf{RM}^{\mathrm{alt}}(1, j+1)$ (for $j = 1$ these are co-Büchi languages),
- $\mathbf{\Delta}_j^{\mathrm{alt}} \stackrel{\mathrm{def}}{=} \mathbf{\Pi}_j^{\mathrm{alt}} \cap \mathbf{\Sigma}_j^{\mathrm{alt}}$,
- $\mathbf{Comp}(\mathbf{\Pi}_j^{\mathrm{alt}})$ be the class of regular tree languages recognised by Comp$(0, j)$-automata.

The above classes are naturally ordered by inclusion, as depicted on Fig. 1.4.

Similarly, one can consider non-deterministic automata instead of alternating ones, i.e. define $\mathbf{RM}^{\mathrm{non-det}}(i, j)$ as the class of languages recognised by non-deterministic (i, j)-automata. The classes $\mathbf{RM}^{\mathrm{non-det}}(i, j)$ form the *non-deterministic index hierarchy*. The shape of the hierarchy is the same as of the alternating one, except the classes $\mathbf{Comp}(\mathbf{\Pi}_j^{\mathrm{alt}})$ that are not defined in the non-deterministic case.

The expressive power of alternating and non-deterministic tree automata is the same (see Theorem 1.16), therefore both hierarchies contain the same languages. However, particular levels of these hierarchies differ (see [NW05]). As shown in [Niw86, Bra98, Arn99, AS05], both hierarchies are strict, in particular, in the alternating case we have

$$\mathbf{Comp}(\mathbf{\Pi}_j^{\mathrm{alt}}) \subsetneq \mathbf{\Delta}_{j+1}^{\mathrm{alt}} \quad \text{for } j > 0. \tag{1.2}$$

A natural question is to compute exact position of a given language in these hierarchies. It is formalised as the following computational problem.

Problem 1.1 (Alternating (resp. non-deterministic) index problem).

- **Input** An alternating tree automaton $\mathcal{A}$.

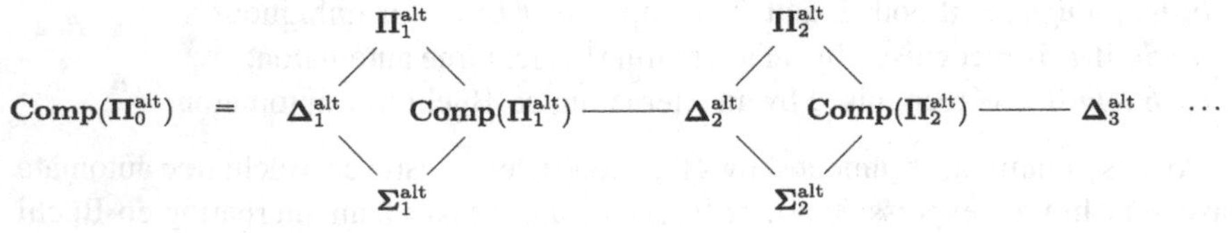

Fig. 1.4 The alternating index hierarchy.

[5] The following assignment of symbols Σ and Π follows the definitions in [AS05, AMN12], however the indices j are shifted by one (also, we use the min-parity condition here). The assignment of the symbols is opposite to the one from [FMS13].

- **Output** The minimal class of the alternating (non-deterministic) index hierarchy that contains $L(\mathcal{A})$.

Both problems were solved for deterministic automata [NW05, NW03], see Sect. 1.7.6. They are both open for general automata. Colcombet and Löding [CL08] have proposed a reduction of the non-deterministic index problem to a boundedness problem for a specific class of tree automata with counters. However, the latter problem is not known to be decidable. The known decidability results regarding these hierarchies are subsumed by the results of [FMS13, CKLV13].

1.7.3 Topological Complexity of Regular Languages

The following results summarize topological complexity of regular languages definable in various ways. In each statement, the given upper bound is optimal from the point of view of the boldface hierarchy. Since there are only countably many regular languages, they cannot fulfil any class of the boldface hierarchy except $\mathbf{\Delta}^0_0$.

Theorem 1.22 (See [TL93]).

- *ω-regular languages are in $\mathcal{BC}(\mathbf{\Sigma}^0_2)$,*
- *WMSO-definable languages of infinite trees lie on the finite levels of the Borel hierarchy,*
- *Büchi-recognisable languages of infinite trees are in $\mathbf{\Sigma}^1_1$,*
- *languages of infinite trees recognisable by deterministic parity automata are in $\mathbf{\Pi}^1_1$,*
- *regular languages of infinite trees are in $\mathbf{\Delta}^1_2$.*

In this thesis the question of *descriptive complexity of a language L* is used in the meaning "is there some simple description of L", for instance:

- is L definable in some weak logic (mainly WMSO logic),
- what is the minimal topological complexity class that contains L?

It should not be confused with the *descriptive complexity* in the meaning of [Imm99].

1.7.4 The languages $W_{i,j}$

The languages $W_{i,j}$ (see [Arn99, AN07]) proved to be convenient tools for studying topological complexity of regular tree languages. As expressed by Theorem 1.23, the language $W_{i,j}$ is complete for the class of languages recognisable by alternating (i, j)-automata.

Definition 1.2. *For $i < j$ consider the following alphabet*

$$A_{i,j} = \{\exists, \forall\} \times \{i, i+1, \dots, j\}.$$

With each $t \in \mathrm{Tr}_{A_{i,j}}$ we associate a parity game $\mathcal{G}_t$ where

- $V = \mathrm{dom}(t)$,
- $E = \{(u, ud) : u \in \mathrm{dom}(t), d \in \{\mathrm{L}, \mathrm{R}\}\}$,
- $v_{\mathrm{I}} = \epsilon$,
- *if $t(u) = (P, n) \in A_{i,j}$ then $\Omega(u) = n$ and $u \in V_P$.*

Let $W_{i,j}$ be the set of all trees over $A_{i,j}$ such that $\exists$ has a winning strategy in $\mathcal{G}_t$.

Theorem 1.23 (Arnold [Arn99]). *The language $W_{i,j}$ can be recognised by a non-deterministic (i, j)-automaton (in particular $W_{i,j} \in \mathbf{RM}^{\text{non-det}}(i, j) \subseteq \mathbf{RM}^{\text{alt}}(i, j)$).*

For every alternating (i, j)-automaton $\mathcal{A}$ there is a canonical continuous function reducing $\mathrm{L}(\mathcal{A})$ to $W_{i,j}$ (i.e. $\mathrm{L}(\mathcal{A}) \leqslant_{\mathrm{W}} W_{i,j}$).

The languages $W_{i,j}$ and the dual $W_{i+1,j+1}$ are incomparable with respect to $\leqslant_{\mathrm{W}}$ (i.e. $W_{i,j} \not\leqslant_{\mathrm{W}} W_{i+1,j+1}$).

Additionally, $W_{0,1}$ is $\mathbf{\Sigma}^1_1$-complete and $W_{1,2}$ is $\mathbf{\Pi}^1_1$-complete.

The following corollary gives an easy way of proving that a particular language does not belong to a given class of the alternating index hierarchy.

Corollary 1.1. *If $W_{i,j} \leqslant_{\mathrm{W}} L$ then $L \notin \mathbf{RM}^{\text{alt}}(i+1, j+1)$ i.e. L cannot be recognised by an alternating $(i+1, j+1)$-automaton.*

In some circumstances one needs to adjust the languages $W_{i,j}$ to current needs. In particular, it is sometimes convenient to add to the alphabet $A_{i,j}$ two additional letters $\top, \bot$ that correspond to an *instant win* in the game $\mathcal{G}_t$. It is expressed by the following remark.

Remark 1.1. All the conditions from Theorem 1.23 are valid for the modification of the languages $W_{i,j}$ by extending the alphabet $A_{i,j}$ with two additional letters $\top, \bot$ of the following semantics: a play π that reaches $\top$ (resp. $\bot$) for the first time in $\mathcal{G}_t$ is winning for $\exists$ (resp. $\forall$) no matter what the priorities occur before and after that.

1.7.5 Separation Property

The notion of separation is an important concept in descriptive set theory and automata theory.

Definition 1.3. *A class of languages $\mathcal{C}$ has the* separation property *with respect to a class $\mathcal{D}$, if the following condition holds:*

> *For every pair of disjoint languages L_1, L_2 from $\mathcal{C}$ there exists a language $L_{\text{sep}} \in \mathcal{D}$ such that*[6]
>
> $$L_1 \subseteq L_{\text{sep}} \quad \text{and} \quad L_2 \subseteq L_{\text{sep}}^{\mathrm{c}}.$$

[6] Recall that X^{c} denotes the complement of a set X.

In that case we say that L_{sep} separates L_1 and L_2. If not stated otherwise, the class $\mathcal{D}$ is taken as $\mathcal{C} \cap \mathcal{C}^c$ — the class of languages L such that both L and L^c belong to $\mathcal{C}$.

Usually, one class from a pair of dual classes $\mathcal{C}$, $\mathcal{C}^c$ has the separation property and the other one does not. Below we recall some known separation-type theorems.

Separation in Topology. The first one is a simple observation about Borel sets.

Theorem 1.24 ([Kec95, Theorem 22.16]). *Let $\eta < \omega_1$. Every two disjoint $\mathbf{\Pi}^0_\eta$ languages can be separated by a language that belongs to $\mathbf{\Pi}^0_\eta \cap \mathbf{\Sigma}^0_\eta$. On the other hand, there exists a pair of disjoint languages in $\mathbf{\Sigma}^0_\eta$ that cannot be separated as above.*

The following theorem is an important extension to the projective hierarchy.

Theorem 1.25 (Lusin (cf. [Kec95, Theorem 14.7, Exercise 28.2])). *If $L_1, L_2 \in \mathbf{\Sigma}^1_1(X)$ are two disjoint analytic subsets of a Polish space X then there exists a Borel set separating them. There exists a pair of disjoint co-analytic (i.e. $\mathbf{\Pi}^1_1$) sets that cannot be separated by any Borel set.*

An important consequence of the above separation result is the following theorem.

Theorem 1.26 (Lusin Souslin [Kec95, Theorem 15.1]). *Assume that $f : X \to Y$ is a continuous function between two Polish spaces and $A \subseteq X$ is Borel. If $f\restriction_A$ is injective then $f(A)$ is Borel.*

Separation in Automata Theory. The following results can be seen as an automata theoretic counterpart of Theorem 1.25.

Theorem 1.27 (Rabin [Rab70]). *If L_1, L_2 are two disjoint Büchi languages of infinite trees then there exists a* WMSO*-definable (i.e.* $\mathbf{Comp}(\mathbf{\Pi}^{\mathrm{alt}}_0)$*) language that separates them.*

This result was extended to higher levels of the non-deterministic index hierarchy, as expressed by the following theorem.

Theorem 1.28 (Arnold Santocanale [AS05]). *Every pair of disjoint languages from $\mathbf{RM}^{\mathrm{non-det}}(0, j)$ (i.e. languages recognised by non-deterministic* min*-parity tree automata of index $(0, j)$) can be separated by a language from* $\mathbf{Comp}(\mathbf{\Pi}^{\mathrm{alt}}_{j-1})$.

Moreover, the construction of an automaton for the separating language is polynomial in the sizes of the given automata from $\mathbf{RM}^{\mathrm{non-det}}(0, j)$.

The following theorem gives negative answers about separability of regular tree languages.

Theorem 1.29 (Hummel Michalewski Niwiński [HMN09], Michalewski Niwiński [MN12], Arnold Michalewski Niwiński [AMN12]). *There exists a pair of disjoint regular tree languages recognised by $(1, 2)$-parity alternating tree automata (i.e. $\mathbf{\Sigma}^{\mathrm{alt}}_1$ languages) that cannot be separated by any Borel set. In particular, these languages cannot be separated by any* WMSO*-definable language.*

For every $j \geqslant 1$ there exists a pair of disjoint regular tree languages from $\mathbf{\Sigma}^{\mathrm{alt}}_j$ that cannot be separated by any $\mathbf{\Delta}^{\mathrm{alt}}_j$ language.

1.7.6 Deterministic Languages

As mentioned earlier, many problems simplify when we restrict to languages recognisable by deterministic automata. Here we collect the decidability results for these languages.

Theorem 1.30 (Niwiński Walukiewicz [NW98]). *The non-deterministic index problem is decidable for deterministic languages.*

The following theorem is often referred to as a *gap property* for deterministic languages.

Theorem 1.31 (Niwiński Walukiewicz [NW03]). *It is decidable if a given regular tree language is deterministic. A deterministic tree language is either:*

- WMSO*-definable and in* $\mathbf{\Pi}^0_3$*,*
- *not* WMSO*-definable and* $\mathbf{\Pi}^1_1$*-complete.*

Moreover, the dichotomy is effective. In particular, a deterministic tree language is either in $\mathbf{Comp}(\mathbf{\Pi}^{\mathrm{alt}}_0)$ *or in* $\mathbf{\Sigma}^{\mathrm{alt}}_1 \setminus \mathbf{Comp}(\mathbf{\Pi}^{\mathrm{alt}}_0)$ *and it is decidable which of the cases holds.*

Finally, the following result of Murlak gives the *ultimate solution* to topological questions about deterministic languages by providing an effective procedure that computes the level in Wadge hierarchy[7] that a given deterministic language occupies.

Theorem 1.32 (Murlak [Mur08]). *The Wadge hierarchy is decidable for deterministic tree languages.*

[7] This hierarchy is not studied in this thesis, it can be seen as a refinement of the Borel hierarchy. In the case of Theorem 1.32, Wadge hierarchy can be seen as the quotient of the class of deterministic tree languages by the order $\leqslant_W$ from Sect. 1.6.2.

Part I

Subclasses of Regular Languages

Chapter 2
Introduction

The fundamental results of Büchi [Büc62] and Rabin [Rab69] state that the monadic second-order (MSO) theory of the ω-chain $(\omega, \leqslant)$ and of the complete binary tree $\left(\{0,1\}^*, \preceq, \leqslant_{\text{lex}}\right)$ is decidable. In both cases the proof relies on a class of finite automata with expressive power equivalent to MSO. Because of effective closure properties and decidability of the emptiness problem, the languages of ω-words and infinite trees definable in MSO are called *regular*. For a broad introduction to the field of regular languages of infinite objects see [Tho96, PP04, TL93].

Since a single ω-word or infinite tree may not have any finite representation, one has to deal with *actual infinity* when studying languages of such objects. In particular, even the set of ω-words over a two-letter alphabet has cardinality continuum. This is the source of strong relationships between properties of regular languages of infinite objects and descriptive set theory. These relationships have a form of synergy: descriptive set theory motivates new problems and methods in automata theory but on the other hand, automata theory introduces natural examples for classical topological concepts.

Recently there has been a number of papers studying these relationships. Properties of regular languages of infinite trees have been studied in [NW03, AN07, ADMN08, Mur08], the Borel complexity of MSO-definable sets of branches of one infinite tree was estimated in [BNR+10], finally the Borel and Wadge complexity of languages of ω-words recognised by various models of computation was estimated in [DFR01, Fin06, CDFM09, DFR13, FS14]. It is worth mentioning that in most of the above cases it turns out that there are languages definable in respective formalisms that are *complete* for the studied topological classes. It shows that these languages are in some sense *representative*. Also, there are some results studying more general set theoretic properties of definable languages. For instance, expressibility of cardinality of sets in MSO was studied in [BKR11], and measurability of regular languages of infinite trees was settled in [GMMS14].

The results of the thesis are based on [HS12, FMS13, BIS13, BS13, Skr14, BGMS14] and the technical report [MS14].

M. Skrzypczak, *Set Theoretic Methods in Automata Theory*, LNCS 9802
DOI: 10.1007/978-3-662-52947-8_2

2.1 Motivations

The following list presents problems studied in the thesis. Most of them have the form of a question about *descriptive complexity* — given a regular language L, is there a description of L that is *simple* in a certain sense.

2.1.1 *Definability in* WMSO

The first question asks how to effectively decide if a given regular language is definable in some logic weaker than MSO. There are two natural candidates for such logics: first-order logic (FO) and weak monadic second-order logic (WMSO) where the set quantification is restricted to finite sets.

In the case of ω-words, definability in FO was solved by Thomas [Tho79] using the methods of Schutzenberger [Sch65] and McNaughton Papert [MP71]. The definability in WMSO trivialises in this case, since every ω-regular language is WMSO-definable.

The problem of definability in WMSO for regular languages of infinite trees is considered as one of the central problems in the area. Recently, there has been some slight progress for various restricted classes of languages. However, the problem in its full generality seems to be out of reach of the currently known methods.

The thesis presents solutions to the problem of WMSO-definability for certain restricted classes of regular languages of infinite trees: for unambiguous Büchi automata in Chap. 3, for general Büchi automata in Chap. 4, for game automata in Chap. 5, and for languages of thin trees in Chap. 6.

2.1.2 *Index Problem*

Another *complexity question* studied in the thesis asks about the *index* of a given regular language of infinite trees L: for a given pair (i, j) is there an alternating top-down parity tree automaton that recognises L and uses only priorities among $\{i, i+1, \ldots, j\}$? It turns out that in the case of languages of infinite trees that are bisimulation-invariant (i.e. definable in μ-calculus, see [JW96]), the index corresponds precisely to the alternation of fixpoints used in the definition of a language [Niw97]. Therefore, the index problem can be seen as a variant of a quantifier alternation question: how many alternations of quantifiers are needed to define a given language.

The decidability of the index problem for general languages of infinite trees is open. As shown in [Bra98, Arn99], the index hierarchy is strict — there are regular languages of infinite trees that cannot be recognised by any automaton of small index. As shown by Rabin [Rab70], the index problem and definability in WMSO are closely related: a regular language of infinite trees is definable in WMSO if and only if both the language and the complement are recognisable by an alternating automaton with Büchi acceptance condition (i.e. condition of the form "infinitely many *accepting* states").

The thesis provides a solution of the index problem for the class of regular languages of infinite trees recognisable by game automata (see Chap. 5). This is the first reasonable class of languages for which the index problem is known to be decidable, that contains languages arbitrarily high in the alternating index hierarchy. Additionally, an effective collapse of index for languages recognisable by unambiguous automata is provided in Chap. 3: it is proved that if an automaton is unambiguous and of certain index then the language recognised by the automaton is lower in the index hierarchy. Although the presented collapse is small, to the author's best knowledge this is the first result that utilizes the fact that a given automaton is unambiguous to give upper bounds on the index of the recognised language.

2.1.3 Bi-unambiguous Languages

One of the difficulties when working with MSO on infinite trees arises from the fact that deterministic automata are too weak to recognise all regular languages. The subclass of regular languages of infinite trees recognisable by deterministic automata seems to be much more tractable [KSV96, NW98, NW03, NW05, Mur08]. Unambiguous automata can be seen as a natural class of automata in-between deterministic and non-deterministic ones. A non-deterministic automaton is *unambiguous* if it has at most one accepting run on every input. As shown by Niwiński and Walukiewicz [NW96], there are regular languages of infinite trees that are inherently ambiguous — there is no unambiguous automaton recognising them. Very little is known about unambiguous languages, for instance it is not known how to decide if a given regular language of infinite trees is recognisable by some unambiguous automaton.

The thesis characterizes the class of bi-unambiguous languages (i.e. languages L such that both L and the complement L^c are unambiguous) as those that can be recognised by finite *prophetic thin algebras*. This theorem constitutes a link between the algebraic framework for thin trees from [Idz12] and languages of general infinite trees. Also, it provides an algebraic way of recognition for a non-trivial class of regular languages of infinite trees.

The following new conjecture has arisen when studying properties of prophetic thin algebras.

Conjecture 2.1. The relation $\varphi'(x, Z)$ expressing that $x \in Z$ and Z is contained in a thin tree does not admit MSO-definable uniformization of the first variable x. In other words, there is no MSO-definable choice function in the class of thin trees.

This conjecture is a strengthening of the theorem of Gurevich and Shelah [GS83] stating that there is no MSO-definable choice function on the complete binary tree. Unfortunately, the conjecture is left open, however some equivalent statements are provided. Also, it is shown that the conjecture implies that it is decidable if a given regular language of infinite trees is bi-unambiguous. Additionally, the conjecture implies that bi-unambiguous languages constitute a very reasonable class (a pseudo-variety from the algebraic point of view).

2.1.4 Borel Languages

The index hierarchy for automata on infinite trees turns out to be closely related to topological hierarchies from descriptive set theory (see for instance [Arn99]). These relations motivate a number of interesting questions, one of them is the following conjecture, stated over 20 years ago.

Conjecture 2.2 (Skurczyński [Sku93]). If a regular language of infinite trees is Borel then it is WMSO-definable.

The converse implication is known to be true: every WMSO-definable language is Borel. Therefore, the conjecture says in fact that a regular language of infinite trees is Borel if and only if it is WMSO-definable. It would mean that if a language is regular and topologically simple then it is also "descriptively" simple. It can also be seen as an automata theoretic counterpart of the relation between the lightface and boldface hierarchies, see [Mos80, Theorem 3E.4].

The conjecture has been proved only in the special case of deterministic languages [NW03]. The thesis provides proofs of the conjecture for wider classes of languages: recognisable by game automata in Chap. 5 and for languages of thin trees in Chap. 6. Additionally, a potential strategy of proving the conjecture for Büchi automata is presented in Chap. 4, unfortunately some additional pumping argument is missing in that case.

2.1.5 Topological Complexity vs. Decidability

In general, there is no direct relationship between decidability of a logic and topological complexity of languages it defines. For instance, the FO theory of the structure of arithmetic $(\omega, \leqslant, +, *)$ is undecidable, while it defines only Borel languages of ω-words. On the other hand one can construct a trivial logic that defines some particular language of very high topological complexity. However, as observed by Shelah [She75] (see also [GS82]) in the case of MSO, the topological complexity and decidability are strongly related: the MSO theory of $(\mathbb{R}, \leqslant)$ is undecidable, however, by Rabin's theorem [Rab69], the theory becomes decidable if we restrict the set quantification to $\mathbf{\Sigma}^0_2$-sets.

These ideas are used in Chaps. 9 and 10 to study decidability of MSO logic equipped with an additional quantifier U (as introduced by Bojańczyk [Boj04] and denoted

MSO+U). Chapter 9 studies topological complexity of languages of ω-words definable in MSO+U. It is shown that the topological complexity of these languages is as high as possible: examples of languages lying arbitrarily high in the projective hierarchy are given. Already this fact implies that there is no *simple* automata model capturing the expressive power of MSO+U on ω-words.

This topological observation is further developed in Chap. 10 to prove that a certain variant of MSO on the Cantor set $\{\mathrm{L}, \mathrm{R}\}^\omega$ (called proj-MSO) can be reduced to the MSO+U theory of the complete binary tree. As shown in [BGMS14], the proj-MSO theory is not decidable in the standard sense (see Theorem 10.88). Therefore, the presented reduction shows that MSO+U is also not decidable in this sense.

The question of decidability of MSO+U on the infinite trees was posed in [Boj04]. The above line of research proves that this question cannot be answered positively. Somehow surprisingly, the technical hearth of the proof relies on purely topological concepts.

2.1.6 Separation Property

The question of separation asks if it is possible to separate every pair of disjoint languages from some class by a *simple* language. A classical example of such property is the following theorem of Lusin: every pair of disjoint analytic (i.e. $\mathbf{\Sigma}^1_1$) sets can be separated by a Borel set.

The separation property has also been studied for certain classes of regular languages, an example is the following result of Rabin: every pair of disjoint regular languages of infinite trees recognisable by Büchi automata can be separated by a language that is WMSO-definable. Recently, the separation turned out to be crucial step in providing a significant result about the decidability of the *dot-depth* hierarchy, see [PZ14].

In Chap. 11 of the thesis the separation property is studied for certain quantitative extensions of ω-regular languages, namely for ωB- and ωS-regular languages introduced by Bojańczyk and Colcombet [BC06]. It is shown that the ωB- and ωS-regular languages have the separation property with respect to ω-regular languages: every pair of disjoint languages recognisable by ωB- (respectively ωS)-automata can be separated by an ω-regular language. This result is somehow surprising as the models of ωB- and ωS-automata are dual: a language is ωB-regular if and only if its complement is ωS-regular. Usually, exactly one class from a pair of dual classes of sets has the separation property.

2.2 Overview of the Parts

The preliminary Chap. 1 introduces basic notions and known results that will be used later. The rest of the thesis is divided into three parts, each part has three chapters.

All the presented results study related problems of *descriptive complexity*. The respective parts group results of similar type. Most of the chapters present results that are technically independent, in particular they can be read separately. The only technical dependencies are: Chaps. 7 and 8 depend on definitions from Chap. 6; results of Chap. 10 depend on Theorem 2.7 from Chap. 9.

A separate chapter (see page 205) presents conclusions of the whole thesis. In particular, some relationships and similarities between the techniques used in the chapters are discussed.

2.2.1 Part I: Subclasses of Regular Languages

The first part of the thesis studies descriptive complexity questions for restricted classes of regular languages of infinite trees: unambiguous automata in Chap. 3, Büchi automata in Chap. 4, and game automata in Chap. 5. Three main theorems of these chapters are the following.

The first theorem shows how to use the fact that a given automaton is unambiguous to derive a collapse in parity index of the language recognised by it.

Theorem 2.1. *If $\mathcal{A}$ is an unambiguous* min*-parity automaton of index* $(0, j)$ *then the language $L(\mathcal{A})$ can be recognised by an alternating* $\mathrm{Comp}(0, j-1)$*-automaton of size polynomial in the size of $\mathcal{A}$.*

In particular, if $\mathcal{A}$ is Büchi and unambiguous then $L(\mathcal{A})$ is WMSO*-definable.*

The second theorem is based on a theory of certain ranks for Büchi automata. Using these ranks, a characterisation of WMSO-definable languages is given.

Theorem 2.2. *It is decidable if the language of infinite trees recognised by a given non-deterministic Büchi tree automaton is* WMSO*-definable.*

The above result was already proved by Kuperberg and Vanden Boom (see for instance [CKLV13]) using the theory of cost functions. However, as discussed in Chap. 4, the methods developed in the presented proof may be of independent interest since they introduce conceptually new techniques based on ranks of well-founded ω-trees.

Finally, the third theorem shows that both index problems are decidable for *game automata* — a class of alternating automata that extends deterministic ones by allowing certain restricted alternation between the players. Two effective procedures that compute the index of the language recognised by a given game automaton are proposed. Then it is shown that the procedures are correct. For this purpose, upper and lower bounds are given. Interestingly, in the case of the alternating index problem, the lower bounds are based on purely topological methods (namely the topological hardness of languages $W_{i,j}$).

Theorem 2.3. *The non-deterministic index problem is decidable for game automata (i.e. if a game automaton is given as the input). The same holds for the alternating index problem.*

2.2.2 Part II: Thin Algebras

The second part is devoted to a study of *thin algebras* and thin trees, i.e. trees having only countably many infinite branches. In Chap. 6 a characterization of languages of thin trees that are WMSO-definable among all infinite trees is given. Chapter 7 is devoted to the recognition of languages of infinite trees by *prophetic thin algebras*. Finally, Chap. 8 studies Conjecture 2.1 and related uniformization problems on thin trees. Three main theorems of these chapters are the following.

The first theorem gives an effective characterisation of regular languages of thin trees that are definable in WMSO among all infinite trees. Additionally, it expresses an upper bound: even if a regular language of thin trees is not WMSO-definable among all infinite trees, it is still topologically simple (i.e. it belongs to $\mathbf{\Pi}^1_1$).

Theorem 2.4. *A regular language of thin trees (i.e. a regular language that contains only thin trees) is either:*

1. $\mathbf{\Pi}^1_1$*-complete among all infinite trees,*
2. WMSO*-definable among all infinite trees (and thus Borel).*

Moreover, it is decidable which of the cases holds.

The second theorem provides an algebraic framework for recognition of a restricted class of regular languages of infinite trees. The idea is to use algebras designed for thin trees to recognise languages of arbitrary infinite trees.

Theorem 2.5. *A language of infinite trees L is recognised by a homomorphism into a finite prophetic thin algebra if and only if L is bi-unambiguous, i.e. both L and the complement L^c can be recognised by unambiguous automata.*

The last theorem consists of three ingredients: an equivalent formulation of Conjecture 2.1, an example of a non-uniformizable relation on thin trees, and an essentially new example of an ambiguous regular language of infinite trees. The non-uniformizable relation uses a concept of *skeleton* — a subset of a thin tree that provides a decomposition of this tree into separate branches.

Theorem 2.6. *Conjecture 2.1 is equivalent to the fact that every finite thin algebra admits some consistent marking on every infinite tree.*

The relation $\varphi(\sigma, t)$ stating that t is a thin tree and σ is a skeleton of t does not admit any MSO*-definable uniformization of σ.*

The language of all thin trees is ambiguous (i.e. it is not recognised by any unambiguous automaton).

Although Conjecture 2.1 is not proved in this thesis, the above non-uniformizability results are of their own interest. In particular, the example about skeletons provides a standalone answer to Rabin's uniformization problem (the problem was solved originally by Gurevich and Shelah in [GS83]).

2.2.3 Part III: Extensions of Regular Languages

The last part of the thesis studies some properties of contemporary *quantitative* developments in automata theory. Topological complexity of MSO+U-definable languages of ω-words is estimated in Chap. 9. Chapter 10 studies consequences of the high topological complexity of MSO+U regarding decidability of this logic on the complete binary tree. Finally, in Chap. 11 the separation property for ωB- and ωS-regular languages is proved. Three main theorems of these chapters are the following.

The first expresses the topological complexity of MSO+U on ω-words.

Theorem 2.7. *There exist languages of ω-words that are definable in* MSO+U *logic and lie arbitrarily high in the projective hierarchy.*

The second theorem uses studies a new variant of MSO (called *proj*-MSO). It is a logic introduced in [BGMS14] where set quantifiers are restricted to projective sets of certain level (fixed explicitly during quantification). For instance, a logic can say "there exists a set X that belongs to $\mathbf{\Sigma}^1_5$ and …".

Theorem 2.8. *The proj-*MSO *theory of* $\{\mathrm{L},\mathrm{R}\}^{\leqslant\omega}$ *with prefix* $\preceq$ *and lexicographic* $\leqslant_{\mathrm{lex}}$ *orders effectively reduces to the* MSO+U *theory of the complete binary tree* $\left(\{\mathrm{L},\mathrm{R}\}^*, \preceq, \leqslant_{\mathrm{lex}}\right)$.

*An algorithm deciding the proj-*MSO *theory of* $\{\mathrm{L},\mathrm{R}\}^{\leqslant\omega}$ *(together with its proof of correctness) would imply that analytic determinacy fails.*

This result was further extended in [BGMS14] using an adaptation of the technique of Shelah [She75]. It is shown there that under a certain set theoretic assumption (namely that V=L, i.e. we work in the *Gödel's constructible universe*) the proj-MSO theory of $\{\mathrm{L},\mathrm{R}\}^{\leqslant\omega}$ is undecidable. Therefore, together with the above theorem, V=L implies that the MSO+U theory of the complete binary tree is undecidable.

Finally, the ninth main theorem of the thesis studies *separation property* for languages of ω-words that are recognised by counter automata introduced by Bojańczyk and Colcombet in [BC06].

Theorem 2.9. *If L_1, L_2 are disjoint languages of ω-words both recognised by ωB-(respectively ωS)-automata then there exists an ω-regular language L_{sep} such that*

$$L_1 \subseteq L_{\mathrm{sep}} \quad \text{and} \quad L_2 \subseteq L_{\mathrm{sep}}^{\mathrm{c}}.$$

Additionally, the construction of L_{sep} is effective.

Chapter 3
Collapse for Unambiguous Automata

A natural class in-between deterministic and non-deterministic automata is the class of *unambiguous* ones — an automaton is unambiguous if it has at most one accepting run on every tree. It seems that an unambiguous automaton represents the structure of the recognised language in a more rigid way than a general non-deterministic automaton. However, as shown in [NW96], there are *ambiguous* regular tree languages — languages that are not recognised by any unambiguous automaton.

In contrast to general regular tree languages, most of the problems are solved in the case of deterministic automata: it is decidable whether a given language is recognisable by a deterministic automaton [NW05], the non-deterministic index problem is decidable [NW03, NW98], as well as the Wadge hierarchy [Mur08].

In comparison, the class of *unambiguous tree languages* (recognisable by unambiguous automata) is still a *terra incognita*. Not only it is unknown how to verify whether a given regular tree language is unambiguous, but also there are no non-trivial upper bounds on the descriptive complexity of unambiguous languages in comparison to all regular tree languages. In particular, it is open whether all unambiguous languages can be recognised by alternating parity automata of a bounded parity index.

There are only two estimations on descriptive complexity of unambiguous languages known. First, a recent result in [Hum12] shows that unambiguous languages are topologically harder than deterministic ones. Second, in [FS09] the authors observe, by a standard descriptive set theoretic argument, that the language recognised by an unambiguous Büchi automaton must be Borel. In this chapter we extend the latter result by showing the following theorem.

Theorem 3.1. *If $\mathcal{A}$ is an unambiguous* min*-parity automaton of index* $(0, j)$ *then the language* $\mathrm{L}(\mathcal{A})$ *can be recognised by an alternating* $\mathrm{Comp}(0, j-1)$*-automaton of size polynomial in the size of $\mathcal{A}$.*

In particular, if $\mathcal{A}$ is Büchi and unambiguous then $\mathrm{L}(\mathcal{A})$ *is* WMSO*-definable.*

This theorem extends the mentioned result from [FS09] in two directions. First, it shows that every unambiguous Büchi automaton recognises a language that is

M. Skrzypczak, *Set Theoretic Methods in Automata Theory*, LNCS 9802
DOI: 10.1007/978-3-662-52947-8_3

WMSO-definable. It is known that every regular tree language definable in WMSO is Borel but the converse is open (see Conjecture 2.2 on page 32). Second, the theorem presented here gives a collapse also for higher priorities.

To the author's best knowledge this is the first result where it is shown how to use the fact that a given automaton is unambiguous to derive upper bounds on the parity index of the recognised language. Therefore, this result should be treated as a first step towards descriptive complexity bounds for unambiguous languages, and generally a better understanding of them.

One should note that in the above theorem the unambiguous-and-Büchi assumptions are put on one automaton. It is still possible for a regular tree language to be both: recognised by an unambiguous automaton and by some (other) Büchi automaton. An example of such a language is the H-language proposed in [Hum12]: "exists a branch containing only a's and turning infinitely many times right". To the author's best knowledge, no non-trivial upper bound is known when the conditions of unambiguity and certain index are put on the language.

The construction presented here can be seen as an automata theoretic adaptation of the proof of the theorem of Lusin and Souslin [Kec95, Theorem 15.1] (see Theorem 1.26 on page 25) stating that if $f\colon X \to Y$ is injective and continuous then the image $f(X)$ is Borel in Y. The proof presented in [Kec95] is based on the Lusin Separation Theorem [Kec95, Theorem 14.7]. Here one can use Rabin's separation result (Theorem 1.27 on page 25) for $j = 1$ and the separation of Arnold and Santocanale (Theorem 1.28 on page 25) for $j > 1$. The idea to use the separation result of Arnold and Santocanale for the case of $j > 1$ was suggested by Henryk Michalewski.

The proof goes as follows. We first observe that if an automaton is unambiguous then the transitions of the automaton have to correspond, is some sense, to disjoint languages. By applying the separation result of Arnold and Santocanale (see Theorem 1.28 on page 25), these disjoint languages can be separated by $\mathbf{Comp}(\mathbf{\Pi}^{\mathrm{alt}}_{j-1})$-languages. This leads to a construction of a unique run ρ_t of a given automaton on a given tree t (Lemma 3.3 in Sect. 3.1).

Then, in Sect. 3.2, we conclude the proof of Theorem 3.1 by providing an effective construction of a Comp$(0, j-1)$-automaton recognising $\mathrm{L}(\mathcal{A})$. This automaton combines the Comp$(0, j-1)$-automata for *transition languages* with an additional game played on the run ρ_t.

3.1 Unique Runs

In this section we prove Lemma 3.3 showing how to define, for a given tree t, a unique run ρ_t of a given unambiguous automaton $\mathcal{A}$ of index $(0, j)$. The crucial property of this construction is that the constrains on ρ_t are $\mathbf{Comp}(\mathbf{\Pi}^{\mathrm{alt}}_{j})$; additionally, ρ_t is accepting if and only if t belongs to the language $\mathrm{L}(\mathcal{A})$.

Let us fix an unambiguous automaton $\mathcal{A}$ of index $(0, j)$. Let Q be the set of states of $\mathcal{A}$ and A be its working alphabet. We will say that a transition $\delta = (q, a, q_{\mathrm{L}}, q_{\mathrm{R}})$ of $\mathcal{A}$ *starts* from (q, a).

A pair $(q, a) \in Q \times A$ is *productive* if it appears in some accepting run: there exists a tree $t \in \mathrm{Tr}_A$ and an accepting run ρ of $\mathcal{A}$ on t such that for some vertex u we have $\rho(u) = q$ and $t(u) = a$. This definition combines two requirements: that there exists an accepting run that leads to the state q and that some tree can be accepted starting from (q, a). Note that if (q, a) is productive then there exists at least one transition starting from (q, a).

For every transition $\delta = (q, a, q_{\mathrm{L}}, q_{\mathrm{R}})$ of $\mathcal{A}$ we define L_δ as the language of trees such that there exists a run ρ of $\mathcal{A}$ on t that is parity-accepting and *uses* δ *in the root of* t:

$$\rho(\epsilon) = q, t(\epsilon) = a, \rho(\mathrm{L}) = q_{\mathrm{L}}, \text{ and } \rho(\mathrm{R}) = q_{\mathrm{R}}.$$

The following lemma is a simple consequence of unambiguity of the given automaton.

Lemma 3.2. *If (q, a) is productive and $\delta_1 \neq \delta_2$ are two transitions starting from (q, a) then the languages L_{δ_1}, L_{δ_2} are disjoint.*

Proof. Assume contrary that there exists a tree $r \in L_{\delta_1} \cap L_{\delta_2}$ with two respective parity-accepting runs ρ_1, ρ_2. Since (q, a) is productive so there exists a tree t and an accepting run ρ on t such that $\rho(u) = q$ and $t(u) = a$ for some vertex u. Consider the tree $t' = t[u \leftarrow r]$ — the tree obtained from t by substituting r as the subtree under u. Since $\rho(u) = q$ and both ρ_1, ρ_2 start from (q, a), we can construct two accepting runs $\rho[u \leftarrow \rho_1]$ and $\rho[u \leftarrow \rho_2]$ on t'. Since these runs differ on the transition used in u, we obtain a contradiction to the fact that $\mathcal{A}$ is unambiguous. ∎

Let (q, a) be a productive pair and $\{\delta_1, \delta_2, \ldots, \delta_n\}$ be the set of transitions of $\mathcal{A}$ starting from (q, a). In that case the languages L_{δ_k} for $k = 1, 2, \ldots, n$ are pairwise disjoint. Theorem 1.28 from page 25 implies that for every pair of transitions $\delta_i \neq \delta_j$ there is an $\mathbf{Comp}(\boldsymbol{\Pi}^{\mathrm{alt}}_{j-1})$-language that separates L_{δ_i} from L_{δ_j}. Since $\mathbf{Comp}(\boldsymbol{\Pi}^{\mathrm{alt}}_{j-1})$-languages are closed under Boolean combinations, we can find Comp$(0, j-1)$-automata $\mathcal{C}_{\delta_k}$ for $k = 1, 2, \ldots, n$ such that:

- for $k = 1, 2, \ldots, n$ we have $L_{\delta_k} \subseteq \mathrm{L}(\mathcal{C}_{\delta_k})$,
- for $k \neq k'$ the languages $\mathrm{L}(\mathcal{C}_{\delta_k})$, $\mathrm{L}(\mathcal{C}_{\delta_{k'}})$ are disjoint,
- the union $\bigcup_{k=1,2,\ldots,n} \mathrm{L}(\mathcal{C}_{\delta_k})$ equals Tr_A.

These automata will be crucial ingredients of the construction.

The following lemma formalizes the notion of the unique runs.

Lemma 3.3. *Let $t \in \mathrm{Tr}_A$ be a tree. There exists a unique maximal partial run ρ_t of $\mathcal{A}$ on t, i.e. a partial function $\rho_t \colon \{\mathrm{L}, \mathrm{R}\}^* \rightharpoonup Q^{\mathcal{A}}$ such that:*

- $\rho_t(\epsilon) = q^{\mathcal{A}}_{\mathrm{I}}$,

– *if* $u \in \mathrm{dom}(\rho_t)$ *and* $(\rho_t(u), t(u))$ *is productive then also* $u\mathrm{L}, u\mathrm{R} \in \mathrm{dom}(\rho_t)$ *and*

$$t\restriction_u \in \mathrm{L}(\mathcal{C}_\delta) \textit{ with } \delta = \big(\rho_t(u), t(u), \rho_t(u\mathrm{L}), \rho_t(u\mathrm{R})\big). \tag{3.1}$$

– $t \in \mathrm{L}(\mathcal{A})$ *if and only if* ρ *is total and accepting.*

Proof. The construction is inductive. We start by putting $\rho_t(\epsilon) = q_\mathrm{I}^\mathcal{A}$. Assume that the value of ρ_t is defined in a vertex $u \in \{\mathrm{L}, \mathrm{R}\}^*$. Let $a = t(u)$ and $q = \rho(u)$. If (q, a) is unproductive we leave the values of ρ on the subtree under u undefined. In that case we call u a *leaf* of ρ_t. Otherwise, the space Tr_A is split into disjoint sets $\mathrm{L}(\mathcal{C}_\delta)$ ranging over transitions δ starting from (q, a). Therefore, there exists exactly one transition $\delta \in \Delta$ starting from (q, a) such that $t\restriction_u \in \mathrm{L}(\mathcal{C}_\delta)$. Let $\delta = (q, a, q_\mathrm{L}, q_\mathrm{R})$ and $\rho(ud) = q_d$ for $d = \mathrm{L}, \mathrm{R}$.

Clearly, the above construction gives a unique maximal partial run ρ satisfying the first two bullets of the statement. If ρ_t is accepting then it is a witness that $t \in \mathrm{L}(\mathcal{A})$. Let ρ be an accepting run of $\mathcal{A}$ on t. We inductively prove that $\rho = \rho_t$. Take a node u of t and define $q = \rho(u)$, $a = t(u)$, $q_\mathrm{L} = \rho_t(u\mathrm{L})$, and $q_\mathrm{R} = \rho_t(u\mathrm{R})$. Observe that ρ is a witness that (q, a) is productive and for $\delta = (q, a, q_\mathrm{L}, q_\mathrm{R})$ we have

$$t \in L_\delta \subseteq \mathrm{L}(\mathcal{C}_\delta).$$

Therefore, $\rho_t(u\mathrm{L}) = \rho(u\mathrm{L})$ and $\rho_t(u\mathrm{R}) = \rho(u\mathrm{R})$. ∎

3.2 Construction of the Automaton

Now we construct an alternating Comp(0, $j-1$)-automaton $\mathcal{R}$ recognising $\mathrm{L}(\mathcal{A})$. It will consist of two sub-automata running in parallel:

1. In the first sub-automaton the role of $\exists$ will be to propose a run ρ on a given tree t. She will be forced to propose precisely the run ρ_t from Lemma 3.3 — at any moment $\forall$ can *challenge* the currently proposed transition and check whether (3.1) in Lemma 3.3 is satisfied. Such a *challenge* will be realised by moving to the initial state of the appropriate automaton $\mathcal{C}_\delta$.
2. In the second sub-automaton the role of $\forall$ will be to prove that the run ρ_t is not accepting. That is, he will find a leaf in ρ_t or an infinite branch of ρ_t that does not satisfy the parity condition. Since he knows the run ρ_t in advance, we can ask him to declare in advance what will be the odd priority n that is the lim inf of priorities of ρ_t on the selected branch.

The automaton $\mathcal{R}$ consists of an *initial component* C described below and of the disjoint union of the automata $\mathcal{C}_{\delta_k}$. States in the initial component C are of the form (q, n) where q is a state of $\mathcal{A}$ and n is either $\perp$ or an odd number between 0 and j. The state q denotes the current state of the run that is being constructed by $\exists$ in the first sub-automaton. The value n (if $\neq \bot$) denotes the odd priority declared by $\forall$ in the second sub-automaton.

The initial state of $\mathcal{R}$ is $(q_I^{\mathcal{A}}, \bot) \in C$. The transitions of $\mathcal{R}$ inside C are built by the following rules. Assume that the label of the current vertex is a and the current state is (q, n):

Step I if the pair (q, a) is not productive, $\exists$ loses,
Step II if $n \neq \bot$ and $\Omega^{\mathcal{A}}(q) < n$ then $\forall$ loses,
Step III $\exists$ declares a transition $\delta = (q, a, q_{\mathrm{L}}, q_{\mathrm{R}})$ of $\mathcal{A}$ that starts from (q, a),
Step IV $\forall$ decides to *challenge* this transition or to *accept* it,
Step V if $\forall$ *challenges* the transition, $\mathcal{R}$ makes an ϵ-transition to the initial state of C_δ (n does not play any role in that case),
Step VI otherwise, if $n = \bot$ then $\forall$ declares a new value n': some odd number between 0 and j, or still $\bot$ (if $n \neq \bot$ then we put $n' = n$),
Step VII finally, $\forall$ selects a direction $d \in \{\mathrm{L}, \mathrm{R}\}$ and the automaton $\mathcal{R}$ makes a d-transition to the state (q_d, n').

Note that for each tree t, each play in the game $\mathcal{G}(\mathcal{R}, t)$ starts in C and either stays in it forever or leaves to some C_δ and stays there forever. Note also that C consists of two parts: C_I with $n = \bot$ and C_F where $n \neq \bot$. Let the priorities of all the states of the form $(q, \bot)$ equal 2. Consider a state (q, n) with $n \neq \bot$. If $\Omega^{\mathcal{A}}(q) = n$ then such a state has priority 1, otherwise (i.e. if $\Omega^{\mathcal{A}}(q) > n$) the priority of (q, n) is 2.

We first argue that if $j > 1$ then the automaton $\mathcal{R}$ is a Comp(0, $j-1$)-automaton. Note that the graph of $\mathcal{R}$ consists of the following strongly-connected components: the components of C_I, C_F, and the components of C_δ for $\delta \in \Delta$. Recall that all the automata C_δ are by the construction Comp(0, $j-1$). By the definition, C_I and C_F are Comp(0, 1)-automata so the whole automaton $\mathcal{R}$ is also Comp(0, $j - 1$).

Consider $j = 1$ (the Büchi case). Observe that the only possible odd value n between 0 and j is $n = 1$. It means that if $\forall$ declares a value $n \neq \bot$ then always $\Omega(q) \leqslant n$, therefore there are no states in C_F of priority 2. It implies that both C_I and C_F are Comp(0, 0)-automata and $\mathcal{R}$ is a Comp(0, 0)-automaton.

Observe that the size of the automaton $\mathcal{R}$ is polynomial in the size of $\mathcal{A}$. The results of the following two sections imply that $\mathrm{L}(\mathcal{R}) = \mathrm{L}(\mathcal{A})$, thus completing the proof of Theorem 3.1.

3.2.1 Soundness

Lemma 3.4. *If $t \in \mathrm{L}(\mathcal{A})$ then $t \in \mathrm{L}(\mathcal{R})$.*

Proof. Fix the accepting run ρ_t of $\mathcal{A}$ on t given by Lemma 3.3. Consider the following strategy $\sigma_\exists$ for $\exists$ in C: always declare δ consistent with ρ_t. Extend it to the winning strategies in C_δ whenever they exist. That is, if the current vertex is u and the state of $\mathcal{R}$ is of the form $(q, n) \in C$ then declare $\delta = (\rho(u), t(u), \rho(u\mathrm{L}), \rho(u\mathrm{R}))$. Whenever the game moves from the component C into one of the automata C_δ in a vertex u, fix some winning strategy in $\mathcal{G}(\mathcal{C}_\delta, t\restriction_u)$ (if exists) and play according to this strategy; if there is no such strategy, play using any strategy.

Take a play consistent with $\sigma_\exists$ in $\mathcal{G}(\mathcal{R}, t)$. First note that $\exists$ does not lose in Step I since all the pairs (q, a) appearing during the play are productive — the run ρ_t is a witness. There are the following cases:

- $\forall$ loses in a finite time in Step II.
- $\forall$ stays forever in C_I never changing the value of n and loses by the parity criterion.
- In some vertex u of the tree $\forall$ *challenges* the transition δ given by $\exists$ and the game proceeds to $\mathcal{C}_\delta$. In that case $t\restriction_u \in L_\delta$ by the definition of L_δ (the run $\rho_t\restriction_u$ is a witness) and therefore $t\restriction_u \in \mathrm{L}(\mathcal{C}_\delta)$. So $\exists$ has a winning strategy in $\mathcal{G}(\mathcal{C}_\delta, t\restriction_u)$ and she wins the rest of the game.
- $\forall$ declares a value $n \neq \perp$ at some point and then *accepts* all successive transitions of $\exists$. In that case the game follows an infinite branch α of t. Since ρ_t is accepting so we know that $k \stackrel{\text{def}}{=} \liminf_{i\to\infty} \Omega^{\mathcal{A}}(\rho_t(\alpha\restriction_i))$ is even. If $k < n$ then $\forall$ loses at some point in Step II. Otherwise $k > n$ and from some point on all the states of $\mathcal{R}$ visited during the game have priority 2, thus $\forall$ loses by the parity criterion in C_F. ■

3.2.2 Completeness

Lemma 3.5. *If $t \notin \mathrm{L}(\mathcal{A})$ then $t \notin \mathrm{L}(\mathcal{R})$.*

Proof. We assume that $t \notin \mathrm{L}(\mathcal{A})$ and give a winning strategy for $\forall$ in the game $\mathcal{G}(\mathcal{R}, t)$. Let us fix the run ρ_t given by Lemma 3.3.

Note that either ρ_t is a partial run: there is a vertex u such that $\rho_t(u) = q$ and $(q, t(u))$ is unproductive, or ρ_t is a total run. Since $t \notin \mathrm{L}(\mathcal{A})$ so ρ_t cannot be a total accepting run. Let α be a finite or infinite branch: either $\alpha \in \{\mathrm{L}, \mathrm{R}\}^*$ and α is a leaf of ρ_t or α is an infinite branch such that $k \stackrel{\text{def}}{=} \liminf_{i\to\infty} \Omega^{\mathcal{A}}(\rho_t(\alpha\restriction_i))$ is odd. If α is finite let us put any odd value between 0 and j as k.

Consider the following strategy for $\forall$:

- $\forall$ keeps $n = \perp$ until there are no more states of priority greater than k along α in ρ_t. Then he declares $n' = k$.
- $\forall$ *accepts* a transition δ given by $\exists$ in a vertex u if and only if it is *consistent with* ρ_t in u (i.e. if $\delta = (\rho_t(u), t(u), \rho_t(u\mathrm{L}), \rho_t(u\mathrm{R}))$).
- $\forall$ always follows α: in vertex $u \in \{\mathrm{L}, \mathrm{R}\}^*$ he chooses the direction d in such a way that $ud \preceq \alpha$.

As before, we extend this strategy to strategies on $\mathcal{C}_\delta$ whenever they exist: if the game moves from the component C into one of the automata $\mathcal{C}_\delta$ in a vertex u then $\forall$ uses some winning strategy in the game $\mathcal{G}(\mathcal{C}_\delta, t\restriction_u)$ (if it exists); if there is no such strategy, $\forall$ plays using any strategy.

Consider any play π consistent with $\sigma_\forall$. Note that if α is a finite word and the play π reaches the vertex α in a state (q, n) in C then $q = \rho_t(\alpha)$ and $\forall$ wins in Step I as $(\rho_t(\alpha), t(\alpha))$ is not productive. Similarly, by the definition of the strategy $\sigma_\forall$, $\forall$ never loses in Step II — if he declared $n \neq \bot$ then the play will never reach a state of priority smaller than n.

Let us consider the remaining cases. First assume that at some vertex u player $\forall$ *challenged* a transition δ declared by $\exists$. It means that there is another transition $\delta' \neq \delta$ consistent with ρ_t in u. By the definition of ρ_t we know that $t\restriction_u \in \mathrm{L}_{\delta'}$ in particular $t\restriction_u \in \mathrm{L}(\mathcal{C}_{\delta'})$. Since the languages $\mathcal{C}_{\delta'}, \mathcal{C}_\delta$ are disjoint, $t\restriction_u \notin \mathcal{C}_\delta$ and $\forall$ has a winning strategy in $\mathcal{G}(\mathcal{C}_\delta, t\restriction_u)$ and wins in that case.

Consider the remaining case: $\forall$ *accepted* all the transitions declared by $\exists$ and the play is infinite. Then, for every $i \in \mathbb{N}$ the game reached the vertex $\alpha\restriction_i$ in a state (q, n) satisfying $q = \rho_t(\alpha\restriction_i)$. In that case there is some vertex u along α where $\forall$ declared $n = k$. Therefore, infinitely many times $\Omega^{\mathcal{A}}(q) = n$ in π so $\forall$ wins that play by the parity criterion. ■

This concludes the proof of Theorem 3.1.

3.3 Conclusions

The results presented in this chapter provide a way of using the fact that a given automaton $\mathcal{A}$ is unambiguous to prove some upper bounds on the index of the language $\mathrm{L}(\mathcal{A})$. Therefore, they can be seen as an attempt to solve the following open problem.

Question 3.1. Does there exist a number n such that every unambiguous tree language belongs to $\mathbf{\Pi}^{\mathrm{alt}}_n$?

As proved in [BIS13], every regular language of thin trees can be recognised by a non-deterministic automaton of index $(1, 3)$. The results of Chap. 7 suggest that there is a strong relationship between bi-unambiguous languages and languages of thin trees (namely that every bi-unambiguous language can be recognised by a homomorphism into a finite prophetic thin algebra). These observations suggest the following conjecture that would give a partial solution to the above question in the case of bi-unambiguous languages.

Conjecture 3.3. If both L and the complement L^{c} are recognisable by unambiguous automata (i.e. L is bi-unambiguous) then $L \in \mathbf{\Delta}^{\mathrm{alt}}_2$.

The best known lower bounds are given by Hummel in [Hum12] where examples of bi-unambiguous languages in $\mathbf{Comp}(\mathbf{\Pi}^{\mathrm{alt}}_1) \setminus \left(\mathbf{\Pi}^{\mathrm{alt}}_1 \cup \mathbf{\Sigma}^{\mathrm{alt}}_1\right)$ are provided.

This chapter is based on the technical report [MS14].

Chapter 4
When a Büchi Language Is Definable in WMSO

A natural subclass of regular tree languages are those that can be defined in weak monadic second-order logic (WMSO). As shown by Rabin (see Theorem 1.18 on page 20), a language L is WMSO-definable if and only if both L and the complement can be recognised by non-deterministic (equivalently alternating, see Theorem 1.19 on page 21) Büchi automata. Therefore, the following decision problem can be seen as a special case of the index problems from Sect. 1.7.2 (see Problem 1.1 on page 22).

Problem 4.2 (Definability in WMSO*)*

- **Input** An alternating tree automaton $\mathcal{A}$.
- **Output** Is $\mathrm{L}(\mathcal{A})\mathcal{A}$ definable in WMSO.

The decidability of this decision problem in full generality is open. Therefore, it is natural to ask for solutions for restricted classes of input languages. In this chapter we study the problem when the input automaton is a non-deterministic (equivalently alternating) Büchi automaton.

The main theorem of this section states that this restricted problem is decidable.

Theorem 4.2. *It is decidable if the language of infinite trees recognised by a given non-deterministic Büchi tree automaton is* WMSO*-definable.*

This decidability result was already proved in [CKLV13]. It is shown there that the reduction from [CL08] applied to Büchi automata produces instances of a domination problem for which an effective procedure is known [Van11, KV11]. The whole structure of the proof is rather involved and makes extensive use of the theory of regular cost functions on ω-words [Col13].

The approach presented in this chapter is different. We start by introducing a rank that measures complexity of trees with respect to a given Büchi automaton $\mathcal{B}$. This leads to the definition of an ordinal $\eta(\mathcal{B}) \leqslant \omega_1$. It turns out that this ordinal is strongly related to the descriptive complexity of the language $\mathrm{L}(\mathcal{B})$. In particular, we prove the following two properties of $\eta(\mathcal{B})$:

- $\eta(\mathcal{B}) < \omega_1$ if and only if $\mathrm{L}(\mathcal{B})$ is Borel (see Proposition 4.2),

M. Skrzypczak, *Set Theoretic Methods in Automata Theory*, LNCS 9802
DOI: 10.1007/978-3-662-52947-8_4

– $\eta(\mathcal{B}) < \omega^2$ if and only if $\mathrm{L}(\mathcal{B})$ is WMSO-definable (see Proposition 4.3)

To prove the latter property, we introduce a *finitary* version of $\eta(\mathcal{B})$ represented by languages of K-reach and K-safe trees.

The obtained properties of the rank $\eta(\mathcal{B})$ seem promising, in particular, Conjecture 2.2 on page 32 (every Borel regular tree language is WMSO-definable) would be proved for Büchi automata if one managed to prove the following claim.

Conjecture 4.4. If $\mathcal{B}$ is a non-deterministic Büchi tree automaton then

$$\eta(\mathcal{B}) \geqslant \omega^2 \quad \Longrightarrow \quad \eta(\mathcal{B}) = \omega_1.$$

Unfortunately, the author is unable to prove the above statement. It can be seen as a distant analogue of the study of *closure ordinals* from [Cza10, AL13].

Theorem 2.2 is proved as a consequence of properties of $\eta(\mathcal{B})$ — it is enough to prove that the condition $\eta(\mathcal{B}) < \omega^2$ is decidable. For this purpose, a variant of *domination games* from [Col13] is introduced. Although the motivations come from [Col13], the presented construction is standalone and does not refer to any results about cost functions.

The organisation of the chapter reflects the two parts of the proof. The first part of the proof, which studies properties of $\eta(\mathcal{B})$, is spread across Sects. 4.1, 4.2 and 4.3. In Sect. 4.1 the ordinal $\eta(\mathcal{B})$ is defined and its basic properties are stated. Section 4.2 introduces notions of K-reach and K-safe trees that are designed as finitary approximations of $\eta(\mathcal{B})$. Section 4.3 introduces Comp(0, 0)-automata that recognise languages of K-reach and K-safe trees. These automata show that if $\eta(\mathcal{B}) < \omega^2$ then $\mathrm{L}(\mathcal{B})$ is WMSO-definable.

The second part of the proof, i.e. the effective procedure itself is presented in Sects. 4.4 and 4.5. Section 4.4 introduces a game $\mathcal{G}$ designed to verify if $\eta(\mathcal{B}) < \omega^2$. The game $\mathcal{G}$ has finite arena and the winning condition of $\mathcal{G}$ is ω-regular, therefore it is decidable who wins $\mathcal{G}$. Section 4.5 shows that $\exists$ has a winning strategy in $\mathcal{G}$ if and only if $\eta(\mathcal{B}) < \omega^2$, what finishes the proof of Theorem 2.2.

Finally, Sect. 4.6 concludes the results of this chapter.

4.1 The Ordinal of a Büchi Automaton

Let L be a regular tree language recognised by a non-deterministic Büchi tree automaton $\mathcal{B}$. Our aim is to define a particular continuous reduction T of $L^c = \mathrm{Tr}_A \setminus L$ to the set of well-founded ω-trees WF. Intuitively, $\mathrm{T}(t)$ will reflect to what extent it is possible to construct runs of $\mathcal{B}$ on t that contain *many* accepting states. Formally, $\mathrm{T}(t)$ will consist of *truncated runs* defined in the following subsection. The reduction T will allow us to bind with a tree $t \in L^c$ an ordinal $\mathrm{rank}\big(\mathrm{T}(t)\big)$ measuring the *complexity of t*. Then, we will define an ordinal $\eta(\mathcal{B})$ (the *ordinal of* $\mathcal{B}$) as the supremum of $\mathrm{rank}\big(\mathrm{T}(t)\big)$ over trees $t \in L^c$.

For the rest of this chapter let us fix a non-deterministic Büchi tree automaton $\mathcal{B}$ recognising L. Let us assume that Q is the set of states of $\mathcal{B}$ and A is its working alphabet. Let $F \stackrel{\text{def}}{=} \{q \in Q : \Omega^{\mathcal{B}}(q) = 0\}$. A sequence of states of Q is parity-accepting if it contains infinitely many states in F. For the purpose of this chapter we call the states in F *accepting*.

4.1.1 Truncated Runs

We start with technical definitions of approximations of accepting runs of a Büchi automaton.

For $d \geqslant 0$ a *truncated run* (shortly a *t-run*) of *depth* d *from* $q \in Q^{\mathcal{B}}$ is a function $\gamma : \{\mathrm{L}, \mathrm{R}\}^{\leqslant d} \to Q$ that looks like a prefix of a run of $\mathcal{B}$:

- $\gamma(\epsilon) = q$,
- for every $u \in \{\mathrm{L}, \mathrm{R}\}^{<d}$ there exists a transition of $\mathcal{B}$ of the form

$$\delta = \big(\gamma(u), a, \gamma(u\mathrm{L}), \gamma(u\mathrm{R})\big), \text{ for some } a \in A. \tag{4.1}$$

If the state q is not mentioned explicitly, we assume that $q = q_{\mathrm{I}}^{\mathcal{A}}$. For a tree $t \in \mathrm{Tr}_A$ we say that a t-run γ *fits to* t if the letters in (4.1) agree with t (i.e. we can take a transition δ in (4.1) such that $t(u) = a$).

Let γ be a t-run of depth d and $d_0 < d_1 \leqslant d$. We say that γ is *accepting between* d_0 *and* d_1 if for every $w \in \{\mathrm{L}, \mathrm{R}\}^{d_1}$ there exists $u \preceq w$ such that

$$|u| > d_0 \text{ and } \gamma(u) \in F.$$

It means that every path visits an accepting state at a depth between d_0 and d_1, see Fig. 4.1. The same definition applies when γ is a total run.

The following fact is a standard application of König's lemma.

Fact 4.33. *If ρ is an accepting run of a Büchi automaton then for every $d_0 \geqslant 0$ there exists $d_1 > d_0$ such that ρ is accepting between d_0 and d_1.*

A pair $N = (\boldsymbol{d}, \gamma)$ is a *sliced truncated run* (or shortly an *st-run*) from $q \in Q^{\mathcal{B}}$ if:

- $\boldsymbol{d} = (d_0, \ldots, d_k)$ with $k \geqslant 0$,
- $0 = d_0 < d_1 < \ldots < d_k$,
- γ is a truncated run of depth d from q with $d_{k-1} \leqslant d \leqslant d_k$ (if $k = 0$ then we use $d_{-1} = 0$),
- for every $i = 1, 2, \ldots, k-1$ the truncated run γ is accepting between d_{i-1} and d_i.

As before, by default we take $q = q_{\mathrm{I}}^{\mathcal{B}}$. An st-run $N = (\boldsymbol{d}, \gamma)$ *fits to* t if γ fits to t. The *depth of an st-run* $(\boldsymbol{d}, \gamma)$ is the depth of γ. An st-run $N = \big((d_0, \ldots, d_k), \gamma\big)$ is *completed* if the depth of γ is d_k.

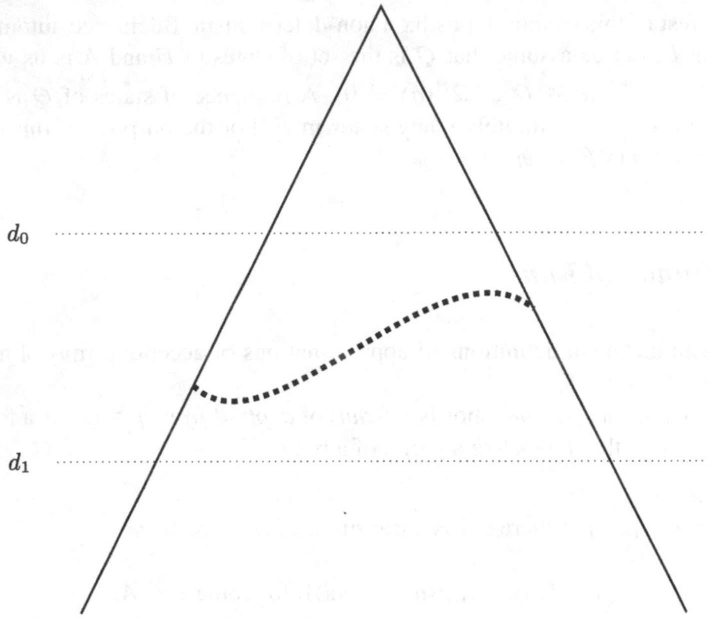

Fig. 4.1 An illustration of a t-run that is accepting between d_0 and d_1: the boldfaced dots mark accepting states that appear on every path between d_0 and d_1.

Let $N = (\boldsymbol{d}, \gamma)$, $N' = (\boldsymbol{d}', \gamma')$ be two st-runs. Assume that the depths of γ, γ' are d, d' respectively. We will define when N' *extends* N (denoted $N \to N'$), there are two cases:

- If N is not completed then we must have $\gamma' \supset \gamma$, $\boldsymbol{d}' = \boldsymbol{d}$, and $d' = d + 1$.
- If N is completed then we must have $\gamma' = \gamma$ and $\boldsymbol{d}' = \boldsymbol{d} \cdot d_{k+1}$ for some $d_{k+1} > d_k$.

Informally, a non-completed st-run can be extended by adding one additional layer to the t-run γ without exceeding the last depth d_k. A completed st-run can be extended by not modifying the t-run γ but declaring a new depth d_{k+1} (in that case the new st-run is not completed).

Fact 4.34. *Let $N_0 = \big((d_0, d_1, \ldots, d_k), \gamma\big)$ be an st-run. Let d be the depth of γ. Then there is no sequence of non-completed st-runs $N_0 \to N_1 \to \ldots \to N_n$ with $n > d_k - d$.*

4.1.2 The Reduction

Now we proceed with a definition of a function, mapping trees $t \in \mathrm{Tr}_A$ to ω-trees $\mathrm{T}(t) \in \omega\mathrm{PTr}$. For the sake of inductive arguments we define one function T_q for each state $q \in Q^{\mathcal{B}}$.

Observe that the set X of all st-runs is countable. Therefore, we can assume that there is a bijection between ω and st-runs: $\omega \ni n \leftrightarrow N^{(n)} \in X$. Assume additionally that $N_q^{(0)} = ((0), \gamma)$ with γ being the unique t-run of depth 0 from q. Modulo the above bijection, a sequence of st-runs $(N_1, N_2, \ldots, N_n)$ can be seen as an element of ω^*. Therefore, we define $\mathtt{T}_q(t) \subseteq \omega^*$ as a set of sequences of st-runs. For a tree $t \in \mathrm{Tr}_A$ let $(N_1, N_2, \ldots, N_n) \in \mathtt{T}_q(t)$ if:

$$N_q^{(0)} \to N_1 \to N_2 \to \ldots \to N_n, \tag{4.2}$$

$$\text{for } i = 1, \ldots, n \text{ the st-run } N_i \text{ fits to } t. \tag{4.3}$$

We define $\mathtt{T}$ as $\mathtt{T}_{q_I^{\mathcal{B}}}$.

Remark 4.2. Assume that N is an st-run from q. Observe that by the definition of $\to$, there is a unique sequence of st-runs $(N_1, N_2, \ldots, N_n)$ satisfying (4.2) with $N_n = N$. This sequence satisfies (4.3) if and only if N fits to t.

In particular, we can identify elements of $\mathtt{T}_q(t)$ with st-runs from q fitting to t. The root of $\mathtt{T}_q(t)$ corresponds to the st-run $N_q^{(0)}$.

Fact 4.35. *The function* $\mathtt{T}_q \colon \mathrm{Tr}_A \to \omega\mathrm{PTr}$ *is continuous.*

Proof. It is enough to observe that for each $\boldsymbol{N} = (N_1, \ldots, N_n)$ the set

$$\left\{t \in \mathrm{Tr}_A : \boldsymbol{N} \in \mathtt{T}_q(t)\right\}$$

is clopen — it depends on the given tree up to the depth of the t-run of N_n. ■

Fact 4.36. *Assume that the vertex of* $\mathtt{T}_q(t)$ *corresponding to an st-run* N *(formally to a sequence* $(N_1, \ldots, N_n = N)$*) is infinitely branching in* $\mathtt{T}_q(t)$*. Then* N *is completed.*

Proof. A non-completed st-run has only finitely many extensions. ■

The following lemma shows that $\mathtt{T}_{q_I^{\mathcal{B}}}$ is a continuous reduction of L^{c} to WF.

Lemma 4.6. *For a tree* $t \in \mathrm{Tr}_A$ *we have*

$$t \in \mathrm{L}(\mathcal{B}) \iff \text{the } \omega\text{-tree } \mathtt{T}_{q_I^{\mathcal{B}}}(t) \text{ is ill-founded (i.e. contains an infinite branch).}$$

Proof. First assume that $t \in \mathrm{L}(\mathcal{B})$. Let ρ be an accepting run of $\mathcal{B}$ on t. Fact 4.33 shows that there is a sequence $0 = d_0 < d_1 < d_2 < \ldots$ such that for every $i > 0$ the pair

$$N_i = \Big((d_0, \ldots, d_i), \rho\restriction_{\{\mathrm{L},\mathrm{R}\}^{\leqslant d_i}}\Big)$$

is a completed st-run that fits to t. The st-runs N_i lay on an infinite branch of $\mathtt{T}_{q_I^{\mathcal{B}}}(t)$.

Now let $N_0 \to N_1 \to \ldots$ be an infinite branch of $\mathtt{T}_{q_I^{\mathcal{B}}}(t)$. Let ρ be the run obtained as the union of the t-runs of these st-runs. By Fact 4.34, this sequence must contain infinitely many completed st-runs. Therefore, ρ is an accepting run of $\mathcal{B}$ on t. ■

4.1.3 Ranks

Now we can define $\eta(\mathcal{B})$ — the ordinal number of main interest in this chapter. Recall that $\mathtt{T}(t)$ stands for $\mathtt{T}_{q_I^{\mathcal{B}}}(t)$. By Lemma 4.6, for every tree $t \notin \mathrm{L}(\mathcal{B})$ the ω-tree $\mathtt{T}(t)$ is well-founded. Let

$$\eta(\mathcal{B}) \stackrel{\text{def}}{=} \sup_{t \notin \mathrm{L}(\mathcal{B})} \mathrm{rank}\big(\mathtt{T}(t)\big). \tag{4.4}$$

The relation between the complexity of $\mathrm{L}(\mathcal{B})$ and $\eta(\mathcal{B})$ is expressed by Propositions 4.2 and 4.3.

Proposition 4.2. *The language* $\mathrm{L}(\mathcal{B})$ *is Borel if and only if* $\eta(\mathcal{B}) < \omega_1$.

Proof. If $\mathrm{L}(\mathcal{B})$ is Borel then

$$\big\{\mathtt{T}(t) : t \notin \mathrm{L}(\mathcal{B})\big\} \subseteq \mathrm{WF} \tag{4.5}$$

is a continuous image of a Borel set, thus an analytic ($\mathbf{\Sigma}_1^1$) set. Therefore, by the boundedness theorem (see Sect. 1.6.4 and Theorem 1.12 on page 17) we have $\eta(\mathcal{B}) < \omega_1$.

Now assume that $\eta(\mathcal{B}) < \omega_1$. Theorem 1.12 implies that the set

$$T_B \stackrel{\text{def}}{=} \{\tau \in \omega\mathrm{Tr} : \mathrm{rank}(\tau) \leqslant \eta(\mathcal{B})\}$$

is Borel. But $\mathrm{Tr}_A \setminus \mathrm{L}(\mathcal{B})$ is the preimage of T_B under the continuous function $\mathtt{T}$, therefore also Borel. ∎

The following proposition constitutes the crucial idea behind the effective characterisation from Theorem 2.2.

Proposition 4.3. *The language* $\mathrm{L}(\mathcal{B})$ *is* WMSO*-definable if and only if* $\eta(\mathcal{B}) < \omega^2$ *(i.e. if there exists* $K \in \omega$ *such that* $\eta(\mathcal{B}) < \omega \cdot K$*).*

The proof of this proposition consists of two lemmas: Lemma 4.7 proved here and Lemma 4.10 from Sect. 4.3.

Lemma 4.7. *If* $\mathrm{L}(\mathcal{B})$ *is* WMSO*-definable then* $\eta(\mathcal{B}) < \omega^2$.

The rest of this section is devoted to proving this lemma. Apart from some technicalities, the reasoning is based on Rabin's pumping lemma from [Rab70].

Assume that $\mathrm{L}(\mathcal{B})$ is WMSO-definable and let $\mathcal{A}$ be a non-deterministic Büchi automaton recognising the complement of $\mathrm{L}(\mathcal{B})$. Let $K = |Q^{\mathcal{A}}| \cdot |Q^{\mathcal{B}}| \cdot |A| + 2$. To arrive to a contradiction assume that $\eta(\mathcal{B}) \geqslant \omega^2$ and let $t \notin \mathrm{L}(\mathcal{B})$ be a tree such that

$$\mathrm{rank}\big(\mathtt{T}(t)\big) \geqslant \omega \cdot K.$$

Since $t \notin \mathrm{L}(\mathcal{B})$ so there exists an accepting run $\rho^{\mathcal{A}}$ of $\mathcal{A}$ on t. Our aim is to construct a t-run γ of $\mathcal{B}$ on t and a sequence of numbers $0 = d_0 < d_1 < \ldots < d_{K-1}$ such that:

$$\text{For every } i < K-1 \text{ both } \gamma \text{ and } \rho^{\mathcal{A}} \text{ are accepting between } d_i \text{ and } d_{i+1}. \qquad (4.6)$$

This will enable us to construct a regular tree t' with accepting runs of both automata $\mathcal{A}$ and $\mathcal{B}$ (see [Rab70]) leading to a contradiction.

Recall that by Remark 4.2 we identify elements (nodes) of $\mathtt{T}(t)$ with st-runs from $q_{\mathrm{I}}{}^{\mathcal{B}}$ fitting to t. The construction is inductive for $i = 1, \ldots, K-1$. The invariant is that N_i is a completed st-run of depth d_i and

$$\mathrm{rank}\big(\mathtt{T}(t)\restriction_{N_i}\big) = \omega \cdot (K - i).$$

Observe that Fact 1.9 from page 17 implies that if $\mathrm{rank}\big(\mathtt{T}(t)\restriction_N\big)$ is a limit ordinal then N is infinitely branching in $\mathtt{T}(t)$. Therefore by Fact 4.36, N is a completed st-run.

We start by fixing N_1 as any node of $\mathtt{T}(t)$ of rank $\omega \cdot (K-1)$ (it exists by Fact 1.10 on page 17) and let d_1 be the depth of N_1.

Assume that a completed st-run $N_{i-1} = (\boldsymbol{d}, \gamma)$ of depth d_{i-1} is defined. Let d' be the depth given by Fact 4.33 such that $\rho^{\mathcal{A}}$ is accepting between d_{i-1} and d'.

Observe that all the st-runs N' in $\mathtt{T}(t)$ such that $N_{i-1} \to N'$ are of the form $(\boldsymbol{d} \cdot d'', \gamma)$ for some d''. In particular, only finitely many of them satisfy $d'' < d'$. Since $\mathrm{rank}\big(\mathtt{T}(t)\restriction_{N_{i-1}}\big) = \omega \cdot (K - i + 1)$ is a limit ordinal, we can find an st-run N' in $\mathtt{T}(t)$ such that:

- $N_{i-1} \to N'$,
- $N' = (\boldsymbol{d} \cdot d_i, \gamma')$ for $d_i \geqslant d'$, and
- $\mathrm{rank}\big(\mathtt{T}(t)\restriction_{N'}\big) \geqslant \omega \cdot (K - i)$.

Now, we use again Fact 1.10 from page 17 to find N_i in $\mathtt{T}(t)$ below N' and satisfying

$$\mathrm{rank}\big(\mathtt{T}(t)\restriction_{N_i}\big) = \omega \cdot (K - i).$$

Now let γ be the t-run of N_{K-1}. Condition 4.6 is clearly satisfied by the construction.

Now it remains to prove the following fact.

Fact 4.37. *There exists a tree $t' \in \mathrm{L}(\mathcal{A}) \cap \mathrm{L}(\mathcal{B})$.*

Proof. We only sketch a proof of this fact, a complete construction is given in [Rab70]. See also [KV99, Theorem 1] for a definition of a *trap* — the sequence $d_0 < d_1 < \ldots < d_{K-1}$ constructed above is a trap for the runs γ and $\rho^{\mathcal{A}}$.

The tree t' (together with the runs of $\mathcal{A}$ and $\mathcal{B}$) is obtained as an unravelling of a finite graph constructed using t. Consider $i \in \{1, \ldots, K-1\}$ and a node $w \in \{\mathtt{L}, \mathtt{R}\}^{d_i}$. If there exists i' such that $0 < i' < i$ and for $u \stackrel{\text{def}}{=} w\restriction_{d_{i'}}$ we have

$$\big(t(u), \gamma(u), \rho^{\mathcal{A}}(u)\big) = \big(t(w), \gamma(w), \rho^{\mathcal{A}}(w)\big)$$

then (for the minimal such i') we remove the edge from the parent of w to w and instead we add an edge from the parent of w to u (preserving the direction $d \in \{\mathtt{L}, \mathtt{R}\}$). In that case we say that w has been *rewired* to u.

Since K is big enough, for every $w \in \{\mathrm{L}, \mathrm{R}\}^{d_{K-1}}$ at least one of the prefixes of w has been rewired. Therefore, none of such vertices w is accessible from ϵ via the edge relation. Let t' be the unravelling of the constructed graph. Clearly, γ and $\rho^{\mathcal{A}}$ are runs of $\mathcal{B}$ and $\mathcal{A}$ on t'. Since both runs are accepting between d_i and d_{i+1} for every i, the respective runs on t' are accepting. ∎

This concludes the proof of Lemma 4.7 finishing the "only if" implication in Proposition 4.3. The "if" implication will be proved in Lemma 4.10 in Sect. 4.3.

4.2 Extending Runs

We now give a more explicit definition expressing the fact that $\eta(\mathcal{B}) \geqslant \omega^2$. It will serve as an intermediate object in a proof of Lemma 4.10. For $K \in \omega$ we will define notions of *K-safe* and *K-reach* trees.

The definitions are designed in such a way to correspond precisely to languages recognised by the alternating automata defined in Sect. 4.3. Because of that, we cannot require here to have exact truncated runs as in Sect. 4.1.1. Therefore, we use a notion of a *partial run* defined as a non-empty finite partial tree $\bar{\rho} \in \mathrm{PTr}_Q$ such that every node $u \in \mathrm{dom}(\bar{\rho})$ is either a leaf of $\bar{\rho}$ or $u\mathrm{L}, u\mathrm{R} \in \mathrm{dom}(\bar{\rho})$ and for some $a \in A$

$$\big(\bar{\rho}(u), a, \bar{\rho}(u\mathrm{L}), \bar{\rho}(u\mathrm{R})\big) \text{ is a transition of } \mathcal{B}.$$

We additionally require that ϵ is not a leaf of $\bar{\rho}$.

A partial run $\bar{\rho}$ is *accepting* if for every leaf $u \in \mathrm{dom}(\bar{\rho})$ of $\bar{\rho}$ we have $\bar{\rho}(u) \in F$ — all the states in the leaves of $\bar{\rho}$ are accepting. A partial run $\bar{\rho}$ is *minimal accepting* if it is accepting and minimal (w.r.t. $\subseteq$) partial tree satisfying the above conditions — $\bar{\rho}$ has a leaf in the first accepting state seen along every branch. This technical assumption will allow us to easily prove Proposition 4.5.

As for t-runs, we say that a partial run $\bar{\rho}$ is *from the state* $\bar{\rho}(\epsilon)$ and it *fits a tree* t if the transitions used in $\bar{\rho}$ use letters of t.

Take a state $q \in Q^{\mathcal{B}}$ and a tree $t \in \mathrm{Tr}_A$. We say that:

- q is always 0-*reach* and 0-*safe* in t.
- q is $(K+1)$-*safe in* t if there exists a total run ρ of $\mathcal{B}$ on t such that $\rho(\epsilon) = q$ and for every $u \in \mathrm{dom}(t)$

$$\rho(u) \text{ is } (K+1)\text{-reach in } t{\restriction}_u.$$

- q is $(K+1)$-*reach in* t if there exists a partial run $\bar{\rho}$ from q such that $\bar{\rho}$ fits t, $\bar{\rho}$ is minimal accepting, and for every leaf u of $\bar{\rho}$ we have

$$\bar{\rho}(u) \text{ is } K\text{-safe in } t{\restriction}_u.$$

In particular, if q is 1-safe in t then there exists a total run ρ of $\mathcal{B}$ on t with $\rho(\epsilon) = q$. In general, the following fact holds.

Fact 4.38. *Assume that $q \in Q^{\mathcal{B}}$ is $(K+1)$-reach in $t \in \mathrm{Tr}_A$. Then, we can find a total run ρ of $\mathcal{B}$ on t and a depth d such that:*

- *$\rho(\epsilon) = q$,*
- *ρ is accepting between 0 and d,*
- *for every $w \in \{\mathrm{L}, \mathrm{R}\}^*$ of length at least d we have*

$$\rho(w) \text{ is } K\text{-reach in } t{\restriction_w}.$$

Directly from the definition, we obtain the following monotonicity property.

Fact 4.39. *Let $K' \geqslant K \geqslant 0$. If q is K'-safe in t then q is K-safe in t. If q is K'-reach in t then q is K-reach in t.*

Also, if q is $(K+1)$-reach in t then q is K-safe in t and if q is $(K+1)$-safe in t then q is $(K+1)$-reach in t.

Proposition 4.4. *The following conditions are equivalent:*

1. *for every K there exists a tree $t \notin \mathrm{L}(\mathcal{B})$ such that $q_{\mathrm{I}}^{\mathcal{B}}$ is K-reach in t,*
2. *$\eta(\mathcal{B}) \geqslant \omega^2$.*

The proof of this proposition is split across the following two subsections. The following remark follows easily from the definition of K-safe, however we will not prove it directly, instead we will use automata defined in Sect. 4.3. It implies, together with the above proposition, that if $\eta(\mathcal{B}) < \omega^2$ then the language $\mathrm{L}(\mathcal{B})$ is WMSO-definable.

Remark 4.3. For every K there exists a WMSO formula φ_K such that

$$\mathrm{L}(\varphi_K) = \{t : q_{\mathrm{I}}^{\mathcal{B}} \text{ is } K\text{-reach in } t\}.$$

4.2.1 K-reach Implies Big Rank

In this subsection we prove one of the estimations needed for Proposition 4.4: if for every K there is a tree $t \notin \mathrm{L}(\mathcal{B})$ such that ${q_{\mathrm{I}}}^{\mathcal{B}}$ is K-reach in t then $\eta(\mathcal{B}) \geqslant \omega^2$. The proof goes by induction, as expressed by the following lemma.

Lemma 4.8. *Let $N = (d, \gamma)$ be a completed st-run of depth d from q. Assume that N fits to a tree $t \in \mathrm{Tr}_A$ and for every $u \in \{\mathrm{L}, \mathrm{R}\}^d$ we know that $\gamma(u)$ is K-reach in $t{\restriction_u}$. Then*

$$\mathrm{rank}\big(\mathrm{T}_q(t){\restriction_N}\big) \geqslant \omega \cdot K.$$

Observe that by putting $N = N_q^{(0)}$ (the unique st-run of depth 0) above, we obtain that if q is K-reach in t then $\mathrm{rank}\big(\mathtt{T}_q(t)\big) \geqslant \omega \cdot K$.

Proof. The proof is inductive in K. For $K = 0$ the thesis holds. Assume that the thesis holds for $K \geqslant 0$ and every $q \in Q^{\mathcal{B}}$, $t \in \mathrm{Tr}_A$. Take an st-run N as in the statement and assume that for every $u \in \{\mathtt{L}, \mathtt{R}\}^d$ we know that $\gamma(u)$ is $(K+1)$-reach in $t{\restriction_u}$.

For every $u \in \{\mathtt{L}, \mathtt{R}\}^d$ we can apply Fact 4.38 to $q = \gamma(u)$ and $t = t{\restriction_u}$ obtaining a total run ρ^u of $\mathcal{B}$ on $t{\restriction_u}$ with $\rho^u(\epsilon) = \gamma(u)$ and a depth d^u. Let us put:

$$d' = \max_{u \in \{\mathtt{L},\mathtt{R}\}^d} d^u, \quad \rho = \gamma\Big[u \leftarrow \rho^u\Big]_{u \in \{\mathtt{L},\mathtt{R}\}^d}.$$

By the construction in Fact 4.38 we know that:

- ρ is a total run of $\mathcal{B}$ on t and $\gamma \subseteq \rho$,
- ρ is accepting between d and d',
- for every u of length at least d' we know that $\rho(u)$ is K-reach in $t{\restriction_u}$.

Now take any $d_1 > d'$ and consider the st-node

$$N' \stackrel{\text{def}}{=} \big(\boldsymbol{d} \cdot d_1, \rho{\restriction_{\{\mathtt{L},\mathtt{R}\}^{\leqslant d_1}}}\big).$$

Clearly N' is a completed st-run of depth d_1 from q that fits to t and

$$N \rightarrow^{(d_1-d)} N' \text{ (i.e. } N' \text{ can be obtained by extending } N\ (d_1-d)\text{-times).}$$

Observe that N' satisfies the inductive assumption for K, so

$$\mathrm{rank}\big(\mathtt{T}_q(t){\restriction_{N'}}\big) \geqslant \omega \cdot K.$$

By considering bigger and bigger values of d_1, we can find arbitrarily long paths in $\mathtt{T}_q(t){\restriction_N}$ that lead to vertices of rank at least $\omega \cdot K$. Therefore

$$\mathrm{rank}\big(\mathtt{T}_q(t){\restriction_N}\big) \geqslant \omega \cdot (K+1).$$

■

4.2.2 Big Rank Implies K-reach

Now we prove the opposite estimation from Proposition 4.4: if $\eta(\mathcal{B}) \geqslant \omega^2$ then for every K there exists a tree $t \notin \mathrm{L}(\mathcal{B})$ such that $q_{\mathrm{I}}{}^{\mathcal{B}}$ is K-reach in t. This statement follows from the following lemma.

Lemma 4.9. *Let $N = (d, \gamma)$ be a completed st-run of depth d from q. Assume additionally that N fits to a tree $t \in \mathrm{Tr}_A$.*

1. *If*
$$\mathrm{rank}\big(\mathtt{T}_q(t){\restriction_N}\big) \geqslant \omega \cdot \big(1 + 2 \cdot K\big)$$
then for every $u \in \{\mathtt{L}, \mathtt{R}\}^{\leqslant d}$ (i.e. $u \in \mathrm{dom}(\gamma)$) the state $\gamma(u)$ is K-safe in $t{\restriction_u}$.
2. *If*
$$\mathrm{rank}\big(\mathtt{T}_q(t){\restriction_N}\big) \geqslant \omega \cdot \big(2 \cdot K\big)$$
then for every $u \in \{\mathtt{L}, \mathtt{R}\}^{\leqslant d}$ (i.e. $u \in \mathrm{dom}(\gamma)$) the state $\gamma(u)$ is K-reach in $t{\restriction_u}$.

As before, by putting $N = N_q^{(0)}$ (the unique st-run of depth 0) above, we obtain that if $\mathrm{rank}\big(\mathtt{T}_q(t)\big) \geqslant \omega \cdot \big(2 \cdot K\big)$ then q is K-reach in t.

The rest of this subsection is devoted to proving this lemma. We start with the following observation.

Fact 4.40. *Assume that $N \in \mathtt{T}_q(t)$ is a completed st-run of depth d and* $\mathrm{rank}\big(\mathtt{T}_q(t){\restriction_N}\big) \geqslant \omega \cdot (K+1)$. *Then for every $d' \geqslant d$ there exists a completed st-run $N' \in \mathtt{T}_q(t){\restriction_N}$ of depth at least d' and such that* $\mathrm{rank}\big(\mathtt{T}_q(t){\restriction_{N'}}\big) \geqslant \omega \cdot K$.

Proof. Let $\tau = \mathtt{T}_q(t){\restriction_N}$. Since $\mathrm{rank}(\tau) \geqslant \omega \cdot (K+1)$, there are arbitrarily long paths in τ that lead to vertices of rank at least $\omega \cdot K$. By Fact 1.10 from page 17, under every such vertex there is a vertex $N' \in \tau$ of rank exactly $\omega \cdot K$. Facts 1.9 and 4.36 imply that N' must be completed in that case. Since the path from N to N' is arbitrary long, so is the depth of N'. ■

Now we can prove our lemma, the proof is inductive on K, for $K = 0$ both parts of the thesis are trivial. Assume that both parts of the thesis hold for K and consider a completed st-run N as in the statement.

Item (1) First assume that
$$\mathrm{rank}\big(\mathtt{T}_q(t){\restriction_N}\big) \geqslant \omega \cdot \big(1 + 2 \cdot K\big).$$
Our aim is to prove that for every $u \in \{\mathtt{L}, \mathtt{R}\}^{\leqslant d}$ the state $\gamma(u)$ is K-safe in $t{\restriction_u}$.

Let $(N_i')_{i\in\mathbb{N}}$ be a sequence of completed st-runs of unbounded depths that are given by Fact 4.40. Let γ_i' be the t-run of N_i'. By compactness, there exists a subsequence of $(\gamma_i')_{i\in\mathbb{N}}$ that is point-wise convergent to a total run ρ. Let us restrict the sequences $(N_i')_{i\in\mathbb{N}}, (\gamma_i')_{i\in\mathbb{N}}$ to this convergent sub-sequence (we do not require N_i' to be convergent in any sense). Clearly, $\gamma \subseteq \rho$ and for every $u \in \{\mathtt{L}, \mathtt{R}\}^*$ there is some i such that $\rho(u) = \gamma_i'(u)$.

What remains to prove is that for every $u \in \{\mathtt{L}, \mathtt{R}\}^*$ the state $\rho(u)$ is K-reach in $t{\restriction_u}$. Take such u and consider i such that $\rho(u) = \gamma_i'(u)$. By the construction of N_i' we know that
$$\mathrm{rank}\big(\mathtt{T}_q(t){\restriction_{N_i'}}\big) \geqslant \omega \cdot \big(2 \cdot K\big).$$
Therefore, $\rho(u)$ is K-reach in $t{\restriction_u}$ because of Item (2) of our lemma.

Item (2) Now assume that

$$\mathrm{rank}\big(\mathrm{T}_q(t)\restriction_N\big) \geqslant \omega \cdot \big(2 \cdot (K+1)\big)$$

and take $u \in \{\mathrm{L}, \mathrm{R}\}^{\leqslant d}$. Our aim is to prove that the state $\gamma(u)$ is $(K+1)$-reach in $t\restriction_u$.

By applying Fact 4.40 to N and any depth greater than d we obtain a completed run N' such that $N' \in \mathrm{T}_q(t)\restriction_N$ and

$$\mathrm{rank}\big(\mathrm{T}_q(t)\restriction_{N'}\big) \geqslant \omega \cdot \big(1 + 2 \cdot K\big). \tag{4.7}$$

Let γ' be the t-run of N' and d' be the depth of γ'. Since both N and N' are completed, γ' is accepting between d and d'

Let $\bar{\rho}$ be the restriction of $\gamma'\restriction_u$ to its initial fragment before the first accepting state:

$$\begin{aligned}\mathrm{dom}(\bar{\rho}) \stackrel{\text{def}}{=} \big\{w : \; & uw \in \mathrm{dom}(\gamma'),\\ & \gamma'(uw) \in F, \text{ and}\\ & \text{for every } \epsilon \prec w' \prec w \text{ we have } \gamma'(uw') \notin F\big\}.\end{aligned}$$

By the definition $\bar{\rho}$ is a partial run and $\bar{\rho}$ is minimal accepting. Observe that by (4.7) and the inductive assumption, for every w that is a leaf of $\bar{\rho}$ we know that the state $\bar{\rho}(w)$ is K-safe in $t\restriction_{uw}$. Therefore, $\bar{\rho}$ is a witness that $\gamma(u)$ is $(K+1)$-reach in $t\restriction_u$.

This concludes the proof of Lemma 4.9 and therefore the proof of Proposition 4.4.

4.3 Automata for K-safe Trees

In this section we define a sequence of automata (defined in a uniform way) that recognise languages of K-safe trees, as expressed in Proposition 4.5 below. The primary goal of this construction will be a proof of Lemma 4.10 (i.e. that if $\eta(\mathcal{B}) < \omega^2$ then $\mathrm{L}(\mathcal{B})$ is WMSO-definable) which completes the proof of Proposition 4.3. Furthermore, in Sect. 4.4 we will define a game based on the automata constructed here; the aim of this game will be verifying if $\eta(\mathcal{B}) < \omega^2$.

The automata constructed here are very similar to the counter automaton $\mathcal{B}$ defined in [CKLV13, Sect. 4.3], however both notions were developed independently basing on the idea of *traps* in [KV99].

Lemma 4.10. *If for some $K \in \omega$ and every $t \notin \mathrm{L}(\mathcal{B})$ we have $\mathrm{rank}\big(\mathrm{T}(t)\big) < \omega \cdot K$ then $\mathrm{L}(\mathcal{B})$ is WMSO-definable.*

We take a number $K \geqslant 0$ and construct an automaton $\mathcal{C}[K]$ over the alphabet A. Let the states of $\mathcal{C}[K]$ be $Q^{\mathcal{B}} \times \{\text{safe}, \text{reach}\} \times \{0, 1, \ldots, K\}$. The initial state is $(q_I^{\mathcal{B}}, \text{safe}, K)$. Let the states of the form (q, safe, i) have priority 0 and the other states have priority 1.

Let us define the transitions of the automaton $\mathcal{C}[K]$. All the states of the form $(q, \text{safe}, 0)$ and $(q, \text{reach}, 0)$ have only trivial transition $\top$ — they accept everything. First we give a formal definition of the form of the transitions, then we explain it informally. Assume that the current state is of the form (q, z, i) with $z \in \{\text{safe}, \text{reach}\}$ and $i > 0$; and a letter a is given. The transition of $\mathcal{C}[K]$ consists of the following choices of the players:

$\forall$ chooses an element $z' \in \{z, \text{reach}\}$ (if $z = \text{reach}$ then $\forall$ has no choice here)
$\exists$ chooses a transition $\delta = (q, a, q_{\mathtt{L}}, q_{\mathtt{R}})$ of $\mathcal{B}$
$\forall$ chooses a direction $d \in \{\mathtt{L}, \mathtt{R}\}$

When these choices are done, the automaton $\mathcal{C}[K]$ moves in direction d to the successive state defined according to the following cases:

- $z' = \text{safe}$ then the successive state is (q_d, safe, i),
- $z' = \text{reach}$ and $q_d \notin F$ then the successive state is (q_d, reach, i),
- $z' = \text{reach}$ and $q_d \in F$ then the successive state is $(q_d, \text{safe}, i - 1)$.

Informally, from each state (q, safe, i) the player $\forall$ can request to jump to the state (q, reach, i) without moving in the tree. Assume that he made his choice and the state of $\mathcal{C}[K]$ is (q, z', i). Now $\exists$ declares a transition δ and $\forall$ picks a direction d. If $z' = \text{safe}$ then they just continue in the state (q_d, safe, i). If $z' = \text{reach}$ then $\mathcal{C}[K]$ waits for an accepting state. If q_d is accepting then $\mathcal{C}[K]$ moves to $(q_d, \text{safe}, i - 1)$, otherwise $\mathcal{C}[K]$ stays in (q_d, safe, i).

By the definition of the transitions of $\mathcal{C}[K]$ we obtain the following fact.

Fact 4.41. *For every $K \geqslant 0$ the automaton $\mathcal{C}[K]$ is* Comp(0, 0).

The following proposition expresses a relation between the notions of K-safe trees and acceptance by the automata $\mathcal{C}[K]$.

Proposition 4.5. *For $K \geqslant 0$ and a tree $t \in \mathrm{Tr}_A$:*

$$t \in \mathrm{L}\big(\mathcal{C}[K], (q, \text{safe}, i)\big) \iff q \text{ is } i\text{-safe in } t,$$
$$t \in \mathrm{L}\big(\mathcal{C}[K], (q, \text{reach}, i)\big) \iff q \text{ is } i\text{-reach in } t.$$

Proof. The proof is inductive in i. For the induction step it is enough to observe that there is a 1–1 correspondence between winning strategies of $\exists$ in the component $Q^{\mathcal{B}} \times \{\text{safe}\} \times \{i\}$ of $\mathcal{C}[K]$ and runs ρ witnessing the i-safety (similarly for the component $Q^{\mathcal{B}} \times \{\text{reach}\} \times \{i\}$ and partial runs $\bar{\rho}$ witnessing i-reachability). ■

Now, all the properties of the ordinal $\eta(\mathcal{B})$ from Sect. 4.1 have been proved. What remains in the following sections is to give an effective procedure deciding if $\eta(\mathcal{B}) < \omega^2$.

4.4 Boundedness Game

In this section we construct a finite game $\mathcal{G}$ with an ω-regular winning condition that satisfies the following proposition.

Proposition 4.6. *The following conditions are equivalent:*

1. $\exists$ *has a winning strategy in* $\mathcal{G}$,
2. $\eta(\mathcal{B}) \geqslant \omega^2$.

Since the winner of $\mathcal{G}$ can be effectively computed (see Theorem 1.21 on page 21), Theorem 2.2 will follow from Proposition 4.3. The game $\mathcal{G}$ is highly motivated by *domination games* from [Col13], however the construction presented here does not depend on any external results about cost functions.

In this section we construct the game $\mathcal{G}$, a proof of Proposition 4.6 is given in Sect. 4.5.

Let us fix a non-deterministic tree automaton $\mathcal{A}$ recognising the complement of $\mathrm{L}(\mathcal{B})$ ($\mathcal{A}$ can have arbitrary index). We will construct $\mathcal{G}$ from $\mathcal{A}$ and $\mathcal{B}$. Intuitively, $\mathcal{G}$ will require the following declarations from the players:

- $\exists$ will be constructing a tree t and a run $\rho^{\mathcal{A}}$ of $\mathcal{A}$ on t,
- $\forall$ will be selecting successive directions constructing an infinite branch α of t, aiming to show that the run $\rho^{\mathcal{A}}$ proposed by $\exists$ is not accepting,
- at the same time both players will simulate (in the *history-deterministic* way in the sense of [Col13]) the game $\mathcal{G}(\mathcal{C}[K], t)$ for an "unknown but big" K.

The set of positions of $\mathcal{G}$ is

$$V \stackrel{\text{def}}{=} \mathsf{P}\left(Q^{\mathcal{B}} \times \{\text{safe}, \text{reach}\}\right) \times Q^{\mathcal{A}} \times \{0, 1, 2, 3\}.$$

A position $(S, p, r) \in V$ of $\mathcal{G}$ consists of a set $S \subseteq Q^{\mathcal{B}} \times \{\text{safe}, \text{reach}\}$ of *active states*, a state $p \in Q^{\mathcal{A}}$, and a *sub-round number* $r \in \{0, \ldots, 3\}$.

The initial position of $\mathcal{G}$ is $(\{(q_{\mathrm{I}}^{\mathcal{B}}, \text{safe})\}, q_{\mathrm{I}}^{\mathcal{A}}, 0)$.

The edges of $\mathcal{G}$ will have an additional structure (i.e. an edge will be more than just a pair of positions $(v, v') \in V \times V$). This richer structure will be used to define the winning condition of $\mathcal{G}$ that will refer to a sequence of edges. From our definition it will be easy to see how to transform such a game into a standard two player game in the sense of Sect. 1.3 (see page 6). To underline that edges have additional structure we refer to them as *multi-transitions*.

A *multi-transition* μ from $(S, p, r) \in V$ to $(S', p', r') \in V$ contains:

- the *pre-state* (S, p, r),
- the *post-state* (S', p', r') with $r' = r + 1 \pmod 4$,
- a set $e \subseteq S \times S'$ of *edges* between the active states S and S',
- a set $\bar{e} \subseteq e$ of *boldfaced edges*, satisfying

$$\text{for every } s' \in S' \text{ exactly one edge to } s' \text{ is boldfaced (i.e. } |\{s : (s, s') \in \bar{e}\}| = 1). \tag{4.8}$$

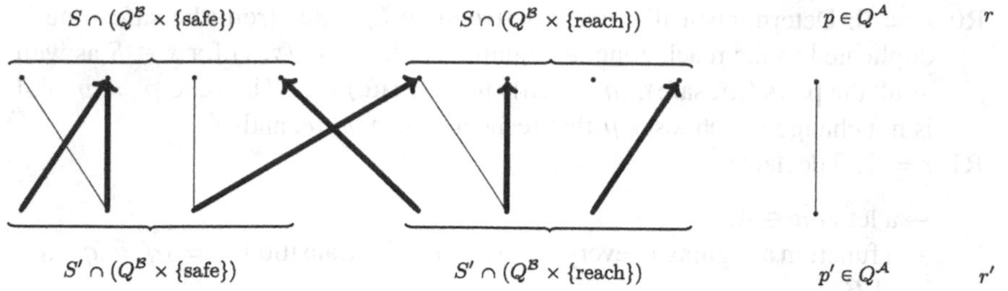

Fig. 4.2 An example of a multi-transition μ.

Observe that by the definition, there is only finitely many multi-transitions. The exact rules how the multi-transitions are selected by the players are given in Sect. 4.4.1.

An active state (q, safe) is said to be *in the* safe *zone* and an active state (q, reach) is said to be *in the* reach *zone*. We say that a pair $(s, s') \in e$ with $s = (q, z)$ and $s' = (q', z')$ *changes zone* if $z \neq z'$, it *changes zone from* safe *to* reach if $z = \text{safe}$ and $z' = \text{reach}$, it *changes zone from* reach *to* safe if $z = \text{reach}$ and $z' = \text{safe}$.

An example multi-transition is depicted on Fig. 4.2. The convention is that all the active states from the safe zone are drawn on the left, then all the active states from the reach zone are drawn in the middle, and finally the state of $\mathcal{A}$ and the sub-round number are drawn on the right. For the purpose of layout, we additionally draw an edge between the states p and p' of $\mathcal{A}$ (this edge does not belong to e). Boldfaced edges are boldfaced.

4.4.1 Rules of the Game

In this section we describe the rules for choosing multi-transitions in $\mathcal{G}$. A multi-transition from a position $(S, p, r) \in V$ will be constructed by first selecting a set of edges $e \subseteq S \times (Q^{\mathcal{B}} \times \{\text{safe}, \text{reach}\})$ and $p' \in Q^{\mathcal{A}}$ according to the rules given below; and then by allowing $\forall$ to choose any multi-transition μ that *respects* (S, p, r), e, *and* p' in the following sense:

- the pre-state of μ is (S, p, r),
- the post-state of μ is (S', p', r') with $S' = \{s' : (s, s') \in e\}$ and $r' = r + 1 \pmod 4$,
- the edges of μ are e,
- the boldfaced edges $\bar{e}$ of μ are chosen arbitrarily by $\forall$ according to Condition 4.8.

That is, the only freedom $\forall$ has when selecting a multi-transition that respects (S, p, r), e, and p' is when choosing the boldfaced edges $\bar{e}$.

Assume that the current position in $\mathcal{G}$ is (S, p, r) and consider the following cases for the number of sub-round r. In all the cases players construct a multi-transition μ that leads to a post-state (S', p', r'):

R0 $r = 0$: Deterministically, every active state (q, safe) from the safe zone is duplicated to the reach zone: e contains all the pairs (s, s) for $s \in S$ as well as all the pairs $((q, \text{safe}), (q, \text{reach}))$ for $(q, \text{safe}) \in S$. The state $p' = p$ of $\mathcal{A}$ is not changed. $\forall$ chooses μ that respects (S, p, r), e, and p'.

R1 $r = 1$: $\exists$ declares:

- a letter $a \in A$,
- a function assigning to every $s = (q, z) \in S$ a transition $\delta_s = (q, a, q^s_{\text{L}}, q^s_{\text{R}})$ of $\mathcal{B}$,
- a transition $\delta = (p, a, p'_{\text{L}}, p'_{\text{R}})$ of $\mathcal{A}$.

If $\exists$ is unable to do such a declaration, she loses.
$\forall$ responds by selecting a direction $d \in \{\text{L}, \text{R}\}$. Then $p' = p'_d$ and e contains all the pairs of the form $((q, z), (q^s_d, z))$ for $s = (q, z) \in S$. $\forall$ chooses μ that respects (S, p, r), e, and p'.

R2 $r = 2$: Deterministically, every active state (q, reach) in the reach zone with $q \in F$ is moved to the safe zone. Formally, e contains:

- all the pairs $((q, \text{safe}), (q, \text{safe}))$ for $(q, \text{safe}) \in S$,
- all the pairs $((q, \text{reach}), (q, \text{reach}))$ for $(q, \text{reach}) \in S$ and $q \notin F$,
- all the pairs $((q, \text{reach}), (q, \text{safe}))$ for $(q, \text{reach}) \in S$ and $q \in F$.

The state $p' = p$ of $\mathcal{A}$ is not changed. $\forall$ chooses μ that respects (S, p, r), e, and p'.

R3 $r = 3$: $\forall$ may remove some active states in S by selecting $e \subseteq \{(s, s) : s \in S\}$. The state $p' = p$ of $\mathcal{A}$ is not changed. $\forall$ chooses μ that respects (S, p, r), e, and p'.

Figure 4.3 presents a round of $\mathcal{G}$ (i.e. four consecutive sub-rounds with $r = 0, 1, 2, 3$).

By the definition of the sub-rounds of the game, we obtain the following fact.

Fact 4.42. *Let μ be a multi-transition constructed in the game $\mathcal{G}$ and $s = (q, z) \in S$ be an active state in the pre-state (S, p, r) of μ. Then one of the following cases holds:*

- $z = \text{safe}$ *and there is precisely one q' such that* $(s, (q', \text{safe})) \in e$,
- $z = \text{reach}$ *and $q \notin F$ and there is precisely one q' such that* $(s, (q', \text{reach})) \in e$,
- *in R2 if* $z = \text{reach}$ *and $q \in F$ then there is no q' such that* $(s, (q', \text{reach})) \in e$,
- *there is no s' such that $(s, s') \in e$ (it may happen only in R3 if $\forall$ removes s).*

The state q' in the first two cases above is called the μ-successor of (q, z). Similarly, for a sequence of multi-transitions $\mu_0, \ldots, \mu_k$ we have the notion of $(\mu_0, \ldots, \mu_k)$-successor. Note that a priori the μ-successors of (q, safe) and (q, reach) may be distinct. For an element $s' \in S'$, the unique s such that $(s, s') \in \bar{e}$ is called the *μ-predecessor* of s'.

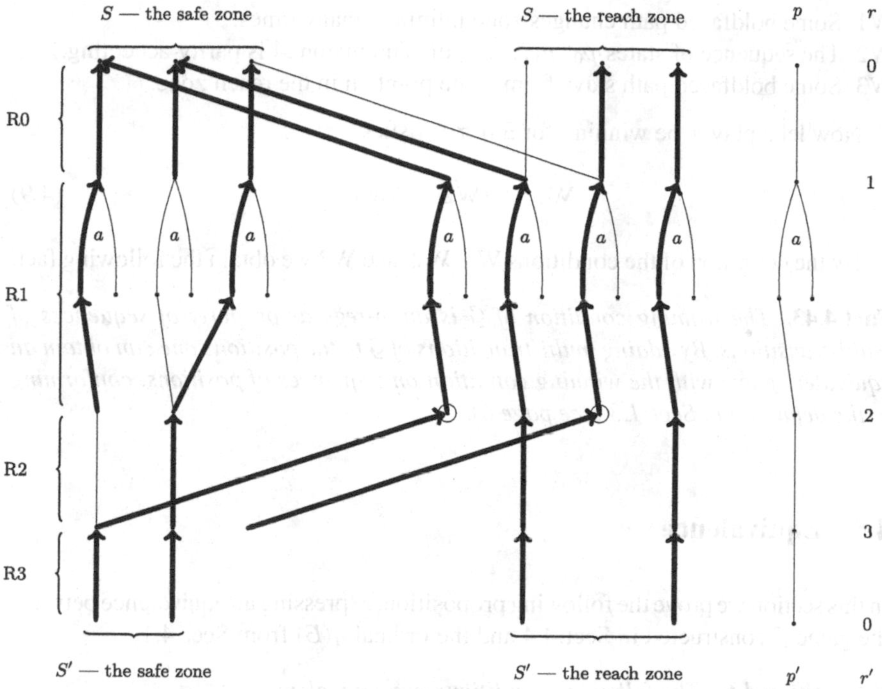

Fig. 4.3 An example round of the game $\mathcal{G}$ consisting of the four sub-rounds. The nodes in circles correspond to accepting states. At sub-round R3 ∀ decides to remove one active state from the safe zone.

4.4.2 Winning Condition

Now we will define the winning condition for ∃ in $\mathcal{G}$. Recall that it will refer to the sequence of multi-transitions on the play.

Let $\pi = \mu_0\mu_1\ldots$ be the infinite sequence of multi-transitions that were played in $\mathcal{G}$. We will refer to the pre-state of μ_n as (S_n, p_n, r_n). Analogously, we will use (S'_n, p'_n, r_n) for the post-state, e_n for the edges, and $\bar{e}_n$ for the boldfaced edges of μ_n, respectively. Since π is a play, $(S'_n, p'_n, r'_n) = (S_{n+1}, p_{n+1}, r_{n+1})$ and $r_n \equiv n \pmod 4$.

Observe that every $s \in S'_n$ has a unique *boldfaced history in* π: a unique sequence $s_0, s_1, \ldots, s_n = s$ such that $(s_i, s_{i+1}) \in \bar{e}_i$ for $i < n$. A *path* in π is a sequence $\alpha = s_0, s_1, \ldots$ such that $(s_i, s_{i+1}) \in e_i$ of all i. A path is *boldfaced* if $(s_i, s_{i+1}) \in \bar{e}_i$ for all i. In particular, every finite prefix of a boldfaced path is a boldfaced history.

Intuitively, we would like to count how many times the boldfaced history of an active state $s \in S'_n$ has changed zone from reach to safe, this number will be denoted $\mathrm{val}(s)$ and will be defined formally in Equation (4.10). The main purpose of $\mathcal{G}$ is to avoid measuring this quantity and to use an ω-regular winning condition instead.

For a play $\pi = \mu_0\mu_1\ldots$ define the following properties:

W1 Some boldfaced path changes zone infinitely many times.
W2 The sequence of states $p_0, p_1, \ldots$ of the automaton $\mathcal{A}$ is parity-accepting.
W3 Some boldfaced path stays from some point on in the reach zone.

Now let a play π be winning for $\exists$ if π satisfies

$$\text{W1} \ \vee \ (\text{W2} \wedge \neg\text{W3}). \tag{4.9}$$

By the definition of the conditions W1, W2, and W3 we obtain the following fact.

Fact 4.43. *The winning condition of $\mathcal{G}$ is an ω-regular property of sequences of multi-transitions. By adding multi-transitions of $\mathcal{G}$ to the positions one can obtain an equivalent game with the winning condition on sequences of positions, conforming to the definition in Sect. 1.3 (see page 6).*

4.5 Equivalence

In this section we prove the following proposition, expressing an equivalence between the game $\mathcal{G}$ constructed in Sect. 4.4 and the ordinal $\eta(\mathcal{B})$ from Sect. 4.1.

Proposition 4.6. *The following conditions are equivalent:*

1. *$\exists$ has a winning strategy in $\mathcal{G}$,*
2. *$\eta(\mathcal{B}) \geqslant \omega^2$.*

4.5.1 Implication (1) ⇒ (2)

In this subsection we assume that $\exists$ has a winning strategy $\sigma_\exists$ in the game $\mathcal{G}$ and prove Item (2) in Proposition 4.6, i.e. that $\eta(\mathcal{B}) \geqslant \omega^2$. For this purpose we take any number $K \in \mathbb{N}$ and we will construct a tree $t \notin \mathrm{L}(\mathcal{B})$ such that $q_{\mathrm{I}}^{\mathcal{B}}$ is K-safe in t. Proposition 4.4 will imply that $\eta(\mathcal{B}) \geqslant \omega^2$.

The main idea behind the game $\mathcal{G}$ is that although the winning condition of $\mathcal{G}$ is ω-regular, the structure of $\mathcal{G}$ allows to keep track of real *values* of active states. These *values* will correspond to the numbers stored in the states of $\mathcal{C}[K]$. We start by formally defining these values for a play in $\mathcal{G}$.

Consider a finite or infinite play $\pi = \mu_0\mu_1 \ldots$ and an active state $s \in S'_n$ with the boldfaced history $s_0, s_1, \ldots, s_n = s$. Let

$$\mathrm{val}(s, n, \pi) \stackrel{\text{def}}{=} \left|\left\{i : s_i \in Q^{\mathcal{B}} \times \{\text{reach}\} \text{ and } s_{i+1} \in Q^{\mathcal{B}} \times \{\text{safe}\}\right\}\right|. \tag{4.10}$$

We usually skip n and π above and write just $\mathrm{val}(s)$ if the current history of the play is known from the context.

Now, given a value K we can consider *genuine strategies of* $\forall$ — strategies that keep track of the values of active states. It will turn out that such strategies allow us to simulate plays in $\mathcal{C}[K]$. We start by formally defining these strategies.
K-genuine strategies of $\forall$. A strategy $\sigma_\forall$ of $\forall$ is called *K-genuine* if it satisfies the three conditions defined below: *genuine-removal*, *val-monotonicity*, and *tie-breaking*.

A strategy $\sigma_\forall$ satisfies *genuine-removal* if in the sub-round R3 it removes an active state $s \in S$ if and only if $\mathrm{val}(s) \geqslant K$.

A strategy $\sigma_\forall$ satisfies *val-monotonicity* if whenever $\forall$ defines boldfaced edges, he does it in such a way to minimize $\mathrm{val}(s)$ — he puts (s, s') into $\bar{e}$ if s has a minimal value $\mathrm{val}(s)$ among all $\{s : (s, s') \in e\}$. In other words, every pair $(s, s') \in \bar{e}$ has to satisfy

$$\forall_{(s_0,s')\in e}\ \mathrm{val}(s) \leqslant \mathrm{val}(s_0). \tag{4.11}$$

Already the two above conditions guarantee the following fact.

Fact 4.44. *If π is an infinite play of $\mathcal{G}$ consistent with a K-genuine strategy of $\forall$ then π does not satisfy W1 (no boldfaced path changes side infinitely many times).*

The last condition, namely the *tie-breaking*, says what to do when defining $\bar{e}$ if there are two possible active states s with the minimal value $\mathrm{val}(s)$, i.e. both satisfying (4.11). The only purpose of this condition is to guarantee the following fact.

Fact 4.45. *Let π be an infinite play of $\mathcal{G}$ that is consistent with a K-genuine strategy of $\forall$. If π contains an infinite path α that from some point on stays in the* reach *zone then this path is eventually boldfaced (i.e. there exists an infinite boldfaced path α' that differs from α on finitely many positions, so α' satisfies W3).*

To express the condition of *tie-breaking* let us assume that during a play the player $\forall$ keeps track of a linear order on the active states: along with the position (S, p, r) he stores an order $\leqslant$ on S. This order is a simplified variant of Latest Appearance Record, see [GH82] and [Büc83b]. When he chooses a multi-transition μ, the new order $\leqslant'$ on S' is defined according to the following rules:

- for an active state $s' \in S'$ that is in the reach zone let us define $\mathrm{pre}(s') = \{s \in Q^{\mathcal{B}} \times \{\mathrm{reach}\} : (s, s') \in e\}$ — the set of e-predecessors of s' that are in the reach zone,
- for s'_0, s'_1 in the reach zone such that both sets $\mathrm{pre}(s'_0)$, $\mathrm{pre}(s'_1)$ are non-empty we put
$$s'_0 \leqslant' s'_1 \quad \text{if} \quad \sup\nolimits^{\leqslant} \mathrm{pre}(s'_0) \leqslant \sup\nolimits^{\leqslant} \mathrm{pre}(s'_1),$$
- all the active states s' in the reach zone such that $\mathrm{pre}(s') = \emptyset$ are added to $\leqslant'$ below all the existing elements (i.e. $s' <' s'_0$ when $\mathrm{pre}(s') = \emptyset$ and $\mathrm{pre}(s'_0) \neq \emptyset$),
- all the active states in the safe zone are added below all the active states in the reach zone (i.e. $(q, \mathrm{safe}) <' (q', \mathrm{reach})$).

– when the above rules do not determine the order, some fixed order on $Q^{\mathcal{B}}$ is used.

Intuitively, the order $\leqslant$ measures, for a given active state s, how long history (possibly not boldfaced) this active state has in the reach zone — the longer history, the $\leqslant$-bigger is s.

Now, a strategy $\sigma_\forall$ satisfies the condition of *tie-breaking* if among all active states s satisfying (4.11) it selects the $\leqslant$-maximal one: if $(s, s') \in \bar{e}$ then

$$\forall_{(s_0,s')\in e} \operatorname{val}(s_0) = \operatorname{val}(s) \Rightarrow s_0 \leqslant s.$$

Proof of Fact 4.45. Let $\pi = \mu_0\mu_1 \ldots$ and consider a path α in π as in the statement (α stays from some point on in the reach zone). Observe that from some point on the value $\operatorname{val}(s)$ for the active states on the path α must stabilize — the values of active states along a path not changing zone can only decrease. Therefore, from some point on, the boldfaced edges to active states on α were chosen using the condition of *tie-breaking*.

For the purpose of this proof, let the *grade* of an active state s in an order $\leqslant$ be the number of elements greater than s in $\leqslant$ — the smaller the grade is the $\leqslant$-bigger the element is. By Fact 4.42 an active state s in the reach zone has at most one e-successor. Therefore, the grades of the active states on the path α are from some point on decreasing. Let n be the moment when both the values and the grades of the active states on α stabilize.

Consider a multi-transition $\mu_{n'}$ in π that is later than n (i.e. $n' \geqslant n$). Let s, s' be the active states on α just before and just after $\mu_{n'}$. The values are already stabilized so $\operatorname{val}(s) = \operatorname{val}(s')$. Since the grades of s and s' are the same, s is $\leqslant$-maximal in $\operatorname{pre}(s')$. Therefore, the edge (s, s') has to be boldfaced in $\mu_{n'}$. ■

The following remark shows how to define a K-genuine strategy.

Remark 4.4. Observe that all the choices of $\forall$ except the directions d are uniquely determined in a K-genuine strategy. Therefore, to define a K-genuine strategy it is enough to say what will be the directions proposed by $\forall$ in the sub-rounds R1.

From a strategy in $\mathcal{G}$ to a K-safe tree. Assume that $\exists$ has a winning strategy $\sigma_\exists$ in $\mathcal{G}$ and $K \in \mathbb{N}$. Our aim is to construct a tree $t \notin \mathrm{L}(\mathcal{B})$ such that $Q^{\mathcal{B}}_{\mathrm{I}}$ is K-safe in t. The requirement that $t \notin \mathrm{L}(\mathcal{B})$ will be ensured by constructing an accepting run ρ of $\mathcal{A}$ on t. It will finish the proof of Item 2 in Proposition 4.6 (i.e. that $\eta(\mathcal{B}) \geqslant \omega^2$).

We define a tree t and a run ρ of $\mathcal{A}$ on t inductively. Let us take $u \in \{\mathtt{L}, \mathtt{R}\}^*$. Consider the play π of $\mathcal{G}$ resulting from $\exists$ playing $\sigma_\exists$ and $\forall$ playing a K-genuine strategy such that the first $|u|$ directions proposed by $\forall$ are $u(0), \ldots, u(|u| - 1)$. Let a, p be the letter and the state of $\mathcal{A}$ from the sub-round R1 of the $|u|$'th round of π. Let us put $t(u) = a$ and $\rho(u) = p$.

Let α be any infinite branch of t. By $\pi(K, \alpha)$ we denote the play resulting from $\exists$ playing $\sigma_\exists$ and $\forall$ playing the K-genuine strategy with consecutive directions $\alpha(0), \alpha(1), \ldots$ By Fact 4.44, the play $\pi(K, \alpha)$ does not satisfy W1. Since $\sigma_\exists$ is winning, $\pi(K, \alpha)$ satisfies W2 and $\neg$ W3. In particular, W2 implies that the run

ρ determined by $\sigma_\exists$ is parity-accepting on α. Since the choice of α is arbitrary, ρ is accepting so $t \notin \mathrm{L}(\mathcal{B})$.

It remains to prove that $q_{\mathrm{I}}^{\mathcal{B}}$ is K-safe in t. It is expressed in an inductive fashion by the following lemma. We assume that the sequence of multi-transitions during $\pi(K, \alpha)$ is $\mu_0\mu_1\ldots$. Note that the four multi-transitions played in the sub-rounds of an n'th round of the play $\pi(K, \alpha)$ are μ_{4n}, μ_{4n+1}, μ_{4n+2}, and μ_{4n+3}.

Lemma 4.11. *Consider the play $\pi(K, \alpha)$ for an infinite branch α. Assume that an n'th round of this play started in the vertex $u = \alpha{\restriction}_n$ of the tree t. Take any active state $s = (q, z) \in S_{4n}$ or S'_{4n} (we allow active states before and after the sub-round R0). For every $i \leqslant K - \mathrm{val}(s)$:*

$$\textit{if } z = \text{reach} \textit{ then } q \textit{ is } i\textit{-reach in } t{\restriction}_u,$$
$$\textit{if } z = \text{safe} \textit{ then } q \textit{ is } i\textit{-safe in } t{\restriction}_u.$$

Note that the above lemma for $n = 0$, $s = (q_{\mathrm{I}}^{\mathcal{B}}, \text{safe})$, and $i = K$ gives us that $q_{\mathrm{I}}^{\mathcal{B}}$ is K-safe in t.

Proof. The proof goes by induction on i. The thesis is trivial for $i = 0$. Assume that we have proved the thesis for $i-1$ (for all n and s). Consider a vertex $u = \alpha{\restriction}_n$ and an active state s as in the statement.

The $z = $ reach case. First consider the case of $z = $ reach. We need to show that q is i-reach in $t{\restriction}_u$.

We will construct a partial tree $\bar{\rho} \in \mathrm{PTr}_{Q^{\mathcal{B}}}$ that will be a partial run witnessing that q is i-reach in $t{\restriction}_u$. The construction of $\bar{\rho}(w)$ is inductive on the length of $w \in \{\mathtt{L}, \mathtt{R}\}^*$. With every w during the construction we bind a prefix of a play in $\mathcal{G}$ that is consistent with the strategy $\sigma_\exists$. The invariant is that $s' = (\bar{\rho}(w), \text{reach})$ is an active state and $\mathrm{val}(s') \leqslant K - i$. We start with $w = \epsilon$, the prefix $\mu_0 \ldots \mu_{4n}$, and $s' = s$.

Assume we reached a vertex w during the construction with the prefix of the play being $\mu_0 \ldots \mu_{4n-1}\mu'_{4n}\mu'_{4n+1} \cdots \mu'_{4n'}$ (here $n' - n = |w|$). Assume that $s_0 = (\bar{\rho}(w), \text{reach})$ is an active state and $\mathrm{val}(s_0) \leqslant K - i$. We need to show how to extend the construction to wd for $d = \mathtt{L}, \mathtt{R}$. Consider such d and let us play the remaining three sub-rounds of the (n')'th round. Let $\exists$ play using $\sigma_\exists$ and let $\forall$ play in this round using a K-genuine strategy with the proposed direction d, the three multi-transitions constructed are $\mu'_{4n'+1}, \ldots, \mu'_{4n'+3}$. Now let us play the first sub-round R0 of the successive round, what gives us $\mu'_{4n'+4}$ — it does not influence the reach zone.

Let $q' = q_d^{s_0}$ — the state from the transition proposed by $\exists$ for s_0. First assume that $q' \notin F$. In that case, by Fact 4.42, the active state (q', reach) is the unique $(\mu'_{4n'+1}, \ldots, \mu'_{4n'+4})$-successor of (q, reach) — since $i > 0$ and $\mathrm{val}(s_0) \leqslant K - i$ so $\forall$ does not remove the active state (q', reach) in $\mu'_{4n'+3}$. In particular, $\mathrm{val}(q', \text{reach}) \leqslant \mathrm{val}(q, \text{reach}) \leqslant K - i$. We define $\bar{\rho}(wd) = q'$, and proceed with $w = wd$, $s_0 = (q', \text{reach})$, and the prefix of a play $\mu_0 \ldots \mu_{4n-1}\mu'_{4n}\mu'_{4n+1} \cdots \mu'_{4n'+4}$.

Now consider the case that $q' \in F$. In that case we finish the inductive constriction by letting w be a leaf of $\bar{\rho}$. Note that in that case in the multi-transition $\mu'_{4n'+2}$ there is an edge $((q', \text{reach}), (q', \text{safe}))$. Therefore, we obtain

$$\text{val}(q', \text{safe}) \leqslant 1 + \text{val}(q', \text{reach}) \leqslant 1 + K - i = K - (i - 1).$$

As before, $\forall$ does not remove (q', safe) in $\mu'_{4n'+3}$. By the inductive assumption for $i - 1 \leqslant K - \text{val}(q', \text{safe})$ we know that q' is $(i-1)$-safe in $t\restriction_{uw}$. It means that $\bar{\rho}$ is a partial run witnessing that the original state q was i-reach in $t\restriction_u$ if and only if $\bar{\rho}$ is a finite tree (does not have any infinite branch).

It remains to prove that $\bar{\rho}$ is finite. Assume contrary that there exists an infinite branch β such that for every $w \prec \beta$ the above construction gave a state $q' \notin F$. It means that there exists a path in the play $\pi(K, u\beta)$ that is from some moment on in the reach zone. By Fact 4.45 it means that W3 is satisfied what contradicts the assumption that $\sigma_\exists$ is winning.

The $z =$ safe *case.* Assume that $z =$ safe. We need to show that q is i-safe in $t\restriction_u$. Similarly as above, we construct a total run ρ of $\mathcal{B}$ on $t\restriction_u$ with $\rho(\epsilon) = q$. We will argue that for every w we know that $\rho(w)$ is i-reach in $t\restriction_{uw}$.

The construction of $\rho(w)$ is inductive on the length of $w \in \{\text{L}, \text{R}\}^*$. With every w during the construction we bind a prefix of a play in $\mathcal{G}$ that is consistent with the strategy $\sigma_\exists$. The invariant is that $s' = (\rho(w), \text{safe})$ is an active state and $\text{val}(s') \leqslant K - i$. During the step in which we define $\rho(wd)$ we additionally argue that $\rho(w)$ is i-reach in $t\restriction_{uw}$. We start with $w = \epsilon$, the prefix $\mu_0 \ldots \mu_{4n-1}$ and $s' = s$.

Assume that we reached a vertex w during the construction with the prefix of the play being $\mu_0 \ldots \mu_{4n-1}\mu'_{4n}\mu'_{4n+1} \cdots \mu'_{4n'-1}$ (here $n' - n = |w|$). Assume that $s_0 = (\bar{\rho}(w), \text{safe})$ is an active state and $\text{val}(s_0) \leqslant K - i$. We need to show how to extend the construction to wd for $d = \text{L}, \text{R}$. Consider such d and let us play the four sub-rounds of the (n')'th round. Let $\exists$ play using $\sigma_\exists$ and let $\forall$ play in this round using a K-genuine strategy with the proposed direction d, the four multi-transitions constructed are $\mu'_{4n'}, \ldots, \mu'_{4n'+3}$.

Let $q' = q_d^{s_0}$ — the state from the transition proposed by $\exists$ for s_0. By Fact 4.42, the active state (q', safe) is the unique $(\mu'_{4n'}, \ldots, \mu'_{4n'+3})$-successor of (q, safe) — since $i > 0$ and $\text{val}(s_0) \leqslant K - i$, $\forall$ does not remove the active state (q', safe) in $\mu'_{4n'+3}$. In particular, $\text{val}(q', \text{safe}) \leqslant \text{val}(q, \text{safe}) \leqslant K - i$. We define $\rho(wd) = q'$, and proceed with $w = wd$, $s_0 = (q', \text{safe})$, and the prefix of a play $\mu_0 \ldots \mu_{4n-1}\mu'_{4n}\mu'_{4n+1} \cdots \mu'_{4n'+3}$.

Additionally observe that there is an edge $((q, \text{safe}), (q, \text{reach}))$ in the multi-transition $\mu_{4n'}$. Therefore, by the inductive invariant we know that $\rho(w)$ is i-reach in $t\restriction_{uw}$. ■

This concludes the proof of the implication (1) $\Rightarrow$ (2).

4.5.2 Implication (2) ⇒ (1)

Now assume that $\forall$ has a winning strategy in $\mathcal{G}$. Since the winning condition of $\mathcal{G}$ is ω-regular so we can take as $\sigma_\forall$ a finite memory winning strategy of $\forall$ (see Sect. 1.3.1 and Theorem 1.21 on page 21). Assume that the memory structure of $\sigma_\forall$ is M. We will prove that there exists a number K such that no tree $t \notin L(\mathcal{B})$ is K-reach, thus showing the negation of Item 2 in Proposition 4.6 (i.e. that $\eta(\mathcal{B}) < \omega^2$).

We start with the following fact exploiting the assumption that the strategy $\sigma_\forall$ has finite memory. Recall that in (4.10) we defined the *value* of an active state s in a play (denoted $\mathrm{val}(s)$).

Fact 4.46. *There exists a global bound K such that for every play consistent with $\sigma_\forall$ and every active state s during the play, we have* $\mathrm{val}(s) < K$. *The bound K can be computed effectively basing on $\mathcal{B}$.*

Proof. Assume contrary and let us take a play $\pi = \mu_0\mu_1\ldots\mu_{4n}$ such that for some active state s we have $\mathrm{val}(s) \geqslant |\mathcal{G}| \cdot |M| \cdot |Q^{\mathcal{B}}| \cdot 2$. A standard pumping technique (see e.g. [AS05]) shows that in that case there exists a loop $\mu_{4i}\mu_{4i+1}\ldots\mu_{4i'}$ in the graph $\mathcal{G} \times M$ and an active state $s \in Q^{\mathcal{B}} \times \{\mathrm{safe}, \mathrm{reach}\}$ such that:

- $s \in S_{4i}$,
- $s \in S'_{4i'+3}$,
- the boldfaced history of s in $\mu_{4i}\mu_{4i+1}\ldots\mu_{4i'}$ reaches s in S_{4i},
- the above boldfaced history changes zone from reach to safe.

Consider the play

$$\pi' = \mu_0\mu_1\ldots\mu_{4i-1}\left(\mu_{4i}\ldots\mu_{4i'+3}\right)^{\infty}.$$

This play is consistent with the strategy $\sigma_\forall$ and satisfies W1. Therefore, π' is winning for $\exists$ what contradicts the fact that $\sigma_\forall$ is a winning strategy of $\forall$. ■

Let us fix the bound K from Fact 4.46. Assume for the contradiction that $\eta(\mathcal{B}) \geqslant \omega^2$. Proposition 4.4 implies that there exists a tree $t \notin L(\mathcal{B})$ such that $q_{\mathrm{I}}^{\mathcal{B}}$ is K-safe in t. Let σ be a winning strategy of $\exists$ in $\mathcal{G}(\mathcal{C}[K], t)$.

We will construct a strategy $\sigma_\exists$ of $\exists$ in $\mathcal{G}$ that will simulate the strategy σ. Then we will show that the play of $\mathcal{G}$ resulting from $\exists$ playing $\sigma_\exists$ and $\forall$ playing $\sigma_\forall$ is winning for $\exists$ what contradicts the assumption that $\sigma_\forall$ is winning.

Let $\rho^{\mathcal{A}}$ be an accepting run of $\mathcal{A}$ on t. The strategy $\sigma_\exists$ will simulate during a play of $\mathcal{G}$ a set of plays of $\mathcal{G}(\mathcal{C}[K], t)$ (by following the boldfaced edges) and play $\rho^{\mathcal{A}}$ as the transitions of $\mathcal{A}$. That is, for every active state s the player $\exists$ will keep track of an *s-play in $\mathcal{G}(\mathcal{C}[K], t)$* defined below. The invariant will be:

$$\text{If } s = (q, z) \in S \text{ at the beginning of a round in } \mathcal{G} \text{ then the } s\text{-play in } \mathcal{G}(\mathcal{C}[K], t) \text{ reached the state } \big(q, z, K - \mathrm{val}(s)\big). \tag{4.12}$$

Let us define the s-play in $\mathcal{G}(\mathcal{C}[K], t)$ more formally. At the beginning of the game $\mathcal{G}$ the only active state is $(q_{\mathrm{I}}^{\mathcal{B}}, \mathrm{safe})$ and $(q_{\mathrm{I}}^{\mathcal{B}}, \mathrm{safe}, K)$ is the initial state of $\mathcal{C}[K]$. We will consider the four sub-rounds of a round in $\mathcal{G}$. Whenever a new multi-transition μ is played in $\mathcal{G}$, the $s' \in S'$-play in $\mathcal{G}(\mathcal{C}[K], t)$ is the continuation of the s-play in $\mathcal{G}(\mathcal{C}[K], t)$ for s being the μ-predecessor of s'. Now consider the successive sub-rounds:

- In the sub-round R0 it is possible that the edge (s, s') changes zone from safe to reach. In that case $\exists$ simulates $\forall$ playing $z' = \text{reach}$ in $\mathcal{G}(\mathcal{C}[K], t)$, otherwise she simulates $z' = \text{safe}$.
- The transition δ_s played by $\exists$ in $\mathcal{G}$ in the sub-round R1 is the transition δ from $\mathcal{G}(\mathcal{C}[K], t)$ played in the s-play in $\mathcal{G}(\mathcal{C}[K], t)$. The transition δ of $\mathcal{A}$ played by $\exists$ is the transition from $\rho^{\mathcal{A}}$. The direction d played by $\forall$ in $\mathcal{G}(\mathcal{C}[K], t)$ is the direction from the sub-round R1.
- In the sub-round R2 it is possible that the edge (s, s') changes zone from reach to safe. In that case $s = (q, \text{reach})$ with $q \in F$ and the s'-play in $\mathcal{G}(\mathcal{C}[K], t)$ moves to the state $(q', \text{safe}, i - 1)$.
- If $\forall$ decides to remove some active states s in the sub-round R3 of $\mathcal{G}$ then they are not longer active after this sub-round. For active states s that are not removed, the s-play is not changed.

Observe that after such a round the invariant (4.12) is satisfied.

Let π be the play resulting from $\exists$ playing $\sigma_\exists$ and $\forall$ playing $\sigma_\forall$ in $\mathcal{G}$. Fact 4.46 implies that π does not satisfy W1. Since the strategy σ of $\exists$ in $\mathcal{G}(\mathcal{C}[K], t)$ is winning, W3 is not satisfied by π. The run $\rho^{\mathcal{A}}$ is accepting so π satisfies W2. Therefore, π is winning for $\exists$ what contradicts the assumption that $\sigma_\forall$ is winning.

This concludes the proof of Implication (2) $\Rightarrow$ (1) and the proof of Proposition 4.6.

4.6 Conclusions

The results presented in this chapter relate descriptive complexity of the language recognised by a Büchi automaton $\mathcal{B}$ with the rank $\eta(\mathcal{B})$. In particular, Conjecture 4.4 stated in this chapter would imply that if a Büchi language is Borel then it is WMSO-definable (i.e. a special case of Conjecture 2.2 for Büchi languages). Unfortunately, Conjecture 4.4 remains open as an appropriate pumping argument is missing.

The study of the ordinal $\eta(\mathcal{B})$ is motivated by the boundedness theorem (see Theorem 1.12 on page 17), saying that if an analytic (i.e. $\mathbf{\Sigma}^1_1$) set X is contained in a ranked set (e.g. well-founded ω-trees) then there is a bound on ranks that are realised in X. This theorem is the crucial tool for proving Proposition 4.2 that relates Borel languages and the rank $\eta(\mathcal{B})$.

Since every Büchi language is analytic, this may suggest to use the boundedness theorem for deciding if a given language is Büchi. However, one should bear in mind the following example. It implies that among $\mathbf{\Sigma}^1_1$-sets there are some Büchi languages and some regular languages that are not Büchi. Therefore, topological methods are not enough to distinguish between the two cases.

Example 4.2. The regular tree language $L_{\neq 1}$ containing these trees $t \in \mathrm{Tr}_{\{a,b\}}$ that have 0 or at least 2 infinite branches labelled by infinitely many letters a is an analytic language (i.e. $\mathbf{\Sigma}^1_1$) but it cannot be recognised by a Büchi automaton.

Sketch of the proof. The fact that $L_{\neq 1}$ is analytic follows from [Kec95, Exercise 33.1]. The fact that $L_{\neq 1}$ is not a Büchi language follows from the standard pumping argument showing that the set of trees where every branch contains only finitely many a is not Büchi. ■

However, there is a hope that some more involved ranks may still be useful for deciding higher levels of the index hierarchy.

Chapter 5
Index Problems for Game Automata

One of the main difficulties when working with regular languages of infinite trees is the lack of a convenient notion of recognition. In particular, since deterministic automata are too weak, one has to deal with an inherent non-determinism. On the other hand, many problems simplify when we restrict to languages recognisable by deterministic automata (called *deterministic languages*), see Sect. 1.7.6 on page 26. The crucial technique standing behind these results is the so-called *pattern method* — the properties of a deterministic language are reflected by certain patterns in the graph of a deterministic automaton recognising it.

The pattern method cannot be applied to non-deterministic nor alternating automata; the reason is that both these classes are closed under union and union is not an operation that preserves the index of languages. However, it turns out that if we avoid closure under union, we can extend the pattern method well-beyond deterministic automata, to so-called game automata.

In this chapter we study game automata that can be seen as a combination of deterministic and co-deterministic ones. They were introduced in [DFM11] as the largest subclass of alternating tree automata extending the deterministic ones, closed under complementation and composition, and for which the latter operation preserves natural equivalence relations on recognised languages, like the topological equivalence, or having the same index. As game automata recognise the languages $W_{i,j}$ from [Arn99] (see Sect. 1.7.4, page 23) the alternating index problem does not trivialise, unlike for deterministic automata.

Recall that an alternating tree automaton $\mathcal{A}$ is *deterministic* if its transitions are of the form $(q_{\mathrm{L}}, \mathtt{L}) \wedge (q_{\mathrm{R}}, \mathtt{R})$. For such automata, all the positions in the induced game $\mathcal{G}(\mathcal{A}, t)$ on a tree t belong to the universal player $\forall$— his aim is to indicate a branch on which the run is rejecting. In the case of game automata we allow dual transitions where $\exists$ is in charge of selecting the direction. More formally, an alternating tree automaton $\mathcal{A}$ is a *game automaton* if every transition of $\mathcal{A}$ is of one of the following forms:

$$\top\,,\quad \bot\,,\quad (q_d, d)\,,\quad (q_{\mathrm{L}}, \mathtt{L}) \vee (q_{\mathrm{R}}, \mathtt{R})\,,\quad (q_{\mathrm{L}}, \mathtt{L}) \wedge (q_{\mathrm{R}}, \mathtt{R})$$

M. Skrzypczak, *Set Theoretic Methods in Automata Theory*, LNCS 9802
DOI: 10.1007/978-3-662-52947-8_5

for $d \in \{\text{L}, \text{R}\}$ and $q_{\text{L}}, q_{\text{R}} \in Q^{\mathcal{A}}$. If $\mathcal{A}$ is a game automaton and t is a tree then both players are allowed to make decisions in the game $\mathcal{G}(\mathcal{A}, t)$. However, for every direction d in the tree, there is at most one successive state that can be reached by moving in this direction.

The following theorem summarizes the results of this chapter.

Theorem 5.3. *The non-deterministic index problem is decidable for game automata (i.e. if a game automaton is given as the input). The same holds for the alternating index problem.*

Let L be a language recognised by a game automaton. If $L \in \mathbf{\Delta}^{\text{alt}}_j$ then $L \in$ $\mathbf{Comp}(\mathbf{\Pi}^{\text{alt}}_{j-1})$. *If L is Borel then L is* WMSO*-definable.*

Additionally, it is shown in [FMS13] that it is decidable if a given regular tree language is recognisable by a game automaton. This characterisation is not presented in this thesis, it follows similar lines as in the deterministic case [NW98]. It implies that the decidability results from Theorem 2.3 hold for the class of languages recognisable by game automata: there exists an algorithm that inputs a representation (possibly a non-game automaton) of a regular tree language, verifies if the language can be recognised by a game automaton and if it can then computes the index of the language.

At this point game automata form the widest class of automata for which both index problems are known to be decidable. It seems that game automata is the frontier of the pattern method — to move further one needs a new insight into the structure of regular tree languages.

The symbols $\mathbf{\Pi}^{\text{alt}}_j$ and $\mathbf{\Sigma}^{\text{alt}}_j$ are used in this thesis in the opposite meaning when compared to [FMS13]. This is to keep consistency with the notions from [AS05, AMN12].

The chapter is organised as follows. In Sect. 5.1 we introduce and study a notion of the *run* of a game automaton on a tree. In Sect. 5.2 we give an easy argument for decidability of the non-deterministic index problem for game automata. Section 5.3 builds some technical tools that will allow to give a solution for the alternating index problem for game automata in Sect. 5.4. In Sect. 5.5 we conclude.

5.1 Runs of Game Automata

The main similarity between game automata and deterministic automata is that their acceptance can be expressed in terms of *runs*, which are unique labellings of input trees. The notion of a run of a game automaton will be used in subsequent sections of this chapter.

For a game automaton $\mathcal{A}$ and a state $q_0 \in Q^{\mathcal{A}}$, with each tree $t \in \text{Tr}_{A^{\mathcal{A}}}$ one can associate the *run*

$$\rho = \rho(\mathcal{A}, t, q_0)\colon \text{dom}(t) \to Q^{\mathcal{A}} \sqcup \{\top, \bot, \star\}$$

such that $\rho(\epsilon) = q_0$ and for all $u \in \mathrm{dom}(t)$, if $\rho(u) = q$, $\delta(q, t(u)) = b_u$ then

- if b_u is $(q_{\mathtt{L}}, \mathtt{L}) \vee (q_{\mathtt{R}}, \mathtt{R})$ or $(q_{\mathtt{L}}, \mathtt{L}) \wedge (q_{\mathtt{R}}, \mathtt{R})$ then $\rho(ud) = q_d$ for $d = \mathtt{L}, \mathtt{R}$;
- if $b_u = (q_d, d)$ for some $d \in \{\mathtt{L}, \mathtt{R}\}$ then $\rho(ud) = q_d$ and $\rho(u\bar{d}) = \star$;
- if $b_u = \bot$ then $\rho(u\mathtt{L}) = \rho(u\mathtt{R}) = \bot$, and dually for $\top$;

and if $\rho(u) \in \{\top, \bot, \star\}$ then $\rho(u\mathtt{L}) = \rho(u\mathtt{R}) = \star$.

The run $\rho = \rho(\mathcal{A}, t, q_0)$ for a tree t is naturally interpreted as a game $\mathbf{G}_\rho(\mathcal{A}, t, q_0)$ with:

- positions $\mathrm{dom}(t) \setminus \rho^{-1}(\star)$,
- where edges follow the child relation and loop on those positions u where $\rho(u) \in \{\top, \bot\}$,
- the priority of u is $\Omega^{\mathcal{A}}(\rho(u))$ with $\Omega(\bot) = 1$, $\Omega(\top) = 0$,
- the owner of u being $\exists$ if and only if $\delta(\rho(u), t(u)) = (q_{\mathtt{L}}, \mathtt{L}) \vee (q_{\mathtt{R}}, \mathtt{R})$ for some $q_{\mathtt{L}}, q_{\mathtt{R}} \in Q^{\mathcal{A}}$.

Note that the symbol $\star$ in ρ denotes the vertices that cannot be visited during the game $\mathbf{G}_\rho(\mathcal{A}, t, q_0)$.

Recall that the game $\mathbf{G}(\mathcal{A}, t, q_0)$ (see Sect. 1.4, page 7) is defined similarly to $\mathbf{G}_\rho(\mathcal{A}, t, q_0)$ but is more complicated: a play in $\mathbf{G}(\mathcal{A}, t, q_0)$ explicitly operates on transitions of $\mathcal{A}$. For instance, one edge in the game $\mathbf{G}_\rho(\mathcal{A}, t, q_0)$ may correspond to three edges in $\mathbf{G}(\mathcal{A}, t, q_0)$:

- from (u, b_u) to (u, b_d) where b_d is an atomic transition that is a sub-formula of b_u,
- from (u, b_d) to (ud, q_d) for an atomic transition $b_d = (q_d, d)$,
- from (ud, q_d) to $\big(ud, \delta(q_d, t(ud))\big)$ where $\delta(q_d, t(ud)) = b_{ud}$.

Therefore, $\mathbf{G}_\rho(\mathcal{A}, t, q_0)$ can be seen as a projection of $\mathbf{G}(\mathcal{A}, t, q_0)$), the advantage of $\mathbf{G}_\rho(\mathcal{A}, t, q_0)$ is that this game explicitly reflects the input tree — the set of positions of $\mathbf{G}_\rho(\mathcal{A}, t, q_0)$ is contained in $\mathrm{dom}(t)$. By the definition, $t \in \mathrm{L}(\mathcal{A}, q_0)$ if and only if $\exists$ has a winning strategy in $\mathbf{G}_\rho(\mathcal{A}, t, q_0)$.

For simplicity we write $\rho(\mathcal{A}, t)$ for $\rho(\mathcal{A}, t, q_I^{\mathcal{A}})$ and $\mathbf{G}_\rho(\mathcal{A}, t)$ for $\mathbf{G}_\rho(\mathcal{A}, t, q_I^{\mathcal{A}})$.

It will be important in this chapter that we assume that every state q of a game automaton $\mathcal{A}$ recognises a non-trivial language, i.e. $\mathrm{L}(\mathcal{A}, q)$ is neither $\varnothing$ nor $\mathrm{Tr}_{A^{\mathcal{A}}}$. This can be achieved for every game automaton recognising a non-trivial language by removing trivial states and simplifying transitions, see Fact 1.4 on page 9 (it is easy to observe that the proposed method produces a game automaton).

The following remark subsumes the crucial property of runs of game automata.

Remark 5.5. Let $\mathcal{A}$ be a game automaton and $t \in \mathrm{Tr}_{A^{\mathcal{A}}}$ be a tree. Assume that $u \in \mathrm{dom}(t)$ is a vertex such that $\rho(\mathcal{A}, t)(u) = q \in Q^{\mathcal{A}}$ (i.e. $\rho(\mathcal{A}, t)(u)$ is not in $\{\top, \bot, \star\}$). Let

$$L' = \{t' \in \mathrm{Tr}_{A^{\mathcal{A}}} : t[u \leftarrow t'] \in \mathrm{L}(\mathcal{A})\}.$$

Then either:

- $L' = \varnothing$,
- $L' = \mathrm{Tr}_{A^{\mathcal{A}}}$,
- $L' = \mathrm{L}(\mathcal{A}, q)$.

Additionally, since all the states of $\mathcal{A}$ recognise non-trivial languages, the above disjunction is exclusive.

Proof. Consider the following cases:

- One of the players $P \in \{\exists, \forall\}$ has a winning strategy σ in $\mathbf{G}_\rho(\mathcal{A}, t)$ (we treat σ as a set of nodes of t) such that $u \notin \sigma$. In that case the same strategy is a winning strategy in $\mathbf{G}_\rho(\mathcal{A}, t[u \leftarrow t'])$, so $L' = \varnothing$ or $L' = \mathrm{Tr}_{A^\mathcal{A}}$ depending whether $P = \forall$ or $\exists$.
- Whenever σ is a winning strategy of a player P in t then $u \in \sigma$. We want to show that $L' = \mathrm{L}(\mathcal{A}, q)$. Consider any tree t' and assume that a player P has a winning strategy σ in $\mathbf{G}_\rho(\mathcal{A}, t[u \leftarrow t'])$. By our assumption $u \in \sigma$ — otherwise σ would be a winning strategy of P in $\mathbf{G}_\rho(\mathcal{A}, t)$ that does not contain u. Note that since $u \in \sigma$, σ induces a winning strategy of P in $\mathbf{G}_\rho(\mathcal{A}, t', q)$. Therefore, $t' \in L'$ if and only if $P = \exists$ if and only if $t' \in \mathrm{L}(\mathcal{A}, q)$. ■

5.2 Non-deterministic Index Problem

In this section we prove the first part of Theorem 2.3: the non-deterministic index problem is decidable for languages recognisable by game automata. It follows directly from the decidability of the non-deterministic index problem for deterministic tree languages [NW05] and the following proposition.

Proposition 5.7. *For each game automaton $\mathcal{A}$ one can effectively construct a deterministic automaton $\mathcal{D}$, such that* $\mathrm{L}(\mathcal{A})$ *is recognised by a non-deterministic automaton of index (i, j) if and only if so is* $\mathrm{L}(\mathcal{D})$.

Proof. Essentially, $\mathcal{D}$ recognises the set of winning strategies for $\exists$ in the games induced by the runs of $\mathcal{A}$. Let $W^\exists_\mathcal{A}$ be the set of all trees $t \otimes s$ over the alphabet $A^\mathcal{A} \times \{\mathrm{L}, \mathrm{R}, \mathrm{LR}\}$ such that s encodes a winning strategy for $\exists$ in the game $\mathbf{G}_\rho(\mathcal{A}, t)$ in the following sense: if $s(u) \in \{\mathrm{L}, \mathrm{R}\}$, $\exists$ should move from u to $u \cdot s(u)$, and $s(u) = \mathrm{LR}$ means that $\exists$ has no choice in u. It is easy to see that $W^\exists_\mathcal{A}$ can be recognised by a deterministic automaton $\mathcal{D}$: it inherits the state-space, the initial state, and the priority function from $\mathcal{A}$. The transitions of $\mathcal{D}$ are defined as follows: for all $q \in Q$, $a \in A^\mathcal{A}$, $d \in \{\mathrm{L}, \mathrm{R}\}$, if $\delta^\mathcal{A}(q, a) = (q_\mathrm{L}, \mathrm{L}) \vee (q_\mathrm{R}, \mathrm{R})$ for some $q_\mathrm{L}, q_\mathrm{R}$, then

$$\delta^\mathcal{D}(q, (a, d)) = (q_d, d) \quad \delta^\mathcal{D}(q, (a, \mathrm{LR})) = \bot$$

otherwise,

$$\delta^\mathcal{D}(q, (a, d)) = \bot \quad \delta^\mathcal{D}(q, (a, \mathrm{LR})) = \delta^\mathcal{A}(q, a).$$

It is easy to check that $\mathrm{L}(\mathcal{D}) = W^\exists_\mathcal{A}$.

Note that

$$\mathrm{L}(\mathcal{A}) = \left\{t \in \mathrm{Tr}_{A^\mathcal{A}} : \exists s \in \mathrm{Tr}_{\{\mathrm{L},\mathrm{R},\mathrm{LR}\}}.\ t \otimes s \in W^\exists_\mathcal{A}\right\}.$$

Hence, if $W^{\exists}_{\mathcal{A}} = \mathrm{L}(\mathcal{B})$ for some non-deterministic automaton $\mathcal{B}$ of index (i, j) then $\mathrm{L}(\mathcal{A}) = \mathrm{L}(\mathcal{B}')$, where $\mathcal{B}'$ is the standard projection of $\mathcal{B}$ on the alphabet $A^{\mathcal{A}}$: for all $q \in Q^{\mathcal{A}}$ and $a \in A^{\mathcal{A}}$, $\delta^{\mathcal{B}'}(q, a) = \delta^{\mathcal{B}}(q, (a, \mathrm{L})) \vee \delta^{\mathcal{B}}(q, (a, \mathrm{R})) \vee \delta^{\mathcal{B}}(q, (a, \mathrm{LR}))$. The projection does not influence the index.

For the other direction, the proof is based on the following observation. For $t \in \mathrm{Tr}_{A^{\mathcal{A}}}$ and $s \in \mathrm{Tr}_{\{\mathrm{L},\mathrm{R},\mathrm{LR}\}}$ let $\mathrm{force}(t, s) \in \mathrm{Tr}_{A^{\mathcal{A}}}$ be the tree obtained from t by the following operation: for each u, if $\rho(\mathcal{A}, t)(u) = q$, $\delta(q, t(u)) = (q_{\mathrm{L}}, \mathrm{L}) \vee (q_{\mathrm{R}}, \mathrm{R})$, and $s(u) = \mathrm{L}$ then replace the subtree of t rooted in $u\mathrm{R}$ by some fixed regular tree in the complement of $\mathrm{L}(\mathcal{A}, q_{\mathrm{R}})$; dually for $s(u) = \mathrm{R}$. (Recall that $\mathcal{A}$ has only non-trivial states, so $\mathrm{L}(\mathcal{A}, q_{\mathrm{R}}) \subsetneq \mathrm{Tr}_{A^{\mathcal{A}}}$.) If s encodes a strategy σ_s for $\exists$ in $\mathbf{G}_\rho(\mathcal{A}, t)$ then σ_s is winning if and only if $\mathrm{force}(t, s) \in \mathrm{L}(\mathcal{A})$. Hence

$$t \otimes s \in W^{\exists}_{\mathcal{A}} \iff s \text{encodes a strategy for } \exists \text{ in } \mathbf{G}_\rho(\mathcal{A}, t) \text{ and } \mathrm{force}(t, s) \in \mathrm{L}(\mathcal{A}). \tag{5.1}$$

What remains is to show that if $\mathrm{L}(\mathcal{A}) = \mathrm{L}(\mathcal{B})$ for some non-deterministic automaton $\mathcal{B}$ of index (i, j) then we can construct a non-deterministic automaton $\mathcal{C}$ of index at most (i, j) recognising $W^{\exists}_{\mathcal{A}}$. The automaton $\mathcal{C}$ simply checks for the input tree $t \otimes s$ if the right-hand side of (5.1) holds: whether s encodes a strategy for $\exists$ in the parity game associated with $\rho(\mathcal{A}, t)$ and if $\mathrm{force}(t, s) \in \mathrm{L}(\mathcal{B})$.

Now we provide a more formal description of the automaton $\mathcal{C}$.

By Rabin's theorem (see Theorem 1.17 on page 20), for each $q \in Q^{\mathcal{A}}$ there exists a regular tree $t_q \notin \mathrm{L}(\mathcal{A}, q)$. We define a sequence of regular languages and then we argue that they can be recognised by non-deterministic automata of indices at most (i, j):

$$\mathrm{St} = \Big\{t \otimes s : \quad s \text{ encodes a strategy for } \exists \text{ in } \mathbf{G}_\rho(\mathcal{A}, t)\Big\},$$

$$\mathrm{StE} = \Big\{t \otimes s \otimes t' t \mathpunct{:} \otimes s \in \mathrm{St} \ \wedge \ \mathrm{force}(t, s) = t'\Big\},$$

$$\mathrm{StEW} = \Big\{t \otimes s \otimes t' t \mathpunct{:} \otimes s \otimes t' \in \mathrm{StE} \ \wedge \ t' \in \mathrm{L}(\mathcal{B}) = \mathrm{L}(\mathcal{A})\Big\},$$

$$\mathrm{StW} = \Big\{t \otimes s : \quad t \otimes s \in \mathrm{St} \ \wedge \ \mathrm{force}(t, s) \in \mathrm{L}(\mathcal{A})\Big\}.$$

where:

- The language St corresponds to a *safety condition* of the form "in every vertex ...". This condition can be verified by a Comp(0, 0)-deterministic automaton,
- The language StE additionally enforces that the respective subtrees are equal t_q where t_q are regular. It can be verified by a Comp(0, 0)-deterministic automaton,
- The language StEW can be recognised by a product of the automata recognising StE and $\mathcal{B}$—the resulting non-deterministic automaton can be constructed in such a way that its index equals (i, j),
- StW is obtained as the projection of StEW onto the first two coordinates, as such can also be recognised by a non-deterministic (i, j)-automaton.

What remains to show is the following equation

$$W^{\exists}_{\mathcal{A}} = \text{StW} \tag{5.2}$$

First assume that $t \otimes s \in W^{\exists}_{\mathcal{A}}$. In that case s encodes a winning strategy σ for $\exists$ in $\mathbf{G}_\rho(\mathcal{A}, t)$. We treat σ as a subset of $\text{dom}(t)$. Note that if $u \in \sigma$ then $t(u) = t'(u)$, so also $\rho(\mathcal{A}, t)(u) = \rho(\mathcal{A}, t')(u)$. Therefore, the strategy σ is also winning in $\mathbf{G}_\rho(\mathcal{A}, t')$. So $t' \in \text{L}(\mathcal{A})$ what implies that $t \otimes s \otimes t' \in \text{StEW}$ and $t \otimes s \in \text{StW}$.

Now assume that $t \otimes s \in \text{StW}$. Let $t' = \text{force}(t, s)$ and σ be the strategy for $\exists$ in $\mathbf{G}_\rho(\mathcal{A}, t)$ encoded by s. By the definition of StEW we obtain that $t' \in \text{L}(\mathcal{A})$ so there exists a winning strategy σ' for $\exists$ in $\mathbf{G}_\rho(\mathcal{A}, t')$.

If $\sigma' \not\subseteq \sigma$ then there exists a minimal (w.r.t. the prefix order) vertex $u \in \sigma' \setminus \sigma$. By the definition of $\text{force}(t, s)$ we obtain that $t'\restriction_u$ is t_q for $q = \rho(\mathcal{A}, t)(u)$. Therefore, since $t_q \notin \text{L}(\mathcal{A}, q)$, there is no winning strategy for $\exists$ in $\mathbf{G}_\rho(\mathcal{A}, t_q, q)$ and we obtain a contradiction. Therefore $\sigma' \subseteq \sigma$ and for every $u \in \sigma'$ we have $\rho(\mathcal{A}, t)(u) = \rho(\mathcal{A}, t')(u)$, so σ' is also a strategy in $\mathbf{G}_\rho(\mathcal{A}, t')$. Since strategies form an anti-chain with respect to inclusion, we know that $\sigma = \sigma'$, $t' \in \text{L}(\mathcal{A})$, and $t \otimes s \in W^{\exists}_{\mathcal{A}}$. ■

5.3 Partial Objects

In this section we build some technical tools that will be used in solving the alternating index problem for game automata.

The proofs in the alternating case will be inductive over the structure of a given game automaton. Therefore, we introduce here definitions that allow partial objects: partial trees have holes, partial automata have exits (where computation stops), and partial games have final positions (where the play stops and no player wins). The definitions become standard when restricted to *total* objects.

5.3.1 *Trees*

It will be convenient in this chapter to work with partial trees PTr_A, as defined in Sect. 1.1.2 (see page 3). A partial tree that is not complete contains *holes*. A *hole* of a partial tree t is a minimal sequence $u \in \{\text{L}, \text{R}\}^*$ that does not belong to $\text{dom}(t)$ (a hole is *off* t in the sense of Sect. 1.1). By $\text{holes}(t) \subseteq \{\text{L}, \text{R}\}^*$ we denote the set of holes of a tree t. If u is a hole of a tree $t \in \text{PTr}_A$ and $t' \in \text{PTr}_A$ we define the partial tree $t[u \leftarrow t']$ obtained by putting the root of t' into the hole u of t.

Let $\mathcal{A}$ be a game automaton and $q_0 \in Q^{\mathcal{A}}$. Recall the inductive definition of a run ρ of $\mathcal{A}$ on a tree t (see Sect. 5.1). Note that the value $\rho(u)$ is uniquely determined by the labels of t on the path leading from the root to u (except u). Therefore, the value $\rho(\mathcal{A}, t, q_0)(u)$ is well-defined even for a partial tree $t \in \text{PTr}_{A^{\mathcal{A}}}$ and

$u \in \mathrm{dom}(t) \sqcup \mathrm{holes}(t)$. In other words, if $t \in \mathrm{PTr}_{A^{\mathcal{A}}}$ then $\rho(\mathcal{A}, t, q_0)$ is a function of the type

$$\mathrm{dom}(t) \sqcup \mathrm{holes}(t) \longrightarrow Q^{\mathcal{A}}.$$

5.3.2 *Games*

A *partial parity game* is a tuple $\langle V = V_\exists \sqcup V_\forall, v_I, F, E, \Omega\rangle$ as in Sect. 1.3 (see page 6) with an additional set F of *final positions*, $F \cap V = \emptyset$. We assume that $E \subseteq V \times (V \sqcup F)$ is the transition relation — there are transitions from positions in V to positions in V and from positions in V to final positions in F.

A *play* in a partial parity game $\mathcal{G}$ may be a finite sequence $\pi = v_I v_1 \ldots v_n$ of positions with v_n being a final position (i.e. $v_n \in F$). In that case v_n is called the *final position of* π. A finite play is not winning for any of the players.

Strategies are defined in the standard way, see Sect. 1.3: a strategy σ is *winning* if all the infinite plays consistent with σ are winning — the finite plays are irrelevant. Theoretically, both players may have a winning strategy in a partial parity game. For a winning strategy σ we define the *guarantee of* σ as the set of all final positions that can be reached in finite plays consistent with σ.

To operate with partial trees, we extend the definition of the parity game $\mathcal{G}_t$ from Sect. 1.7.4 (see page 23) to the case when $t \in \mathrm{PTr}_{A_{i,j}}$. Whole Definition 1.2 from page 23 is unchanged, the only difference is that we additionally put $F = \mathrm{holes}(t)$ — each hole of t is treated as a final position of the game $\mathcal{G}_t$. As defined in Sect. 1.7.4, the language $W_{i,j}$ is the set of complete trees over $A_{i,j}$ such that $\exists$ has a winning strategy in $\mathcal{G}_t$.

5.3.3 *Automata*

A *partial alternating automaton* $\mathcal{A}$ is defined as a tuple $\langle A, Q, F, \delta, \Omega\rangle$ as in Sect. 1.4 (see page 7) with an additional finite set F of *exits*. We assume that F is disjoint from Q and we allow atomic transitions of the form (f, d) for $f \in F$ and $d \in \{\mathrm{L}, \mathrm{R}\}$ — a transition can lead to an exit but there are no transitions from exits, i.e. the domain of δ is $Q \times A$. Note that a partial automaton does not have an initial state.

An automaton $\mathcal{A}$ is *total* if $F = \emptyset$. In that case the presented definitions take the form from Sect. 1.4.

For a partial alternating automaton $\mathcal{A}$, a state $q_0 \in Q$, and a partial tree $t \in \mathrm{PTr}_A$ we define the partial parity game $\mathcal{G}(\mathcal{A}, t, q_0)$ similarly as in Sect. 1.4:

$$V = \mathrm{dom}(t) \times (S_\delta \sqcup Q),$$
$$F = \big(\mathrm{holes}(t) \times (Q \sqcup F)\big) \sqcup \mathrm{dom}(t) \times F,$$

where S_δ is the set of all sub-formulae of formulae in $\mathrm{rg}(\delta)$; all positions of the form $(u, b_1 \vee b_2)$ belong to $\exists$ and the remaining ones to $\forall$. The edges E follow the transition relation.

Note that the set of final positions of $\mathcal{G}(\mathcal{A}, t, q_0)$ can be split into two disjoint parts: positions in the holes of t, visited in a state or an exit of $\mathcal{A}$, and positions inside t visited in an exit $f \in F$ of $\mathcal{A}$.

5.3.4 Composing Automata

Let $\mathcal{A} = \langle A^{\mathcal{A}}, Q^{\mathcal{A}}, F^{\mathcal{A}}, \delta^{\mathcal{A}}, \Omega^{\mathcal{A}} \rangle$ be a partial alternating automaton and $Q' \subseteq Q^{\mathcal{A}}$ be a set of states. By $\mathcal{A}\restriction_{Q'}$ we denote the *restriction of* $\mathcal{A}$ *to* Q' obtained by replacing the set of states by Q', the set of exits by $F^{\mathcal{A}} \sqcup (Q^{\mathcal{A}} \setminus Q')$, the priority function by $\Omega^{\mathcal{A}}\restriction_{Q'}$, and the transition function by $\delta^{\mathcal{A}}\restriction_{Q' \times A^{\mathcal{A}}}$. We say that $\mathcal{B}$ is a *sub-automaton of* $\mathcal{A}$ (denoted $\mathcal{B} \subseteq \mathcal{A}$) if $Q^{\mathcal{B}} \subseteq Q^{\mathcal{A}}$ and $\mathcal{B} = \mathcal{A}\restriction_{Q^{\mathcal{B}}}$.

For two partial alternating automata $\mathcal{A}, \mathcal{B}$ over an alphabet A with $Q^{\mathcal{A}} \cap Q^{\mathcal{B}} = \emptyset$, we define the composition $\mathcal{A} \cdot \mathcal{B}$ as the automaton over A, with states $Q = Q^{\mathcal{A}} \sqcup Q^{\mathcal{B}}$, exits $(F^{\mathcal{A}} \cup F^{\mathcal{B}}) \setminus Q$, transitions $\delta^{\mathcal{A}} \cup \delta^{\mathcal{B}}$, and priorities $\Omega^{\mathcal{A}} \cup \Omega^{\mathcal{B}}$. What is very important is that some exits of $\mathcal{A}$ may be states of $\mathcal{B}$ and vice versa.

Fact 5.47. *If $\mathcal{A}$ is a partial alternating automaton and $Q^{\mathcal{A}} = Q_1 \sqcup Q_2$ is a partition of the states of $\mathcal{A}$ then $\mathcal{A}\restriction_{Q_1} \cdot \mathcal{A}\restriction_{Q_2} = \mathcal{A}$.*

5.3.5 Resolving

Let $t \in \mathrm{PTr}_A$ be a partial tree and $\rho = \rho(\mathcal{A}, t, q_0)$ be the run of a total game automaton $\mathcal{A}$ on t from a state q_0. We say that t *resolves* $\mathcal{A}$ *from* $q_0 \in Q$ if $\rho(w) \neq \star$ for each hole w of t and for every $u \in \mathrm{dom}(t)$ if $t\restriction_{ud}$ is the only total tree in $\{t\restriction_{u\mathrm{L}}, t\restriction_{u\mathrm{R}}\}$, either $\rho(ud) = \star$ or ud is losing for the owner of u in $\mathbf{G}_\rho(\mathcal{A}, t, q_0)$.

The following fact shows the crucial property of trees that resolve game automata. It can be seen as an extension of Remark 5.5.

Fact 5.48. *Assume that t resolves $\mathcal{A}$ from q_0 and $\rho = \rho(\mathcal{A}, t, q_0)$ assigns states to all the holes of t. If t has a single hole u then for every $s \in \mathrm{Tr}_A$ we have*

$$t[u \leftarrow s] \in \mathrm{L}(\mathcal{A}, q_0) \iff s \in \mathrm{L}(\mathcal{A}, \rho(u)).$$

If t has two holes u, u', whose closest common ancestor w satisfies $\delta_{\mathcal{A}}(\rho(w), t(w)) = (q_\mathrm{L}, \mathtt{L}) \wedge (q_\mathrm{R}, \mathtt{R})$ for some q_L, q_R then for all s, s'

$$t[u \leftarrow s, u' \leftarrow s'] \in \mathrm{L}(\mathcal{A}, q_0) \iff \big(s \in \mathrm{L}(\mathcal{A}, \rho(u)) \text{ and } s' \in \mathrm{L}(\mathcal{A}, \rho(u'))\big);$$

dually for $(q_\mathrm{L}, \mathtt{L}) \vee (q_\mathrm{R}, \mathtt{R})$ with or *on the right-hand side.*

Proof. The proof of the first claim is exactly the same as in Remark 5.5.

For the second claim, it follows easily that in this case the trees $t{\restriction}_{u\mathrm{L}}$, $t{\restriction}_{u\mathrm{R}}$ and the tree obtained by putting a hole in t instead of u, resolve $\mathcal{A}$ from q_{L}, q_{R}, and q_0, respectively. We obtain the second claim by applying the first claim three times. ■

5.4 Alternating Index Problem

In this section we prove the second part of Theorem 2.3: the alternating index problem is decidable for game automata. As a consequence of our characterisation, in the case of languages recognisable by game automata the respective classes $\mathbf{Comp}(\mathbf{\Pi}_i^{\mathrm{alt}})$ and $\mathbf{\Delta}_{i+1}^{\mathrm{alt}}$ coincide for all levels. All these properties are summarized by the following proposition.

Proposition 5.8. *For each game automaton $\mathcal{A}$, the language $\mathrm{L}(\mathcal{A})$ belongs to exactly one of the classes:*

$$\mathbf{Comp}(\mathbf{\Pi}_0^{\mathrm{alt}}),\quad \mathbf{\Pi}_i^{\mathrm{alt}} \setminus \mathbf{\Sigma}_i^{\mathrm{alt}},\quad \mathbf{\Sigma}_i^{\mathrm{alt}} \setminus \mathbf{\Pi}_i^{\mathrm{alt}},\quad or\quad \mathbf{Comp}(\mathbf{\Pi}_i^{\mathrm{alt}}) \setminus \left(\mathbf{\Pi}_i^{\mathrm{alt}} \cup \mathbf{\Sigma}_i^{\mathrm{alt}}\right),$$

for $i > 0$.

Moreover, it can be effectively decided which class it is and an automaton from this class can be constructed.

If a game language L is Borel then it belongs to $\mathbf{Comp}(\mathbf{\Pi}_0^{\mathrm{alt}})$ (i.e. L is WMSO*-definable).*

The rest of this section is devoted to showing this result. Section 5.4.1 describes a recursive procedure to compute the *class* of the given automaton $\mathcal{A}$, i.e. $\mathbf{\Pi}_i^{\mathrm{alt}}$, $\mathbf{\Sigma}_i^{\mathrm{alt}}$, or $\mathbf{Comp}(\mathbf{\Pi}_i^{\mathrm{alt}})$, depending on which of the possibilities holds. Sections 5.4.2 and 5.4.3 show that the procedure is correct. The estimation of Sect. 5.4.2 is in fact an effective construction of an automaton from the respective class. The continuous reductions from Sect. 5.4.3 imply that if $\mathrm{class}(\mathcal{A}) \neq \mathbf{Comp}(\mathbf{\Pi}_0^{\mathrm{alt}})$ then $\mathrm{L}(\mathcal{A})$ is non-Borel.

5.4.1 The Algorithm

Let $\mathcal{A}$ be an alternating automaton of index (i, j). For $n \in \mathbb{N}$ we denote by $\mathcal{A}^{\geqslant n}$ the partial sub-automaton obtained from $\mathcal{A}$ by restricting to states of priority at least n:

$$\mathcal{A}^{\geqslant n} \stackrel{\mathrm{def}}{=} \mathcal{A}{\restriction}_{Q'} \quad \text{for } Q' = \left(\Omega^{\mathcal{A}}\right)^{-1}\left(\{n, n+1, \ldots, j\}\right).$$

Observe that the index of $\mathcal{A}^{\geqslant n}$ is at most (n, j). A partial sub-automaton $\mathcal{B} \subseteq \mathcal{A}$ is an *n-component of* $\mathcal{A}$ if $\mathrm{Graph}(\mathcal{B})$ is a strongly-connected component of $\mathrm{Graph}(\mathcal{A}^{\geqslant n})$ (in particular $\mathcal{B} \subseteq \mathcal{A}^{\geqslant n}$). We say that $\mathcal{B}$ is *non-trivial* if $\mathrm{Graph}(\mathcal{B})$ contains at least one edge. Our algorithm computes the *class* of each n-component $\mathcal{B}$ of $\mathcal{A}$, based on the classes of $(n+1)$-components of $\mathcal{B}$ and transitions between them. (We shall see that for n-components the class does not depend on the initial state.)

We begin with a simple preprocessing. An automaton $\mathcal{A}$ of index (i, j) is *priority-reduced* if for all $n > i$, each n-component of $\mathcal{A}$ is non-trivial and contains a state of priority n.

Lemma 5.12. *Each game automaton $\mathcal{A}$ can be effectively transformed into an equivalent priority-reduced game automaton.*

Proof. We iteratively decrease priorities in n-components of $\mathcal{A}$, for $n > i$. As long as there is an n-component that is not priority-reduced, pick any such n-component, if it is trivial, set all its priorities to $n-1$, if it is non-trivial but does not contain a state of priority n, decrease all its priorities by 2 (this does not influence the recognised language). After finitely many steps the automaton is priority-reduced. Note that no trivial states are introduced and the language of the automaton is preserved. ■

Therefore, we can assume that $\mathcal{A}$ is a priority-reduced automaton of index (i, j). The algorithm starts from $n = j$ and proceeds downward. Let $\mathcal{B}$ be an n-component. We define class($\mathcal{B}$) by considering the following cases.

If $\mathcal{B}$ has only states of priority n then it is an (n, n)-automaton and we can put $\text{class}(\mathcal{B}) = \mathbf{Comp}(\mathbf{\Pi}_0^{\text{alt}})$.

If $\mathcal{B}$ has no states of priority n then, since $\mathcal{A}$ is priority-reduced, it follows that $n = i$ and $\mathcal{B}$ coincides with a single $(n+1)$-component $\mathcal{B}_1$. In that case we put $\text{class}(\mathcal{B}) = \text{class}(\mathcal{B}_1)$.

Otherwise, let $\mathcal{B}_1, \mathcal{B}_2, \dots, \mathcal{B}_k, k \geqslant 1$, be the $(n+1)$-components of $\mathcal{B}$. Assume that n is even (for odd n, the procedure is entirely dual: $\exists$ is replaced with $\forall$, $(q_{\mathrm{L}}, \mathrm{L}) \vee (q_{\mathrm{R}}, \mathrm{R})$ with $(q_{\mathrm{L}}, \mathrm{L}) \wedge (q_{\mathrm{R}}, \mathrm{R})$, and $\mathbf{\Pi}_m^{\text{alt}}$ with $\mathbf{\Sigma}_m^{\text{alt}}$).

For a class K let us define the operation $K^{\exists}$ by the following equation

$$\left(\mathbf{\Pi}_m^{\text{alt}}\right)^{\exists} = \left(\mathbf{\Sigma}_{m-1}^{\text{alt}}\right)^{\exists} = \left(\mathbf{Comp}(\mathbf{\Pi}_{m-1}^{\text{alt}})\right)^{\exists} = \mathbf{\Pi}_m^{\text{alt}}.$$

A component $\mathcal{B}_\ell$ is $\exists$ *-branching* if $\mathcal{B}$ contains a transition

$$\delta(p, a) = (q_{\mathrm{L}}, \mathrm{L}) \vee (q_{\mathrm{R}}, \mathrm{R})$$

with $\left(p, q_{\mathrm{L}} \in Q^{\mathcal{B}_\ell}, q_{\mathrm{R}} \in Q^{\mathcal{B}}\right)$ or $\left(p, q_{\mathrm{R}} \in Q^{\mathcal{B}_\ell}, q_{\mathrm{L}} \in Q^{\mathcal{B}}\right)$. Now, for $\ell = 1, 2, \dots, k$ let us compute a class K_ℓ by considering the following cases:

- if $\mathcal{B}_\ell$ is $\exists$-branching then $K_\ell = \text{class}(\mathcal{B}_\ell)^{\exists}$,
- otherwise $K_\ell = \text{class}(\mathcal{B}_\ell)$.

We set

$$\text{class}(\mathcal{B}) = \bigvee_{\ell=1}^{k} K_\ell\,,$$

i.e. the largest class among $K_1, K_2, \ldots, K_k$ if it exists, or $\mathbf{Comp}(\mathbf{\Pi}^{\mathrm{alt}}_m)$ if among these classes there are two maximal ones, $\mathbf{\Pi}^{\mathrm{alt}}_m$ and $\mathbf{\Sigma}^{\mathrm{alt}}_m$.

Let $\mathrm{class}(\mathcal{A}) = \bigvee_{\ell=1}^{k} \mathcal{A}_\ell$ where $\mathcal{A}_1, \mathcal{A}_2, \ldots, \mathcal{A}_k$ are the i-components of $\mathcal{A}$ reachable from $q_I^{\mathcal{A}}$ in $\mathrm{Graph}(\mathcal{A})$.

The following fact follows directly from the definition. It shows that to reach $\mathrm{class}(\mathcal{B}_\ell)$ higher than $\mathbf{\Pi}^{\mathrm{alt}}_1$ an $\exists$-branching transition has to occur.

Fact 5.49. *Using the above notions, if* $K_\ell \geqslant \mathbf{\Pi}^{\mathrm{alt}}_1$ *then* $\mathcal{B}_\ell$ *is* $\exists$*-branching.*

5.4.2 Upper Bounds

In this subsection we show that $L(\mathcal{A})$ can be recognised by a $\mathrm{class}(\mathcal{A})$-automaton. The argument will closely follow the recursive algorithm, pushing through an invariant guaranteeing that each n-component $\mathcal{B}$ of $\mathcal{A}$ can be replaced with an "equivalent" $\mathrm{class}(\mathcal{B})$-automaton. The notion of equivalence for non-total automata is formalised by simulations, see Definition 5.4.

Recall from Sect. 5.3.3 that if t is a total tree and $\mathcal{A}$ is a partial alternating automaton then the final positions of $\mathbf{G}(\mathcal{A}, t)$ are of the form (u, f) where $u \in \{\mathrm{L}, \mathrm{R}\}^*$ and f is an exit of $\mathcal{A}$. Similarly, for every $u \in \{\mathrm{L}, \mathrm{R}\}^*$ and $q \in Q^{\mathcal{A}}$ there is a position of the form (u, q) in $\mathbf{G}(\mathcal{A}, t)$ (in may not be reachable from the initial position).

Definition 5.4. *Assume that* $\mathcal{S}$ *is a partial alternating automaton and* $\mathcal{A}$ *is a partial game automaton, both over the same alphabet* A. *We say that* $\mathcal{S}$ simulates $\mathcal{A}$ *if* $F^{\mathcal{S}} \subseteq F^{\mathcal{A}}$ *and there exists an embedding* $\iota\colon Q^{\mathcal{A}} \to Q^{\mathcal{S}}$ *(usually* $Q^{\mathcal{A}} \subseteq Q^{\mathcal{S}}$*) such that for all* $t \in \mathrm{Tr}_A$, $q_0 \in Q^{\mathcal{A}}$, *and for each winning strategy* σ *for a player* $P \in \{\exists, \forall\}$ *in* $\mathbf{G}(\mathcal{A}, t, q_0)$ *there is a winning strategy* σ^S *for* P *in* $\mathbf{G}(\mathcal{S}, t, \iota(q_0))$ *such that the guarantee of* σ^S *is contained in the guarantee of* σ.

Note that if $\mathcal{A}$ and $\mathcal{S}$ are total and $\mathcal{S}$ simulates $\mathcal{A}$ then $\mathrm{L}(\mathcal{A}) = \mathrm{L}(\mathcal{S}, \iota(q_I^{\mathcal{A}}))$. The following lemma formalises the inductive invariant that we will prove.

Lemma 5.13. *For every* n*-component* $\mathcal{B}$ *of a game automaton* $\mathcal{A}$, $\mathcal{B}$ *can be simulated by a* $\mathrm{class}(\mathcal{B})$*-automaton.*

From this lemma it follows easily that $\mathrm{L}(\mathcal{A})$ can be recognised by a $\mathrm{class}(\mathcal{A})$-automaton: the automaton can be obtained as a loop-less composition of the $\mathrm{class}(\mathcal{A}_\ell)$-automata simulating the i-components $\mathcal{A}_\ell$ of $\mathcal{A}$ reachable from $q_I^{\mathcal{A}}$. In other words, the upper bounds computed by the algorithm in Sect. 5.4.1 are correct.

The rest of this section is devoted to a proof of this lemma. Assume that the index of $\mathcal{A}$ is (i, j). We proceed by induction on $n = j, j-1, \ldots, i$. Assume that $\mathcal{B}$ is an n-component of $\mathcal{A}$. If all the states of $\mathcal{B}$ have priority n or all have priority strictly greater than n, the claim is immediate.

Let us assume that neither is the case and let $\mathcal{B}_1, \mathcal{B}_2, \ldots, \mathcal{B}_k$ be the $(n+1)$-components of $\mathcal{B}$. By the inductive hypothesis we get a $\mathrm{class}(\mathcal{B}_\ell)$-automaton $\mathcal{B}_\ell^S$, simulating $\mathcal{B}_\ell$. We shall construct a class($\mathcal{B}$)-automaton $\mathcal{B}^S$ that simulates $\mathcal{B}$ by combining the automata $\mathcal{B}_\ell^S$. By symmetry it is enough to give the construction for even n. Examining the algorithm we see that for each ℓ, either $K_\ell = \text{class}(\mathcal{B}_\ell)^\exists = \mathbf{\Pi}_{m_\ell}^{\text{alt}}$ for some m_ℓ, or $K_\ell = \text{class}(\mathcal{B}_\ell) \leqslant \mathbf{\Sigma}_1^{\text{alt}}$ and $\mathcal{B}_\ell$ is not $\exists$-branching.

First assume that class($\mathcal{B}$) > $\mathbf{Comp}(\mathbf{\Pi}_1^{\text{alt}})$. In that case class($\mathcal{B}$) $= \bigvee_\ell K_\ell = \mathbf{\Pi}_m^{\text{alt}}$ for some $m \geqslant 2$, and each $\mathcal{B}_\ell^S$ can be assumed to be an $(n, n{+}m)$-automaton. Hence, we can put

$$B^S = \mathcal{B}{\restriction}_{\Omega^{-1}(n)} \cdot \mathcal{B}_1^S \cdot \mathcal{B}_2^S \cdot \ldots \cdot \mathcal{B}_k^S \tag{5.3}$$

to get an $(n, n{+}m)$-automaton. We need to show that $\mathcal{B}^S$ simulates $\mathcal{B}$. Let ι be defined by inductive assumption on automata $\mathcal{B}_\ell$ and as the identity on $\mathcal{B}{\restriction}_{\Omega^{-1}(n)}$. Clearly the exits of $\mathcal{B}^S$ are contained in the exits of $\mathcal{B}$. Assume that $t \in \text{Tr}_A$, $q_0 \in Q^{\mathcal{B}}$, and σ is a winning strategy of a player $P \in \{\exists, \forall\}$ in $\mathbf{G}(\mathcal{B}, t, q_0)$. Consider a strategy σ^S in $\mathbf{G}(\mathcal{B}^S, t, \iota(q_0))$ that repeats the decisions of σ in $\mathcal{B}{\restriction}_{\Omega^{-1}(n)}$ and uses the inductive assumption to play on the components $\mathcal{B}_\ell^S$.

Consider any finite or infinite play π^S consistent with σ^S in $\mathbf{G}(\mathcal{B}^S, t, \iota(q_0))$. Observe that this play can be split into a sequence (finite or infinite) of plays $\pi_0^S \cdot \pi_1^S \cdot \ldots$ corresponding to the elements of the product (5.3) — after every prefix $\pi_0^S \ldots \pi_k^S$ an exit of the current sub-automaton is visited and the play moves to another sub-automaton in (5.3). By the inductive assumption about the containment of the guarantees we know that the same sequence of sub-automata (using the same exits) can be visited by a play π in $\mathbf{G}(\mathcal{B}, t, q_0)$. If π^S is finite then π is also finite and ends in the same final position (u, f). Therefore, the guarantee of σ^S is contained in the guarantee of σ. Now assume that π^S is infinite. By the definition of n-components, we know that either π^S visits infinitely many times a state in $\mathcal{B}{\restriction}_{\Omega^{-1}(n)}$ (in that case both π^S and π are winning for $\exists$), or π^S stays from some point on in one of the sub-automata $\mathcal{B}_\ell^S$. In that case, by the inductive assumption we know that π^S is winning for P. Therefore, σ^S is winning for P.

Now assume that class($\mathcal{B}$) $\leqslant \mathbf{Comp}(\mathbf{\Pi}_1^{\text{alt}})$. We will repeat the above construction by taking special care to obtain a class($\mathcal{B}$)-automaton. We call a component $\mathcal{B}_\ell$ *problematic* if $\mathcal{B}_\ell$ is not $\exists$-branching. For such components we replace $\mathcal{B}_\ell^S$ in (5.3) by $\mathcal{B}_\ell^R \cdot \mathcal{B}_\ell^T$, where

- $\mathcal{B}_\ell^T$ is $\mathcal{B}_\ell^S$ with each transition leading to an exit of $\mathcal{B}_\ell$ that is not an exit of $\mathcal{B}$ replaced with a transition to $\top$ (losing for $\forall$);
- $\mathcal{B}_\ell^R$ is $\mathcal{B}_\ell$ with all priorities set to n and additional ϵ-transitions (which can be eliminated in the usual way): for each state q of $\mathcal{B}_\ell^R$ allow $\forall$ to decide to stay in q or to move to the respective state $\iota(q)$ in $\mathcal{B}_\ell^T$ (such a move is treated as an exit of $\mathcal{B}_\ell^R$).

As in (5.3), $\mathcal{B}^S$ is the composition of $\mathcal{B}{\restriction}_{\Omega^{-1}(n)}$ and the appropriate automata $\mathcal{B}_\ell^S$, $\mathcal{B}_\ell^R$, $\mathcal{B}_\ell^T$. This composition gives a class($\mathcal{B}$)-automaton: each problematic $\mathcal{B}_\ell$ was replaced with an (n, n)-automaton $\mathcal{B}_\ell^R$ that is further composed with class($\mathcal{B}_\ell$)-automata $\mathcal{B}_\ell^T$ in a loop-less way.

What remains is to show that $\mathcal{B}^S$ simulates $\mathcal{B}$. Let ι be defined as before for non-problematic components and on a problematic component $\mathcal{B}_\ell$ as the identity $Q^{\mathcal{B}_\ell} \to Q^{\mathcal{B}_\ell^R}$. Consider a tree $t \in \mathrm{Tr}_A$, a state q_0 of $\mathcal{B}$, and games $\mathbf{G}(\mathcal{B}, t, q_0)$ and $\mathbf{G}(\mathcal{B}^S, t, \iota(q_0))$.

Firstly assume that σ is a winning strategy of $\exists$ in $\mathbf{G}(\mathcal{B}, t, q_0)$. Since $\exists$ has no additional choices in $\mathcal{B}^S$ comparing to the above case and all the changes of priorities in $\mathcal{B}_\ell^R$, $\mathcal{B}_\ell^T$ are favourable to her, the previous construction gives a strategy σ^S that simulates σ.

Now assume that σ is a winning strategy for $\forall$ in $\mathbf{G}(\mathcal{B}, t, q_0)$. Let us define a strategy σ^S for $\forall$ in $\mathbf{G}(\mathcal{B}^S, t, \iota(q_0))$ as follows:

- in positions corresponding to states of priority n in $\mathcal{B}$ as well as in the components $\mathcal{B}_\ell^R$ the strategy σ^S follows the decisions of σ;
- $\forall$ immediately moves from $\mathcal{B}_\ell^R$ to $\mathcal{B}_\ell^T$ whenever each extension of the current play, conforming to σ, stays forever in $\mathcal{B}_\ell$ or reaches an exit that is also an exit of $\mathcal{B}$;
- in components $\mathcal{B}_\ell^S$ and $\mathcal{B}_\ell^T$ the strategy σ^S simulates σ using the inductive assumption.

As before the guarantee of σ^S is contained in the guarantee of σ. It remains to prove that σ^S is winning for $\forall$. Let π^S be a play consistent with σ^S. It is enough to exclude the following cases (in other cases we know that π^S is winning because σ was a winning strategy):

1. π^S stays from some point on in $\mathcal{B}_\ell^R$ (and therefore is losing for $\forall$ by the parity criterion),
2. π^S reaches the transition $\top$ in an automaton $\mathcal{B}_\ell^T$ (such transition corresponds to a transition to an exit of $\mathcal{B}_\ell$ that is not an exit of $\mathcal{B}$).

Let $\mathcal{B}_\ell$ be a problematic component (i.e. $\mathcal{B}_\ell$ is not $\exists$-branching in $\mathcal{B}$).

Consider the first case above. By the definition of σ^S it means that there is a play π that is consistent with σ and that from some point on in $\mathcal{B}_\ell$. We can assume that π starts in $\mathcal{B}_\ell$ and never leaves it. By the assumption that $\mathcal{B}_\ell$ is not $\exists$-branching in $\mathcal{B}$ we know that whenever $\exists$ has a choice during π exactly one of the successive states is an exit of $\mathcal{B}$. Therefore, the strategy σ^S moves from $\mathcal{B}_\ell^R$ to $\mathcal{B}_\ell^T$ what contradicts the assumption that π^S stays forever in $\mathcal{B}_\ell^R$.

Now consider the second case above: the transition $\top$ is reached in $\mathcal{B}_\ell^T$. Again we can assume that the moment when $\forall$ decided to move from $\mathcal{B}_\ell^R$ to $\mathcal{B}_\ell^T$ was at the initial position of the game. By the inductive assumption about $\mathcal{B}_\ell^S$ it means that it is possible to visit an exit of $\mathcal{B}_\ell$ that is not an exit of $\mathcal{B}$ by a play consistent with σ. But this contradicts the definition of σ^S — the only case when $\forall$ moves to $\mathcal{B}_\ell^T$ is when he knows that the strategy σ will never reach any exit of $\mathcal{B}_\ell$ that is not an exit of $\mathcal{B}$.

This concludes the proof of Lemma 5.13.

5.4.3 Lower Bounds

It remains to see that $L(\mathcal{A})$ cannot be recognised by an alternating automaton of index lower than $\mathrm{class}(\mathcal{A})$. For this purpose we will use the pre-order $\leqslant_W$ from Sect. 1.6.2 and the $W_{i,j}$ languages from Sect. 1.7.4, page 23.

By Corollary 1.1 from page 24, in order to show that the index bound computed by the algorithm from Sect. 5.4.1 is tight, it suffices to show that if $\mathbf{RM}^{\mathrm{alt}}(i,j) \leqslant \mathrm{class}(\mathcal{A})$ then $W_{i,j} \leqslant_W L(\mathcal{A})$. Therefore, our aim will be to construct a continuous reduction from $W_{i,j}$ to $L(\mathcal{A})$.

We construct the reduction in three steps:

1. we show that if the class computed by the algorithm is at least $\mathbf{RM}^{\mathrm{alt}}(i,j)$ then this is witnessed with a certain hard subgraph in the graph of the automaton, called (i,j)-edelweiss;
2. we introduce intermediate languages $\widehat{W}_{i,j}$, whose internal structure corresponds precisely to (i,j)-edelweisses, and show that $\widehat{W}_{i,j} \leqslant_W L(\mathcal{A})$ if only $\mathcal{A}$ contains an (i,j)-edelweiss reachable from $q_I^{\mathcal{A}}$;
3. we prove that $W_{i,j} \leqslant_W \widehat{W}_{i,j}$.

The combinatorial core of the argument is the last step.

Definition 5.5. *We say that in a game automaton $\mathcal{B}$ there is an* i-loop rooted in p *if there exists a word u such that on the path $p \xrightarrow{u} p$ in* $\mathrm{Graph}(\mathcal{B})$ *the minimal priority is i.*

An automaton $\mathcal{B}$ contains an (i,j)-loop for $\exists$ rooted in p *if there exist states q, q_L, q_R of $\mathcal{B}$, a letter a, and words u, u_L, u_R such that:*

- $\delta(q,a) = (q_L, \mathrm{L}) \vee (q_R, \mathrm{R})$;
- $p \xrightarrow{u} q$; $q_L \xrightarrow{u_L} p$; $q_R \xrightarrow{u_R} p$;
- *on one of the paths $p \xrightarrow{u(a,\mathrm{L})u_L} p$, $p \xrightarrow{u(a,\mathrm{R})u_R} p$ the minimal priority is i and on the other it is j.*

For $\forall$ dually, with $\vee$ replaced with $\wedge$.

For an even $j > i$, $\mathcal{B}$ *contains an* (i,j)-edelweiss rooted in p *(see Fig. 5.1) if for some even n it contains:*

- *$(n+k)$-loops for $k = i, i+1, \ldots, j-3$,*
- *$(n+j-2, n+j-1)$-loop for $\exists$, if $i \leqslant j-2$,*
- *$(n+j-1, n+j)$-loop for $\forall$*

all rooted in p. For an odd j swap $\forall$ and $\exists$ but keep n even.

Lemma 5.14. *Let $\mathcal{A}$ be a game automaton. If* $\mathrm{class}(\mathcal{A}) \geqslant \mathbf{RM}^{\mathrm{alt}}(i,j)$ *then $\mathcal{A}$ contains an (i,j)-edelweiss rooted in a state reachable from $q_I^{\mathcal{A}}$.*

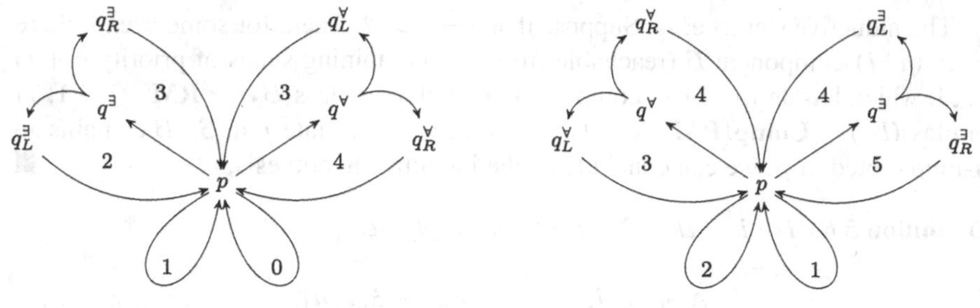

Fig. 5.1 (0, 4)-edelweiss and (1, 5)-edelweiss.

Proof. Let us first assume that $(i, j) = (0, 1)$. Analysing the algorithm we see that the only case when class($\mathcal{A}$) jumps to $\mathbf{RM}^{\text{alt}}(0, 1)$ is when for some even n there is an n-component $\mathcal{B}$ in $\mathcal{A}$, reachable from $q_I^{\mathcal{A}}$, and containing states of priority n, such that some $(n+1)$-component $\mathcal{B}_\ell$ of $\mathcal{B}$ is $\exists$-branching in $\mathcal{B}$, i.e. $\mathcal{B}$ contains a transition of the form

$$\delta(p, a) = (q_{\mathrm{L}}, \mathrm{L}) \vee (q_{\mathrm{R}}, \mathrm{R})$$

with $p, q_{\mathrm{L}} \in Q^{\mathcal{B}_\ell}$, $q_{\mathrm{R}} \in Q^{\mathcal{B}}$ (or symmetrically, $p, q_{\mathrm{R}} \in Q^{\mathcal{B}_\ell}$, $q_{\mathrm{L}} \in Q^{\mathcal{B}}$). Since $\mathcal{A}$ is priority-reduced, p is reachable from q_{L} within $\mathcal{B}_\ell$ via a state of priority $n+1$, and from q_{R} within $\mathcal{B}$ via a state of priority n. This gives an $(n, n+1)$-loop for $\exists$ (a (0, 1)-edelweiss) rooted in a state reachable from $q_I^{\mathcal{A}}$. The argument for (1, 2) is entirely dual.

Next, assume that $(i, j) = (0, 2)$. It follows immediately from the algorithm that $\mathcal{A}$ contains an n-component $\mathcal{B}$ (reachable from $q_I^{\mathcal{A}}$, containing states of priority n) such that n is even and there exists an $(n+1)$-component $\mathcal{B}_\ell$ such that

1. class($\mathcal{B}_\ell$) $= \mathbf{\Sigma}_1^{\text{alt}}$ and $\mathcal{B}_\ell$ is $\exists$-branching in $\mathcal{B}$; or
2. class($\mathcal{B}_\ell$) $= \mathbf{Comp}(\mathbf{\Pi}_1^{\text{alt}})$.

In the first case, by the claim for (1, 2), $\mathcal{B}_\ell$ contains an $(n', n'+1)$-loop for $\forall$, for some odd $n' \geqslant n$. Since $\mathcal{A}$ is priority-reduced, for each state q in $\mathcal{B}_\ell$ and each r between n and $\Omega(q)$, there is a loop from q to q with the lowest priority r. Hence, the $(n', n'+1)$-loop can be turned into an $(n+1, n+2)$-loop. Thus, $\mathcal{B}_\ell$ contains an $(n+1, n+2)$-loop for $\forall$, rooted in a state p. We claim that $\mathcal{B}$ contains an $(n, n+1)$-loop for $\exists$, also rooted in p (giving a (0, 2)-edelweiss rooted in p).

Indeed, since $\mathcal{B}_\ell$ is $\exists$-branching, arguing like for (0, 1), we obtain an $(n, n+1)$-loop for $\exists$ rooted in a state p' in $\mathcal{B}_\ell$. Since $\mathcal{B}_\ell$ is an $(n+1)$-component, there are paths in $\mathcal{B}_\ell$ from p to p' and back; the lowest priority on these paths is at least $n+1$. Using these paths one easily transforms the $(n, n+1)$-loop rooted in p' into an $(n, n+1)$-loop rooted in p.

In the second case, we also get an $(n+1, n+2)$-loop for $\forall$, rooted in a state p of $\mathcal{B}_\ell$. Moreover, the first claim implies as well that $\mathcal{B}_\ell$ contains an $(n'', n''+1)$-loop for $\exists$, for some even $n'' \geqslant n$. Arguing like in the second case we turn the latter loop into an $(n, n+1)$-loop for $\exists$ rooted in p.

The inductive step is easy. Suppose that $j - i > 2$. Then, for some even n there is an $(n+i)$-component $\mathcal{B}$ (reachable from $q_I^{\mathcal{A}}$, containing states of priority $n + i$) in $\mathcal{A}$, which has an $(n+i+1)$-component $\mathcal{B}_\ell$ such that $\mathrm{class}(\mathcal{B}_\ell) = \mathbf{RM}^{\mathrm{alt}}(i + 1, j)$ or $\mathrm{class}(\mathcal{B}_\ell) = \mathbf{Comp}\big(\mathbf{RM}^{\mathrm{alt}}(i + 1, j)\big)$. Since for each state p in $\mathcal{B}_\ell$, $\mathcal{B}$ contains an n-loop rooted in p, we can conclude by the inductive hypothesis. ■

Definition 5.6. *For $i \leqslant 2k - 2$ consider the alphabet*

$$\widehat{A}_{i,2k} = \{i, i + 1, \ldots, 2k - 3, e, a\}.$$

With each $t \in \mathrm{PTr}_{\widehat{A}_{i,2k}}$ we associate a partial parity game $\widehat{\mathcal{G}}_t$ with positions $\mathrm{dom}(t)$ *and final positions* $\mathrm{holes}(t)$ *such that*

- *if $\epsilon \in \mathrm{dom}(t)$ then $\Omega(\epsilon) = i$,*
- *if $t(u) = a$ then in u the player $\forall$ can choose to go to $u\mathrm{L}$ or to $u\mathrm{R}$, and $\Omega(u\mathrm{L}) = 2k - 1$, $\Omega(u\mathrm{R}) = 2k$,*
- *if $t(u) = e$ then in u the player $\exists$ can choose to go to $u\mathrm{L}$ or to $u\mathrm{R}$, and $\Omega(u\mathrm{L}) = 2k - 2$, $\Omega(u\mathrm{R}) = 2k - 1$,*
- *if $t(u) \in \{i, i + 1, \ldots, 2k - 3\}$, the only move from u is to $u\mathrm{L}$ and $\Omega(u\mathrm{L}) = t(u)$.*

For $i = 2k - 1$, let $\widehat{A}_{i,2k} = \{a, \top\}$, and let $\widehat{\mathcal{G}}_t$ be defined like above, except that if $t(u) = \top$ then $\Omega(u) = 2k$ and the only move from u is back to u.

Let $\widehat{W}_{i,2k} \subseteq \mathrm{Tr}_{\widehat{A}_{i,2k}}$ be the set of all total trees over $\widehat{A}_{i,2k}$ such that $\exists$ has a winning strategy in $\widehat{\mathcal{G}}_t$.

The languages $\widehat{W}_{i,2k+1}$ are defined dually, with e, a and $\exists, \forall$ swapped, and $\top$ replaced with $\bot$.

Observe that the index of the game $\widehat{\mathcal{G}}_t$ is (i, j) for $t \in \mathrm{PTr}_{\widehat{A}_{i,j}}$.

Lemma 5.15. *If a total game automaton $\mathcal{A}$ contains an (i, j)-edelweiss rooted in a state reachable from the initial state $q_I^{\mathcal{A}}$ then $\widehat{W}_{i,j} \leqslant_{\mathrm{W}} \mathrm{L}(\mathcal{A})$.*

Proof. We only give a proof for $(i, j) = (1, 2)$; for other values of (i, j) the argument is entirely analogous. By the definition, $\mathcal{A}$ contains an $(1, 2)$-loop for $\forall$, rooted in a state p reachable from $q_I^{\mathcal{A}}$. Since $\mathcal{A}$ is a game automaton and has no trivial states, it follows that there exist

- a partial tree t_I resolving $\mathcal{A}$ from $q_I^{\mathcal{A}}$, with a single hole h, labelled with p in $\rho(\mathcal{A}, t_I)$;
- a partial tree t_a resolving $\mathcal{A}$ from p with two holes h_1, h_2, such that in $\rho(\mathcal{A}, t_a, p)$ both holes are labelled p, the lowest priority on the path from the root to h_i is i, and the closest common ancestor u' of h_1 and h_2 is labelled with a state q such that $\delta^{\mathcal{A}}(q, t(u')) = (q_{\mathrm{L}}, \mathrm{L}) \wedge (q_{\mathrm{R}}, \mathrm{R})$ for some $q_{\mathrm{L}}, q_{\mathrm{R}}$; and
- a total tree $t_\top \in \mathrm{L}(\mathcal{A}, p)$.

Let us see how to build t_a. The paths $p \xrightarrow{u(a,\mathrm{L})u_{\mathrm{L}}} p$, $p \xrightarrow{u(a,\mathrm{R})u_{\mathrm{R}}} p$ guaranteed by Definition 5.5 give as a partial tree s with a single branching in some node u and two leaves h_1, h_2, which we replace with holes. For $\rho = \rho(\mathcal{A}, s, p)$, $\rho(h_1) = \rho(h_2) = p$ and $\delta^{\mathcal{A}}(\rho(u), t(u)) = (q_{\mathrm{L}}, \mathrm{L}) \wedge (q_{\mathrm{R}}, \mathrm{R})$. At each hole of s, except h_1 and h_2, we substitute a total tree such that the run on the resulting tree with two holes resolves $\mathcal{A}$ from p, e.g. if $w\mathrm{L}$ is a hole and $\delta(s(w), \rho(w)) = (q', \mathrm{L}) \vee (q'', \mathrm{R})$, we substitute at $w\mathrm{L}$ any tree that is not in $\mathrm{L}(\mathcal{A}, q')$, relying on the assumption that $\mathcal{A}$ has no trivial states.

Observe that for $(i, j) = (1, 2)$ the alphabet $\widehat{A}_{i,j}$ equals $\{a, \top\}$. Let us define the reduction $g\colon \mathrm{Tr}_{\{a,\top\}} \to \mathrm{Tr}_{A^{\mathcal{A}}}$. Let $t \in \mathrm{Tr}_{\{a,\top\}}$. For $u \in \mathrm{dom}(t)$, define t_u co-inductively (see Sect. 1.6.5, page 18) as follows: if $t(u) = \top$, set $t_u = t_\top$; if $t(u) = a$ then t_u is obtained by plugging in the holes h_1, h_2 of t_a the trees $t_{u\mathrm{L}}$ and $t_{u\mathrm{R}}$. Let $g(t)$ be obtained by plugging t_ϵ in the hole of t_I. It is easy to check that g continuously reduces $\widehat{W}_{1,2}$ to $\mathrm{L}(\mathcal{A})$. ■

It remains to see that $W_{i,j} \leqslant_{\mathrm{W}} \widehat{W}_{i,j}$. For the lowest level we give a separate proof.

Lemma 5.16. *$W_{0,1} \leqslant_{\mathrm{W}} \widehat{W}_{0,1}$ and $W_{1,2} \leqslant_{\mathrm{W}} \widehat{W}_{1,2}$.*

Proof. By the symmetry it is enough to prove the first claim. Let us take a tree $t \in \mathrm{Tr}_{A_{0,1}}$. By König's lemma, the player $\exists$ has a winning strategy in $\mathcal{G}_t$ if and only if she can produce a sequence of finite strategies $\sigma_0, \sigma_1, \sigma_2, \ldots$ (viewed as subtrees of t, see Sect. 1.3.1 on page 6) such that

1. σ_0 consists of the root only;
2. for each n the strategy σ_{n+1} extends σ_n in such a way that below each leaf of σ_n a non-empty subtree is added, and all the leaves of σ_{n+1} have priority 0.

Using this observation we can define the reduction. Let $(\tau_i)_{i\in\mathbb{N}}$ be the list of all finite subsets of $\{\mathrm{L}, \mathrm{R}\}^*$. Some of these trees naturally induce a strategy for $\exists$ in $\mathcal{G}_t$. For those we define $t_{\tau_i} \in \mathrm{Tr}_{\{e,\bot\}}$ co-inductively, as follows:

- $t_{\tau_i}(\mathrm{R}^j) = e$ for all j;
- if τ_j induces in $\mathcal{G}_t$ a strategy that is a legal extension of the strategy induced by τ_i in the sense of Item 2 above then the subtree of t_{τ_i} rooted at $\mathrm{R}^j\mathrm{L}$ is t_{τ_j};
- otherwise, all nodes in this subtree are labelled with $\bot$.

Let $f(t) = t_{\sigma_0}$. By the initial observation, $t_{\sigma_0} \in \widehat{W}_{0,1}$ if and only if $\exists$ has a winning strategy in $\mathcal{G}_t$. The function f is continuous: to determine the labels in nodes $\mathrm{R}^{n_1}\mathrm{L}\mathrm{R}^{n_2}\mathrm{L}\ldots\mathrm{R}^{n_k}$ and $\mathrm{R}^{n_1}\mathrm{L}\mathrm{R}^{n_2}\mathrm{L}\ldots\mathrm{R}^{n_k}\mathrm{L}$ we only need to know the restriction of t to the union of the domains of $\tau_{n_1}, \tau_{n_2}, \ldots, \tau_{n_k}$. Hence, f continuously reduces $W_{0,1}$ to $\widehat{W}_{0,1}$. ■

Lemma 5.17. *For all i and $j \geqslant i + 2$, $W_{i,j} \leqslant_{\mathrm{W}} \widehat{W}_{i,j}$.*

Proof. By duality we can assume that $j = 2k$. For $t \in \mathrm{Tr}_{A_{i,2k}}$, let us consider a game $\tilde{\mathcal{G}}_t$ defined as follows. The positions are pairs (u, σ), where u is a node of t, and σ is finite strategy from u for $\forall$ (viewed as a subtree of $t{\restriction_u}$). Initially $u = \varepsilon$ is the root of t and $\sigma = \{\varepsilon\}$. In each round, in a position (u, σ), the players make the following moves:

- $\forall$ extends σ under the leaves of priority $2k-1$ to σ' in such a way that on every path leading from a leaf of σ to a leaf of σ' all the nodes have priority $2k$, except the leaf of σ', which has priority at most $2k-1$;
- $\exists$ has the following possibilities:
  - select a leaf u' of σ' with priority at most $2k-2$, and let the next round start with $(u', \{u'\})$, or
  - if σ' has leaves of priority $2k-1$, continue with (u, σ').

A play is won by $\exists$ if she selects a leaf infinitely many times and the least priority of these leaves seen infinitely often is even, or $\forall$ is unable to extend σ in some round. Otherwise, the play is won by $\forall$.

We claim that a player P has a winning strategy in $\mathcal{G}_t$ if and only if P has a winning strategy in $\tilde{\mathcal{G}}_t$.

For a winning strategy $\sigma_\exists$ for $\exists$ in $\mathcal{G}_t$, let $\tilde{\sigma}_\exists$ be the strategy in $\tilde{\mathcal{G}}_t$ in which $\exists$ selects a leaf u' in σ' if and only if $u' \in \sigma_\exists$. Consider an infinite play conforming to $\tilde{\sigma}_\exists$. If in the play $\exists$ selects a leaf infinitely many times, she implicitly defines a path in t conforming to $\sigma_\exists$, and so the play must be winning for $\exists$. Assume that $\exists$ selects a leaf only finitely many times. Then, $\forall$ produces an infinite sequence of finite strategies $\{u\} = \sigma_0 \subset \sigma_1 \subset \ldots$ in $\mathcal{G}_t$. Let σ_∞ be the union of these strategies. Consider the play π in $\mathcal{G}_t$ passing through u and conforming to σ_∞ and $\sigma_\exists$. Observe that for each σ_i, the strategy $\sigma_\exists$ must choose some path; hence, either $\exists$ selects a leaf of σ_i, or this path goes via a leaf of priority $2k-1$. Thus, π is infinite and by the rules of $\tilde{\mathcal{G}}_t$ priorities at most $2k-1$ are visited infinitely often. Since $\exists$ selects a leaf only finitely many times, priorities strictly smaller than $2k-1$ are visited finitely many times in π. Hence, π is won by $\forall$, what contradicts the assumption that $\sigma_\exists$ is winning for $\exists$.

Now, let $\sigma_\forall$ be a winning strategy for $\forall$ in $\mathcal{G}_t$. Then, for each $u \in \sigma_\forall$ there exists a finite sub-strategy σ' of $\sigma_\forall$ from u such that all internal nodes of σ' have priority $2k$ and leaves have priority at most $2k-1$. This shows that for each current strategy $\sigma \subset \sigma_\forall$, $\forall$ is able to produce a legal extension σ' such that $\sigma \subset \sigma' \subset \sigma_\forall$. Let $\tilde{\sigma}_\forall$ be a strategy of $\forall$ in $\tilde{\mathcal{G}}_t$ that extends every given σ by σ' as above. Consider any play conforming to $\tilde{\sigma}_\forall$. By the initial observation, the play is infinite, so priorities strictly smaller than $2k$ are visited infinitely often. If $\exists$ selects a leaf only finitely many times, priorities strictly smaller then $2k-1$ occur only finitely many times and $\forall$ wins. If $\exists$ selects a leaf infinitely many times, then the lowest priority seen infinitely often must be odd, as otherwise $\exists$ would show a losing path in $\sigma_\forall$. Hence, $\forall$ wins in this case as well.

It remains to encode $\tilde{\mathcal{G}}_t$ as a tree $f(t) \in \mathrm{Tr}_{\widehat{A}_{u,2k}}$ in a continuous manner. The argument is similar to the one in Lemma 5.16. Let $(\tau_n)_{n \in \mathbb{N}}$ be the list of all finite subsets of $\{\mathrm{L}, \mathrm{R}\}^*$. For some pairs (u, τ_n), τ_n induces a finite strategy in $\mathcal{G}_t$ from the node u. For such (u, τ_n) we define $t^\forall_{u,\tau_n}$ and $t^\exists_{u,\tau_n}$ co-inductively (see Sect. 1.6.5, page 18), as follows:

- $t^{\forall}_{u,\tau_n}(\mathrm{R}^m) = a$ for all m;
- the subtree of $t^{\forall}_{u,\tau_n}$ rooted at $\mathrm{R}^m\mathrm{L}$ is $t^{\exists}_{u,\tau_m}$ if τ_m induces a strategy from u that is a legal extension of τ_m according to the rules of $\tilde{\mathcal{G}}_t$, and otherwise the whole subtree is labelled with e's (losing choice for $\forall$);
- $t^{\exists}_{u,\tau_n}(\mathrm{R}^m) = e$ for $m = 0, 1, \ldots, \ell$, where $u_0, u_1, \ldots, u_\ell$ are the leaves in the strategy induced by τ_n from u;
- the subtree of $t^{\exists}_{u,\tau_n}$ rooted at $\mathrm{R}^{\ell+1}$ is $t^{\forall}_{u,\tau_m}$ if the strategy induced by τ_m from u has leaves of priority $2k-1$, otherwise the whole subtree is labelled with a's (losing choice for $\exists$);
- for $m \leqslant \ell$, consider the following cases to define the subtree s_m of $t^{\exists}_{u,\tau_n}$ rooted at $\mathrm{R}^m\mathrm{L}$:
  - if $\Omega(u_m) \in \{2k-1, 2k\}$ then s_m is labelled everywhere with a's (losing choice for $\exists$),
  - if $\Omega(u_m) = 2k-2$ then $s_m = t^{\forall}_{u_m,\{u_m\}}$,
  - if $\Omega(u_m) = r < 2k-2$ then $s_m(\varepsilon) = r$, the left subtree of s_m is $t^{\forall}_{u_m,\{u_m\}}$, and the right subtree of s_m is labelled with a's (irrelevant for $\mathcal{G}_t$).

Let $f(t)$ be $t^{\forall}_{\varepsilon,\{\varepsilon\}}$. Checking that f continuously reduces $W_{i,j}$ to $\widehat{W}_{i,j}$ does not pose any difficulties. ■

5.5 Conclusions

The results of this chapter should be treated as an intermediate step to proving decidability of index problems for general regular tree languages. Additionally, edelweisses studied in Sect. 5.4 are new *hard patterns* for alternating automata. The lower bounds proved in Lemma 5.16 seem to be of independent interest — in some cases it is easier to construct a reduction from the language $\widehat{W}_{i,j}$ instead of $W_{i,j}$.

Interestingly, the matching upper and lower bounds in the alternating case are of very different nature. The upper bounds are proved by providing an effective construction of an alternating automaton of certain index, where the lower bounds are obtained using continuous reductions. The structure of this reductions do not seem to be implementable in any *regular* way (e.g. by some kind of MSO interpretation).

The rigid structure of game automata should allow to give more decidability results in future. An instance of such a result is expressed by the following conjecture.

Conjecture 5.5. It is decidable, given $n \in \mathbb{N}$ and a game automaton $\mathcal{B}$, whether[1] $\mathrm{L}(\mathcal{B}) \in \mathbf{\Sigma}^0_n$ (i.e. the level of the Borel hierarchy occupied by a game language can be decided).

This chapter is based on [FMS13].

[1] If $\mathrm{L}(\mathcal{B}) \notin \mathbf{Comp}(\mathbf{\Pi}^{\mathrm{alt}}_0)$ then for all n the answer is no.

Part II

Thin Algebras

Chapter 6
When a Thin Language Is Definable in WMSO

In this chapter, we study *thin trees*, which generalize both finite trees and ω-words, but which are still simpler than arbitrary infinite trees. A tree is *thin* if it contains only countably many infinite branches. It turns out [BIS13] that some problems are more tractable on thin trees than in full generality. Therefore, thin trees can be seen as an intermediate step in understanding regular languages of general infinite trees.

The term *thin trees* comes from [BIS13], in [RR12] they are called *scattered trees*. Also, a tree is thin if it is a *tame tree* in the meaning of [LS98] (the converse is not true as [LS98] deals with trees treated as ordered structures, i.e. a tame tree may have a branch of length ω^2). A language of trees L is called *regular language of thin trees* if L is regular and contains only thin trees.

The notions induced in this chapter (mainly trees over ranked alphabets and thin algebras) are used in the following three chapters.

This chapter contains two main results, summarized by Theorem 2.4: the first result gives an upper bound on the topological complexity of regular languages of thin trees stating that they are all $\mathbf{\Pi}^1_1$ among all trees; the second result can be seen as a dichotomy: a regular language of thin trees is either topologically hard (i.e. $\mathbf{\Pi}^1_1$-hard) or is WMSO-definable among all trees. Additionally, we prove that it is decidable which of the cases holds. The following definition formalizes the notion of definability we use.

Definition 6.7. *Let L be a regular language of thin trees over a ranked alphabet A_{R} and φ be a formula of* WMSO. *We say that φ* defines L among all trees *if $L = \{t \in \mathrm{Tr}_{A_{\mathrm{R}}} \colon t \models \varphi\}$.*

This definition can be seen as a non-standard approach to restricting the class of all trees to thin ones — a standard one would say that L is WMSO-definable if $L = \{t \in \mathrm{Th}_{A_{\mathrm{R}}} : t \models \varphi\}$ for a WMSO formula φ. The requirement in Definition 6.7 for a formula to be satisfied only by thin trees is quite strong, in particular the class of languages definable in WMSO among all trees is not closed under complement with respect to thin trees: the relative complement of the empty language $\varnothing \subseteq \mathrm{Th}_{A_{\mathrm{R}}}$ is $\mathrm{Th}_{A_{\mathrm{R}}}$ which is $\mathbf{\Pi}^1_1$-complete and thus not WMSO-definable among all trees.

M. Skrzypczak, *Set Theoretic Methods in Automata Theory*, LNCS 9802
DOI: 10.1007/978-3-662-52947-8_6

The problem of deciding WMSO-definability among thin trees (i.e. using the standard approach) is open: it is not known how to decide if for a given regular language of thin trees L there exists a WMSO formula φ such that $L = \{t \in \mathrm{Th}_{A_R} : t \models \varphi\}$. Here, contrary to Definition 6.7, we explicitly restrict to trees t that are thin. In particular, there are more languages of thin trees that are WMSO-definable among thin trees (i.e. in the above standard sense) than in the sense of Definition 6.7.

In Proposition 6.10 we show that even in the sense of Definition 6.7 we can define languages as complicated as in the general case. The proof is based on examples from [Sku93] — the proof there is given for general trees but the proposed languages can be seen as regular languages of thin trees.

Now we can state the main result of this chapter as the following dichotomy similar in the spirit to the gap property proved by Niwiński and Walukiewicz [NW03] (see Theorem 1.31 on page 26).

Theorem 6.4. *A regular language of thin trees (i.e. a regular language that contains only thin trees) is either:*

1. $\mathbf{\Pi}^1_1$*-complete among all infinite trees,*
2. WMSO*-definable among all infinite trees (and thus Borel).*

Moreover, it is decidable which of the cases holds.

One of the applications of our characterisation is the following proposition.

Proposition 6.9. *Assume that L is a regular language of trees that is recognized by a non-deterministic (or equivalently alternating) Büchi automaton. Assume additionally that L contains only thin trees. Then L can be defined in* WMSO *among all trees.*

Proof. Since L is recognizable by a Büchi automaton, Theorem 1.22 on page 23 implies that L is an analytic subset of Tr_{A_R}. Therefore, L cannot be $\mathbf{\Pi}^1_1$-hard, thus L is WMSO-definable by Theorem 2.4. ■

The proof of Theorem 2.4 consists of two parts: first we prove in Sect. 6.3 that every regular language of thin trees is in $\mathbf{\Pi}^1_1$ among all trees (i.e. an upper bound). The best upper bound for general regular tree languages in terms of the projective hierarchy is $\mathbf{\Delta}^1_2$. Therefore, the presented result shows that regular languages of thin trees are descriptively simpler than general regular languages of infinite trees. The proof of Theorem 2.4 is concluded in Sect. 6.4 by proving the dichotomy: a regular language of thin trees is either WMSO-definable among all trees or $\mathbf{\Pi}^1_1$-hard (as expressed by Proposition 6.16).

The chapter is organized as follows. In Sect. 6.1 we introduce basic notions, in particular thin trees and tools allowing to inductively decompose them. In Sect. 6.2 we introduce thin algebras that will be used in the successive chapters of this part. Also, these algebras turns out to be convenient in Sect. 6.4. Section 6.3 we prove the upper bounds and in Sect. 6.4 we prove Proposition 6.16. Finally, in Sect. 6.5 we conclude.

6.1 Basic Notions

In the following three chapters we operate on binary trees over ranked alphabets. A *ranked alphabet* is a pair $A_R = (A_{R2}, A_{R0})$ where A_{R2} contains binary symbols and A_{R0} contains nullary symbols (labelling leafs of a tree). We assume that both sets A_{R2} and A_{R0} are finite and that A_{R2} is non-empty.

6.1.1 Thin Trees

We say that t is a *ranked tree* over a ranked alphabet (A_{R2}, A_{R0}) if t is a function from its non-empty prefix-closed domain $\mathrm{dom}(t) \subseteq \{\mathrm{L}, \mathrm{R}\}^*$ into $A_{R2} \cup A_{R0}$ (i.e. an element of $\mathrm{PTr}_{A_{R2} \cup A_{R0}}$ in the meaning of Sect. 1.1, page 1) such that for every node $u \in \mathrm{dom}(t)$ either:

- u is an *internal node of* t (i.e. $u\mathrm{L}, u\mathrm{R} \in \mathrm{dom}(t)$) and $t(u) \in A_{R2}$, or
- u is a *leaf of* t (i.e. $u\mathrm{L}, u\mathrm{R} \notin \mathrm{dom}(t)$) and $t(u) \in A_{R0}$.

A ranked tree containing no leaf is *complete*. The set of all ranked trees over a ranked alphabet A_R is denoted as Tr_{A_R}; in particular if $A_{R0} = \emptyset$ then $\mathrm{Tr}_{(A_{R2}, A_{R0})}$ contains only complete trees and coincides with $\mathrm{Tr}_{A_{R2}}$ as defined in Sect. 1.1.

Definition 6.8. *A ranked tree* $t \in \mathrm{Tr}_{A_R}$ *is* thin *if there are only countably many infinite branches of* t. *The set of all thin trees over a ranked alphabet* A_R *is denoted by* Th_{A_R}. *A ranked tree that is not thin is* thick.

A *context* over a ranked alphabet $A_R = (A_{R2}, A_{R0})$ is a ranked tree $p \in \mathrm{Tr}_{(A_{R2}, A_{R0} \sqcup \{\square\})}$ such that exactly one leaf $u \neq \epsilon$ of p is labelled by $\square$. The leaf u is called the *hole of* p. The set of all contexts over a ranked alphabet A_R is denoted as Con_{A_R}. The set of all contexts over A_R that are thin as trees is denoted by ThCon_{A_R}.

Given a ranked tree $t \in \mathrm{Tr}_{A_R}$ and $u \in \mathrm{dom}(t)$ $(u \neq \epsilon)$ we can construct a context $t[u \leftarrow \square]$ by replacing the subtree of t under u by $\square$: u becomes the hole of the context $t[u \leftarrow \square]$.

Assume that p is a context over a ranked alphabet A_R with the hole u. For every ranked tree $t \in \mathrm{Tr}_{A_R}$ the *composition of* p *and* t, denoted $p(t) \in \mathrm{Tr}_{A_R}$, is defined as $p[u \leftarrow t]$ — we put t in the place of the hole u of p. In particular, if r is a context then $p(r)$ is a new context. If p, r, and t are thin then also $p(t)$ and $p(r)$ are thin.

Let $w_1 \prec w_2$ be two nodes of a given ranked tree t. By $t\restriction_{[w_1, w_2)}$ we denote the ranked context rooted in w_1 with the hole in w_2:

$$t\restriction_{[w_1, w_2)} \stackrel{\text{def}}{=} t\restriction_{w_1}[w_2 \leftarrow \square].$$

Recall that a ranked tree $t' \in \mathrm{Tr}_{A_R'}$ is a *labelling* of a ranked tree $t \in \mathrm{Tr}_{A_R}$ if $\mathrm{dom}(t') = \mathrm{dom}(t)$. In such a case $t \otimes t'$ stands for the ranked tree over the product of ranked alphabets, i.e. an element of $\mathrm{Tr}_{A_R \times A_R'}$ with $A_R \times A_R' = \big(A_{R2} \times A_{R2}', A_{R0} \times A_{R0}'\big)$.

For a pair of ranked contexts $p \in \mathrm{Con}_{A_R}$, $p' \in \mathrm{Con}_{A_R'}$ with the same domain $\mathrm{dom}(p) = \mathrm{dom}(r)$ and the same hole u, by $p \otimes p'$ we denote the ranked context over the product alphabet $A_R \times A_R' = (A_{R2} \times A_{R2}', A_{R0} \times A_{R0}')$ with the hole u:

$$\text{for } w \in \mathrm{dom}(p),\ w \neq u \text{ we have } (p \otimes p')(w) = \big(p(w), p'(w)\big).$$

6.1.2 Automata

For the purpose of the following three chapters we introduce a notion of non-deterministic tree automata working over a ranked alphabet. Again, these notions become standard when we restrict to purely-binary alphabets, i.e. when $A_{R0} = \emptyset$.

A non-deterministic parity tree automaton over a ranked alphabet is a tuple $\mathcal{A} = \langle A_R{}^{\mathcal{A}}, Q^{\mathcal{A}}, I^{\mathcal{A}}, \delta^{\mathcal{A}}, \Omega^{\mathcal{A}} \rangle$ where

- $A_R{}^{\mathcal{A}} = (A_{R2}{}^{\mathcal{A}}, A_{R0}{}^{\mathcal{A}})$ is a ranked alphabet,
- $Q^{\mathcal{A}}$ is a finite set of *states*,
- $I^{\mathcal{A}} \subseteq Q^{\mathcal{A}}$ is a set of *initial states*,
- $\delta^{\mathcal{A}} = \delta_2^{\mathcal{A}} \sqcup \delta_0^{\mathcal{A}}$ is a *transition relation*: $\delta_2^{\mathcal{A}} \subseteq Q^{\mathcal{A}} \times A_{R2}{}^{\mathcal{A}} \times Q^{\mathcal{A}} \times Q^{\mathcal{A}}$ contains *transitions over internal nodes* (q, a, q_L, q_R) and $\delta_0^{\mathcal{A}} \subseteq Q^{\mathcal{A}} \times A_{R0}{}^{\mathcal{A}}$ contains *transitions over leafs* (q, b),
- $\Omega^{\mathcal{A}} \colon Q^{\mathcal{A}} \to \mathbb{N}$ is a *priority function*.

A *run of an automaton* $\mathcal{A}$ *on a ranked tree* $t \in \mathrm{Tr}_{A_R{}^{\mathcal{A}}}$ is a labelling ρ of t over the ranked alphabet $(Q^{\mathcal{A}}, Q^{\mathcal{A}})$ such that for every $u \in \mathrm{dom}(t)$:

- if u is an internal node of t then $\big(\rho(u), t(u), \rho(u\mathrm{L}), \rho(u\mathrm{R})\big) \in \delta_2^{\mathcal{A}}$,
- if u is a leaf of t then $\big(\rho(u), t(u)\big) \in \delta_0^{\mathcal{A}}$.

A *run* ρ *on a ranked context* p is a labelling of p (treated as a tree) by states of $\mathcal{A}$ that obeys the transition relation in all the nodes except the hole u of p. The *value of* ρ *in the hole of* p is $\rho(u)$.

Now we repeat the definitions from Sect. 1.4 (see page 7) in the context of ranked trees:

- A run ρ is *accepting* if it is parity-accepting and $\rho(\epsilon) \in I^{\mathcal{A}}$ (see Sect. 1.4). By the definition we verify the parity condition only on infinite branches of ρ, the finite ones do not influence acceptance.
- A ranked tree $t \in \mathrm{Tr}_{A_R{}^{\mathcal{A}}}$ is *accepted by* $\mathcal{A}$ if there exists an accepting run ρ of $\mathcal{A}$ on t.
- The set of ranked trees accepted by $\mathcal{A}$ is called the *language recognised by* $\mathcal{A}$ and is denoted by $\mathrm{L}(\mathcal{A})$.
- A language $L \subseteq \mathrm{Tr}_{A_R{}^{\mathcal{A}}}$ is *regular* if there exists an automaton recognising L.

By repeating the standard automata constructions over the ranked alphabet, we obtain the following fact.

Fact 6.50. *A language $L \subseteq \mathrm{Tr}_{A_R}$ is regular if and only if it is* MSO*-definable.*

Definition 6.9. *A* regular language of thin trees *is a regular language of ranked trees $L \subseteq \mathrm{Tr}_{A_R}$ such that L contains only thin trees (i.e. $L \subseteq \mathrm{Th}_{A_R}$).*

As we will see later (see Remark 6.6), equivalently one can say that a regular language of thin trees is a language that is the intersection of a regular tree language with Th_{A_R}.

6.1.3 Examples of Skurczyński

In this section we adjust the examples of WMSO-definable languages proposed by Skurczyński [Sku93] to the case of thin trees, as expressed by the following proposition. This can be seen as an argument that there are languages of thin trees that are definable in WMSO among all trees and topologically as complex as general WMSO-definable languages.

Proposition 6.10 (Skurczyński [Sku93]). *For every n there exists a regular language of thin trees $L \subseteq \mathrm{Th}_{A_R}$ that is* WMSO*-definable among all trees and $\mathbf{\Sigma}^0_n(\mathrm{Tr}_{A_R})$-complete.*

Proof. Take $n \in \mathbb{N}$. We will base our construction on languages of trees $W_{i,j}$ (see Sect. 1.7.4, page 23) — we consider trees over a ranked alphabet that encodes parity games of index (i, j) and $W_{i,j}$ contains those trees where $\exists$ has a winning strategy. As observed in Remark 1.1 on page 24, one can extend the alphabet with additional symbols $\top$ and $\bot$ that finish the game indicating that one of the players ($\exists$ or $\forall$ respectively) wins instantly.

Our language L will be obtained as a restriction of a variant of $W_{0,1}$ to thin trees of a particular shape. Consider a ranked alphabet $A_R = (A_{R2}, A_{R0})$ with $A_{R2} = A_{0,1} = \{\exists, \forall\} \times \{0, 1\}$ (see Sect. 1.7.4, page 23) and $A_{R0} = \{\top, \bot\}$ and let $W_{0,1}$ be the set of all trees t over A_R such that $\exists$ has a winning strategy in $\mathcal{G}_t$ (see Definition 1.2 on page 23 and Remark 1.1 on page 24).

Recall that by $\sharp_a(u)$ we denote the number of occurrences of a latter a in a finite word u. Take any $n > 0$ and let $X^n_{[\mathrm{Sku93}]}$ contain all trees $t \in \mathrm{Tr}_{A_R}$ such that (see Fig. 6.1):

$$t(u) = \begin{cases} (\exists, 1) & \text{if } \sharp_{\mathrm{R}}(u) < n \text{ and } \sharp_{\mathrm{R}}(u) \equiv 0 \pmod 2, \\ (\forall, 0) & \text{if } \sharp_{\mathrm{R}}(u) < n \text{ and } \sharp_{\mathrm{R}}(u) \equiv 1 \pmod 2, \\ \top \text{ or } \bot & \text{if } \sharp_{\mathrm{R}}(u) = n. \end{cases}$$

Clearly, for a tree $t \in X^n_{[\mathrm{Sku93}]}$ we have $\mathrm{dom}(t) = \{u \in \{\mathrm{L}, \mathrm{R}\}^* : \sharp_{\mathrm{R}}(u) \leqslant n\}$ so $X^n_{[\mathrm{Sku93}]} \subseteq \mathrm{Th}_{A_R}$. Also, the set $X^n_{[\mathrm{Sku93}]}$ itself is WMSO-definable among all trees.

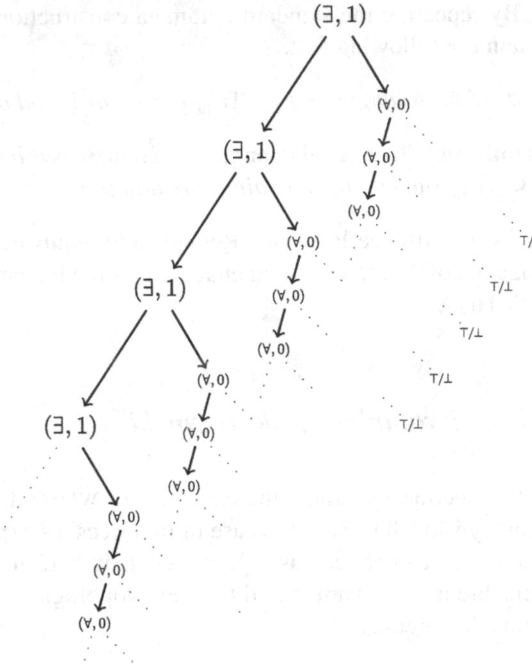

Fig. 6.1 An example of a tree $t \in X^n_{[\text{Sku93}]}$.

By the same argument as in [Sku93], the language $L \overset{\text{def}}{=} W_{0,1} \cap X^n_{[\text{Sku93}]}$ is WMSO-definable among all trees and $\boldsymbol{\Sigma}^0_n(\mathrm{Tr}_{A_\mathrm{R}})$-complete. ■

6.1.4 *Ranks*

The crucial tool in our analysis of thin trees is structural induction — we inductively decompose a given thin tree into *simpler* ones. A measure of complexity of thin trees is called a *rank* — a function that assigns to each thin tree a countable ordinal number. The rank of a thin tree t depends only on the domain of t. During the inductive computation of ranks, we work with partial binary trees (i.e. elements of PTr, see Sect. 1.1, page 1) that may not be ranked trees (e.g. a node may have exactly one child). For the sake of this chapter, we call elements of PTr *tree-shapes*. The set of all tree-shapes that have countably many branches is denoted $\mathrm{PTh} \subseteq \mathrm{PTr}$.

The rank we use is based on the Cantor-Bendixson derivative [Kec95, Chap. 6.C]: we inductively remove *simple* parts of a given tree. Let us fix the set $B_{\mathrm{CB}} \subseteq \mathrm{PTr}$ (the *basis* of the rank) containing all tree-shapes $\tau \in \mathrm{PTr}$ that have only finitely many finite and infinite branches. Equivalently, B_{CB} contains all tree-shapes that contain only finitely many branching nodes.

Fact 6.51. *For every tree-shape $\tau \in \mathrm{PTr}$ we have:*

1. *if no subtree of τ belongs to B_{CB} then τ contains a branching node,*
2. *if τ belongs to B_{CB} then all the subtrees of τ also belong to B_{CB}.*

Consider the following operation on tree-shapes called *derivative*: for a tree-shape $\tau \in \mathrm{PTr}$ we define the tree-shape $\mathrm{Dv}(\tau) \subseteq \tau$ that contains only these nodes $u \in \mathrm{dom}(\tau)$ such that $\tau\restriction_u \notin B_{\mathrm{CB}}$ — we remove from τ those nodes u such that the subtree of τ under u belongs to B_{CB}.

Now we inductively define transfinite compositions of Dv: let $\mathrm{Dv}^0(\tau) = \tau$, $\mathrm{Dv}^{\eta+1}(\tau) = \mathrm{Dv}(\mathrm{Dv}^{\eta}(\tau))$, and if η is a limit ordinal let

$$\mathrm{Dv}^{\eta}(\tau) = \bigcap_{\eta'<\eta} \mathrm{Dv}^{\eta'}(\tau).$$

Fact 6.52. *Let $\tau \in \mathrm{PTr}$ be a tree-shape. The sequence $\mathrm{Dv}^{\eta}(t)$ for $\eta < \omega_1$ is a decreasing sequence of tree-shapes. There exists $\eta_0 < \omega_1$ such that*

$$\mathrm{Dv}^{\eta_0}(\tau) = \mathrm{Dv}^{\eta_0+1}(\tau).$$

The following proposition shows a connection of this iterated derivative and thin trees.

Proposition 6.11. *Let τ be a tree-shape and η be an ordinal such that $\mathrm{Dv}^{\eta}(\tau) = \mathrm{Dv}^{\eta+1}(\tau)$. The tree-shape $\mathrm{Dv}^{\eta}(\tau)$ is empty if and only if τ has only countably many branches. Otherwise τ contains the complete binary tree as a minor*[1].

Proof. Assume that $\mathrm{Dv}^{\eta}(\tau)$ is empty. Observe that every application of the derivative decreases the number of branches of τ by countably many: there are countably many nodes $u \in \mathrm{dom}(\tau)$ and the subtree under a removed node u belongs to the family B_{CB}. Since there are countably many applications of the derivative, the total number of removed branches is also countable.

Assume that $\tau' = \mathrm{Dv}^{\eta}(\tau)$ is non-empty. We show that in that case $\tau' \subseteq \tau$ has uncountably many branches. We construct a Cantor scheme that maps finite sequences $w \in \{\mathtt{L}, \mathtt{R}\}^*$ into nodes $u_w \in \tau'$ in a way monotone with respect to the prefix order $\preceq$ and lexicographic order $\leqslant_{\mathrm{lex}}$. We start with any $u_\epsilon \in \tau'$. Let $w \in \{\mathtt{L}, \mathtt{R}\}^*$ be a sequence such that the node $u_w \in \tau'$ is defined. Observe that there must be a branching node u' under u_w in τ' (since all the subtrees of $\tau'\restriction_{u_w}$ do not belong to B_{CB}, see Fact 6.51). Put $u_{w\mathtt{L}}, u_{w\mathtt{R}}$ as the two children of u' (i.e. $u_{wd} = u'd$ for $d \in \{\mathtt{L}, \mathtt{R}\}$).

The above definition gives us the unique, infinite branch of τ' for every $\beta \in \{\mathtt{L}, \mathtt{R}\}^{\omega}$. Therefore, τ' has uncountably many infinite branches and so does τ. ■

Definition 6.10. *For a thin tree $t \in \mathrm{Th}_{A_{\mathrm{R}}}$ we define the* rank *of t (denoted* $\mathrm{rank}(t)$*) as the smallest ordinal η such that $\mathrm{Dv}^{\eta}(\mathrm{dom}(t)) = \varnothing$.*

---

[1]Formally, it means that there exists an injective function $\iota\colon \{\mathtt{L}, \mathtt{R}\}^* \to \tau$ that preserves the prefix and lexicographic orders.

We extend this definition to $\mathrm{rank}(u,t)$ *(the rank of* u *in* t*) for a node* $u \in \mathrm{dom}(t)$ *in such a way that* $\mathrm{rank}(u,t)$ *is the least* $\eta < \omega_1$ *such that* $u \notin \mathrm{Dv}^{\eta}(\mathrm{dom}(t))$.

For an ordinal $\eta < \omega_1$ *by* $\mathrm{Th}_{A_R}^{\leqslant\eta}$ *we denote the set of thin trees of rank at most* η.

Fact 6.53. *For every thin tree* $t \in \mathrm{Th}_{A_R}$ *and node* $u \in \mathrm{dom}(t)$ *we have* $\mathrm{rank}(u,t) = \mathrm{rank}(t\restriction_u)$.

If t *is a thin tree then* $\mathrm{rank}(t)$ *is not a limit ordinal. In particular the ordinal* $\mathrm{rank}(t){-}1$ *is defined.*

If $u \preceq w$ *are two nodes of a thin tree* t *then* $\mathrm{rank}(u,t) \geqslant \mathrm{rank}(w,t)$.

The crucial way of using ranks is induction: we can decompose a given tree as its *spine* and a number of trees connected to it: the *spine* of a thin tree t is

$$\tau = \mathrm{Dv}^{\mathrm{rank}(t)-1}(\mathrm{dom}(t)) \in \mathrm{PTr}.$$

Since $\mathrm{Dv}(\tau) = \varnothing$ so $\tau \in B_{\mathrm{CB}}$ — the spine has only finitely many branches. Also, if $\mathrm{rank}(t) > 1$ then the spine of t is infinite, otherwise already $\mathrm{Dv}^{\mathrm{rank}(t)-1}(\mathrm{dom}(t))$ would be empty, contradicting minimality of $\mathrm{rank}(t)$.

Intuitively, a thin tree t has rank equal m if t contains m nested levels of infinite branches. In comparison, the rank of well-founded ω-trees from Sect. 1.6.3 (see page 17) counts each node of an ω-tree separately. In particular, a finite ω-tree may have arbitrarily big finite rank in the meaning of Sect. 1.6.3 while a finite thin tree always belongs to B_{CB} and therefore has rank 1.

Figure 6.2 presents a sequence of thin trees of increasing rank. The leftmost branch of each thin tree is its spine.

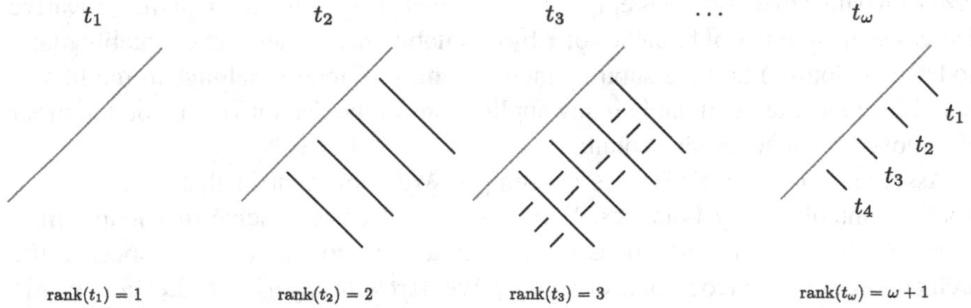

Fig. 6.2 A sequence of thin trees and their spines.

6.1.5 Skeletons

The second tool used to analyse structural properties of thin trees are *skeletons*. A skeleton can be seen as a witness that a given ranked tree is thin. Moreover, a skeleton of a thin tree t represents a structural decomposition of t.

A subset of nodes $\sigma \subseteq \mathrm{dom}(t)$ of a given ranked tree $t \in \mathrm{Tr}_{A_R}$ is a *skeleton of t* if:

- $\epsilon \notin \sigma$,
- for every internal node u of t the set σ contains exactly one of the nodes $u\mathrm{L}$, $u\mathrm{R}$,
- on every infinite branch α of the tree t almost all nodes $u \prec \alpha$ belong to σ.

Observe that we can identify σ with its characteristic function — a labelling of nodes of t by the ranked alphabet $A_R' = (A_{R2}', A_{R0}')$ with $A_{R2}' = A_{R0}' = \{0, 1\}$ so that $\sigma \in \mathrm{Tr}_{A_R'}$.

Assume that σ is a skeleton of a tree t. Take any node $u \in \mathrm{dom}(t)$. The branch α passing through u that follows at every point the skeleton σ is called the *main branch of σ from u*. It can be defined as the unique maximal finite or infinite branch $\alpha \in \{\mathrm{L}, \mathrm{R}\}^{\leqslant\omega}$ such that:

$$u \preceq \alpha \wedge \forall_{w \preceq \alpha} \ (w \preceq u \vee w \in \sigma).$$

Note that the main branch may be finite if it reaches a leaf of the tree. Otherwise it is infinite. By the assumption that a skeleton contains almost all nodes on every branch, we obtain the following fact.

Fact 6.54. *Take a ranked tree $t \in \mathrm{Tr}_{A_R}$ with a skeleton σ and an infinite branch α of t. There exists a node $u \in \mathrm{dom}(t)$ such that α is the main branch of σ from u.*

Proposition 6.12. *A given ranked tree $t \in \mathrm{Tr}_{A_R}$ has a skeleton if and only if t is thin.*

Proof. If a ranked tree has a skeleton then by the above fact every infinite branch of t is from some point on its main branch (from some node of t). So there are at most countably many branches of t.

Now assume that t is a thin tree. We inductively on the rank η of t construct a skeleton of t. The thesis holds for $\eta = 0$ because there is no thin tree of rank 0. Assume the thesis for all thin trees of rank strictly smaller than η. Let t be a thin tree, $\mathrm{rank}(t) = \eta$, and τ be the spine of t. For every $u \in \mathrm{dom}(t)$ that is off τ (i.e. $u \notin \tau$ but the parent of u is a node of τ) we know that $\mathrm{rank}(u, t) < \eta$. Therefore, there exists a skeleton σ_u of the subtree of t under u; we assume that σ_u is a subset of $\mathrm{dom}(t)$, i.e.

$$\sigma_u \subseteq \mathrm{dom}(t) \cap u\{\mathrm{L}, \mathrm{R}\}^*.$$

Let σ_ϵ contain all those elements $u \neq \epsilon$ of τ such that u does not have a sibling in τ. Also, if both $u\mathrm{L}$ and $u\mathrm{R}$ belong to τ let σ_ϵ contain $u\mathrm{L}$. Finally, let σ be the union of σ_ϵ and σ_u for $u \in \mathrm{dom}(t)$ that are off τ. By the construction σ does not contain ϵ and contains exactly one sibling from every pair of siblings in t.

What remains to show is that σ contains almost all nodes on every infinite branch of t. Let α be an infinite branch of t. If α is not an infinite branch of τ then there exists $u \prec \alpha$ that is off τ. Since σ_u is a skeleton so it contains almost all nodes on α. Now assume that α is an infinite branch of τ. Since $\tau \in B_{\mathrm{CB}}$ so it contains only finitely many finite and infinite branches, in particular, almost all nodes $u \prec \alpha$ are not branching in τ. Therefore, σ_ϵ contains almost all nodes on α. ■

The skeleton σ constructed in the above construction is called the *canonical skeleton for t* and is denoted by $\sigma(t)$.

Remark 6.6. Since the conditions on a skeleton are MSO-definable so the family of all thin trees $\mathrm{Th}_{A_R} \subseteq \mathrm{Tr}_{A_R}$ is a regular tree language.

6.2 Thin Algebra

In the following three chapters we use a variant of the *thin forest algebra* as introduced by Bojańczyk and Idziaszek in [BIS13, Idz12] adapted to the case of ranked trees. It can be seen as a natural extension of ω-semigroups and Wilke algebras [Wil93, Wil98] (see Sect. 1.5.2, page 11). The use of thin algebra is this chapter could be avoided, however it seems to be more convenient to use it (thin algebras are used in the proof of Proposition 6.16). Additionally, thin algebras are crucial concepts in the following two chapters.

Let us fix a ranked alphabet $A_R = (A_{R2}, A_{R0})$. A *thin algebra over* A_R is a two-sorted algebra (H, V) where H corresponds to types of trees and V to types of contexts. A thin algebra is equipped with the following operations:

- $s \cdot s' \in V$ for $s, s' \in V$,
- $s \cdot h \in H$ for $s \in V, h \in H$,
- $s^\infty \in H$ for $s \in V$,
- $\prod \colon V^\omega \to H$,
- $\mathrm{Node}(a, d, h) \in V$ for $a \in A_{R2}$, $d \in \{\mathrm{L}, \mathrm{R}\}$, and $h \in H$,
- $\mathrm{Leaf}(b) \in H$ for $b \in A_{R0}$.

Note that the first four operations are the same as in the case of Wilke algebras and ω-semigroups. The last two operations allow to operate on trees. For simplicity, we write $a(\square, h)$ instead of $\mathrm{Node}(a, \mathrm{L}, h)$ and $a(h, \square)$ instead of $\mathrm{Node}(a, \mathrm{R}, h)$. Similarly, $b()$ stands for $\mathrm{Leaf}(b)$ and $a(h_\mathrm{L}, h_\mathrm{R}) \in H$ denotes the result of $a(h_\mathrm{L}, \square) \cdot h_\mathrm{R}$.

The axioms of thin algebra are:

the axioms of Wilke algebra:

$$s \cdot (s' \cdot s'') = (s \cdot s') \cdot s'' \tag{6.1}$$

$$s \cdot (s' \cdot h) = (s \cdot s') \cdot h \tag{6.2}$$

$$(s \cdot s')^\infty = s \cdot (s' \cdot s)^\infty \tag{6.3}$$

$$\forall_{n \geqslant 1} \left(s^n\right)^\infty = s^\infty \tag{6.4}$$

the axioms of ω-semigroups:

$$\prod(s, s, \ldots) = s^\infty \tag{6.5}$$

$$s \cdot \prod(s_0, s_1, \ldots) = \prod(s, s_0, s_1, \ldots) \tag{6.6}$$

$$\prod(s_0 \cdot \ldots \cdot s_{k_1},\ s_{k_1+1} \cdot \ldots \cdot s_{k_2},\ \ldots) = \prod(s_0, s_1, s_2, \ldots) \tag{6.7}$$

and one additional axiom:

$$a(\Box, h_{\mathrm{R}}) \cdot h_{\mathrm{L}} = a(h_{\mathrm{L}}, \Box) \cdot h_{\mathrm{R}}. \tag{6.8}$$

Fact 6.55. *If a finite structure (H, V) satisfies all the axioms of thin algebra except the ones about infinite product:* (6.5), (6.6) *and* (6.7) *then (H, V) can be equipped, in a unique way, with infinite product $\prod$ satisfying axioms* (6.5), (6.6) *and* (6.7).

Proof. The same as in the case of Wilke algebra, see Theorem 1.6 on page 12. ∎

However, as shown in Example 1.1 on page 12, we cannot erase the infinite product $\prod$ from the definition of thin algebra; this operation is important when we consider homomorphisms between thin algebras.

It is easy to verify that the pair $(\mathrm{Tr}_{A_{\mathrm{R}}}, \mathrm{Con}_{A_{\mathrm{R}}})$ has a natural structure of a thin algebra. In particular, the operation $p \mapsto p^\infty$ constructs the ranked tree p^∞ from a ranked context p by *looping* the hole of p to the root of p, that is p^∞ is the unique ranked tree satisfying

$$p(p^\infty) = p^\infty.$$

The subalgebra $(\mathrm{Th}_{A_{\mathrm{R}}}, \mathrm{ThCon}_{A_{\mathrm{R}}}) \subset (\mathrm{Tr}_{A_{\mathrm{R}}}, \mathrm{Con}_{A_{\mathrm{R}}})$ consisting of thin trees and thin contexts is free in the class of thin algebras over the ranked alphabet A_{R}, see [Idz12, Theorem 30] (for more details see Sect. 7.4.1, page 132). The algebra $(\mathrm{Tr}_{A_{\mathrm{R}}}, \mathrm{Con}_{A_{\mathrm{R}}})$ is not free. In Sect. 7.4.1 we will see how thin algebras can be used to recognise languages of general ranked trees (not necessarily thin).

A homomorphism $f\colon (H, V) \to (H', V')$ between two thin algebras over the same alphabet A_{R} is defined in the usual way: f should be a function mapping elements of H into H' and elements of V into V' that preserves all the operations of thin algebra. Such a homomorphism is *surjective* if $f(H) = H'$ and $f(V) = V'$.

Fact 6.56. *Since every context $p \in \mathrm{ThCon}_{A_{\mathrm{R}}}$ can be obtained as a finite combination of trees $t \in \mathrm{Tr}_{A_{\mathrm{R}}}$ using the operation* Node*, if $f_1, f_2\colon (\mathrm{Tr}_{A_{\mathrm{R}}}, \mathrm{Con}_{A_{\mathrm{R}}}) \to (H, V)$ are two homomorphisms that agree on $\mathrm{Tr}_{A_{\mathrm{R}}}$ then $f_1 = f_2$.*

The operations of thin algebra (namely the infinite product $\prod$) imply that homomorphisms have to be *path-wise consistent*, as expressed by the following fact.

Fact 6.57. *Let $t \in \mathrm{Th}_{A_{\mathrm{R}}}$ be a thin tree, $\alpha = d_0 d_1 \ldots$ an infinite branch of t, and $f\colon (\mathrm{Tr}_{A_{\mathrm{R}}}, \mathrm{Con}_{A_{\mathrm{R}}}) \to S$ be a homomorphism into a finite thin algebra S. Let $u_i = d_0 d_1 \ldots d_{i-1} \bar{d}_i$ be the sequence of vertices of t that are off α and $a_i = t(\alpha{\restriction}_i)$ be the i'th letter of t along α. Then*

$$f(t) = \prod_{i \in \mathbb{N}} \mathrm{Node}\big(a_i, d_i, f(t{\restriction}_{u_i})\big).$$

The following fact follows from induction over the rank of a thin tree, see [Idz12, Lemma 34] or the proof of Lemma 8.36 on page 145 in Chap. 8.

Fact 6.58. *Let (H, V) be a thin algebra over a ranked alphabet A_R. Then there exists a unique homomorphism $f\colon (\mathrm{Th}_{A_R}, \mathrm{ThCon}_{A_R}) \to (H, V)$.*

Let $L \subseteq \mathrm{Th}_{A_R}$ be a language of thin trees. We say that a homomorphism $f\colon (\mathrm{Th}_{A_R}, \mathrm{ThCon}_{A_R}) \to (H, V)$ *recognises* L if there is a set $F \subseteq H$ such that $L = f^{-1}(F)$, see Sect. 1.5.3, page 13. We say that (H, V) *recognises* L if there exists a set F as above (Fact 6.58 implies that there exists a unique homomorphism f).

Similarly, $f\colon (\mathrm{Tr}_{A_R}, \mathrm{Con}_{A_R}) \to (H, V)$ *recognises* $L \subseteq \mathrm{Tr}_{A_R}$ if $L = f^{-1}(F)$ for some $F \subseteq H$.

6.2.1 The Automaton Algebra

Every non-deterministic tree automaton $\mathcal{A}$ induces a finite thin algebra $S_\mathcal{A}$ (called the *automaton algebra*) and a homomorphism $f_\mathcal{A}$ from all ranked trees to $S_\mathcal{A}$ (called the *automaton morphism*). The automaton algebra is an example of a finite thin algebra recognising $\mathrm{L}(\mathcal{A}) \subseteq \mathrm{Tr}_{A_R\mathcal{A}}$.

Let $\mathcal{A}$ be a non-deterministic automaton over a ranked alphabet A_R such that $\mathcal{A}$ recognises $L \subseteq \mathrm{Tr}_{A_R}$. Assume that $\mathcal{A}$ has states Q and uses priorities from $\{0, \ldots, k\}$ for some k. Let us define $f_\mathcal{A}(t)$ for a tree $t \in \mathrm{Tr}_{A_R}$ and $f_\mathcal{A}(p)$ for a context $p \in \mathrm{Con}_{A_R}$:

$$f_\mathcal{A}(t) = \Big\{ q : \exists_\rho\ \rho \text{ is a run of } \mathcal{A} \text{ on } t \text{ such that: } \rho \text{ is parity-accepting, } \rho(\epsilon) = q. \Big\} \subseteq Q \tag{6.9}$$

$$f_\mathcal{A}(p) = \Big\{ (q, i, q') : \exists_\rho\ \rho \text{ is a run of } \mathcal{A} \text{ on } p \text{ such that: } \rho \text{ is parity-accepting, } \rho(\epsilon) = q, \rho(u) = q' (\text{where } u \text{ is the hole of } p), \text{ the minimal priority on the path from } \epsilon \text{ to } u \text{ in } \rho \text{ is } i. \Big\} \subseteq Q \times \{0, \ldots, k\} \times Q \tag{6.10}$$

Fact 6.59. *The function $f_\mathcal{A}$ induces uniquely the structure of thin algebra on its image $S_\mathcal{A} \stackrel{\mathrm{def}}{=} (H_\mathcal{A}, V_\mathcal{A}) \subseteq \big(\mathsf{P}(Q), \mathsf{P}(Q \times \{0, \ldots, k\} \times Q)\big)$ in such a way that $f_\mathcal{A}$ becomes a homomorphism of thin algebras.*

Moreover, $f_{\mathcal{A}}$ recognises $\mathrm{L}(\mathcal{A})$*, since*

$$\mathrm{L}(\mathcal{A}) = f_{\mathcal{A}}^{-1}\left(\{h \in H_{\mathcal{A}} : h \cap I^{\mathcal{A}} \neq \varnothing\}\right).$$

For every $h \in H_{\mathcal{A}}$ the language $L_h \stackrel{def}{=} f_{\mathcal{A}}^{-1}(\{h\}) \subseteq \mathrm{Tr}_{A_{\mathrm{R}}}$ is regular.

For the sake of completeness, let us write down the operations of the automaton algebra $S_{\mathcal{A}}$. The formulae are similar to the case of thin forest algebra, see [Idz12, Sect. 4.4.1]. We do not define the infinite product $\prod$, it can be uniquely introduced by Fact 6.55. We implicitly assume that $h \in H_{\mathcal{A}}$, $s, s' \in V_{\mathcal{A}}$, $e \in V_{\mathcal{A}}$ is an idempotent, $a \in A_{\mathrm{R2}}$, $b \in A_{\mathrm{R0}}$, and $d \in \{\mathrm{L}, \mathrm{R}\}$.

$$s \cdot s' = \{(q, \min(j, j'), q'') : (q, j, q') \in s, (q', j', q'') \in s'\}, \tag{6.11a}$$

$$s \cdot h = \{q : (q, j, q') \in s, q' \in h\}, \tag{6.11b}$$

$$s^{\infty} = \left(s^{\sharp}\right)^{\infty} \text{ for } s^{\sharp} \text{ being the idempotent power of } s, \tag{6.11c}$$

$$e^{\infty} = \{q : (q, j, q) \in e, j \equiv 0 \pmod 2\} \text{ for } e \text{ being an idempotent,} \tag{6.11d}$$

$$\mathrm{Node}(a, d, h) = \{(q, \min(q, q_d), q_d) : (q, a, q_{\mathrm{L}}, q_{\mathrm{R}}) \in \delta_2^{\mathcal{A}}, q_{\bar{d}} \in h\}, \tag{6.11e}$$

$$\mathrm{Leaf}(b) = \{q : (q, b) \in \delta_0^{\mathcal{A}}\}. \tag{6.11f}$$

Observe that if $\mathrm{L}(\mathcal{A}) \subseteq \mathrm{Th}_{A_{\mathrm{R}}}$ is a language of thin trees then we can restrict the automaton morphism to $\mathrm{Th}_{A_{\mathrm{R}}}$. After this restriction it recognises $\mathrm{L}(\mathcal{A})$ as a language of thin trees.

The following fact is a direct consequence of the existence of an automaton algebra. It is not used in this thesis, we use only the "only if" part: if a language is regular then it is recognised by a homomorphism into a finite thin algebra.

Fact 6.60. *A language of thin trees $L \subseteq \mathrm{Th}_{A_{\mathrm{R}}}$ is a regular language of thin trees if and only if it is recognised by a homomorphism into a finite thin algebra.*

Sketch of a Proof. If a language is regular then we can take the automaton algebra. The opposite direction follows from the definition of *consistent markings* in Sect. 7.1 and Fact 8.36 — we can define in MSO a consistent marking τ of a given thin tree and check that $\tau(\epsilon) \in F$. ■

6.3 Upper Bounds

In this section we prove an upper bound on descriptive complexity of regular languages of thin trees from Theorem 2.4, as expressed by the following proposition.

Proposition 6.13. *Every regular language of thin trees L is co-analytic as a set of ranked trees.*

Note that despite the fact that the space of thin trees $\mathrm{Th}_{A_{\mathrm{R}}}$ is co-analytic among all trees, it is an uncountable set and contains arbitrarily complicated subsets.

6.3.1 Embeddings and Quasi-skeletons

The definition of a skeleton σ of a tree t is a co-analytic definition — σ has to contain almost all nodes on every branch of t. Our aim in this section is to define objects less rigid than skeletons but definable in an analytic way. For this purpose, we introduce two relations R_{Embed} and R_{QSkel}. Let us fix a ranked alphabet A_{R}.

Proposition 6.14. *There exists an analytic ($\mathbf{\Sigma}^1_1$) relation $R_{\mathrm{Embed}} \subseteq \mathrm{Tr}_{A_{\mathrm{R}}} \times \mathrm{Tr}_{A_{\mathrm{R}}}$ such that for every tree t_1 and every thin tree t_2:*

$$\big(t_1 \text{ } is \text{ } thin \text{ and } \mathrm{rank}(t_1) \leqslant \mathrm{rank}(t_2)\big) \quad if \text{ and only } if \quad (t_1, t_2) \in R_{\mathrm{Embed}}.$$

Intuitively, the relation R_{Embed} is defined by the expression of the form: $(t_1, t_2) \in R_{\mathrm{Embed}}$ if there exists an *embedding* of $\mathrm{dom}(t_1)$ to $\mathrm{dom}(t_2)$. However, to avoid technical difficulties, we do not introduce exact definition of an embedding. Instead, we recall some standard methods from descriptive set theory, see [Kec95, Sect. 34.D], namely the *Borel derivatives*. It will be shown that the derivative Dv from Sect. 6.1.4 is (modulo some technical extension) a *Borel derivative*. We follow here the notions used in [Kec95].

Definition 6.11. *Let X be a countable set and $\mathcal{D} = \mathsf{P}(X)$. A* derivative *on $\mathcal{D}$ is a map $D\colon \mathcal{D} \to \mathcal{D}$ such that $D(A) \subseteq A$ and $D(A) \subseteq D(B)$ for $A \subseteq B$, $A, B \in \mathcal{D}$. For $A \in \mathcal{D}$ we define $D^0(A) = A$, $D^{\eta+1}(A) = D\big(D^{\eta}(A)\big)$ and for a limit ordinal η*

$$D^{\eta}(A) \stackrel{def}{=} \bigcap_{\eta' < \eta} D^{\eta'}(A).$$

Now, let $|A|_D$ for $A \in \mathcal{D}$ be the least ordinal η such that $D^{\eta}(A) = D^{\eta+1}(A)$. Such an ordinal exists by monotonicity of D and since X is countable, $\eta < \omega_1$. We additionally put

$$D^{\infty}(A) \stackrel{def}{=} D^{|A|_D}(A).$$

Now let us state [Kec95, Theorem 34.10] in the case of countable X.

Theorem 6.61 (Theorem 34.10 from [Kec95, Sect. 34.E]). *Let X be a countable set and $\mathcal{D} = \mathsf{P}(X)$. Let $D\colon \mathcal{D} \to \mathcal{D}$ be a derivative that is Borel. Put*

$$\Omega_D = \{F \in \mathcal{D}\colon\ D^{\infty}(F) = \varnothing\}.$$

Then Ω_D is $\mathbf{\Pi}^1_1$ and the map $F \mapsto |F|_D$ is a $\mathbf{\Pi}^1_1$-rank on Ω_D.

Our aim is to present Dv as a Borel derivative in such a way that $\Omega_D = \mathrm{PTh}$ and the map $F \mapsto |F|_D$ is the rank of thin trees in the sense of Sect. 6.1.4. The above theorem will then imply that the rank of thin trees is a $\mathbf{\Pi}^1_1$-rank. Then, by the definition of $\preceq^*_{\mathrm{rank}}$ (see [Kec95, Sect. 34.B]) we obtain that

$$\begin{aligned} R_{\mathrm{Embed}}(t,t') &\overset{\mathrm{def}}{\Leftrightarrow} \mathrm{dom}(t) \preceq^*_{\mathrm{rank}} \mathrm{dom}(t') \\ &\Leftrightarrow \mathrm{dom}(t') \notin \mathrm{PTh} \vee \big(\mathrm{dom}(t), \mathrm{dom}(t') \in \mathrm{PTh} \wedge \\ &\qquad\qquad \wedge\ \mathrm{rank}(\mathrm{dom}(t)) \leq \mathrm{rank}(\mathrm{dom}(t'))\big) \\ &\Leftrightarrow t' \notin \mathrm{Th}_{A_{\mathrm{R}}} \vee (t, t' \in \mathrm{Th}_{A_{\mathrm{R}}} \wedge \mathrm{rank}(t) \leq \mathrm{rank}(t')) \end{aligned}$$

is a $\mathbf{\Sigma}^1_1$-relation.

Fact 6.62. *The rank of thin tree-shapes comes from a Borel derivative, as in the assumptions of Theorem 6.61.*

Proof. Let $X = \{\mathrm{L}, \mathrm{R}\}^*$ and $\mathcal{D} = \mathsf{P}(X)$. Note that in this case $\mathrm{PTr} \subseteq \mathcal{D}$. We will extend the derivative Dv to a function $D\colon \mathcal{D} \to \mathcal{D}$ by defining it also on sets $F \subseteq X$ such that $F \notin \mathrm{PTr}$. Let $F \subseteq X$ and let $\bar{F}$ be the prefix-closure of F:

$$\bar{F} \overset{\mathrm{def}}{=} \{u : \exists_{w \in F}\, u \preceq w\}.$$

Now let $D(F) \overset{\mathrm{def}}{=} \mathrm{Dv}(\bar{F})$.

The function D defined this way is monotone and Borel: the operation $\mathcal{D} \ni F \mapsto \bar{F} \in \mathrm{PTr}$ is Borel and the property that $u \in \mathrm{Dv}(\tau)$ is a Borel property of a tree-shape τ: $u \in \tau$ and $\tau\restriction_u$ does not have a finite number of branches (this property is Borel because our trees are finitely branching). Also, $D^{\infty}(\tau) = \emptyset$ if and only if $\tau \in \mathrm{PTh}$. By applying Theorem 34.10 we obtain that the rank induced by D (that is the rank of thin trees) is a $\mathbf{\Pi}^1_1$-rank. ■

Our second relation R_{QSkel} is intended to witness the existence of a particular skeleton $\tilde{\sigma}$ of a given thin tree t. The trick is that $\tilde{\sigma}$ witnesses a skeleton of t given that t is thin. Otherwise, $\tilde{\sigma}$ does not witness anything interesting. Such a (conditional) skeleton is denoted as a *quasi-skeleton*.

We will encode a subset $\tilde{\sigma} \subseteq \mathrm{dom}(t)$ of nodes of a tree t as its characteristic function — a tree (denoted also $\tilde{\sigma}$) over the ranked alphabet $(\{0,1\}, \{0,1\})$ such that $\mathrm{dom}(t) = \mathrm{dom}(\tilde{\sigma})$. To simplify the notions we will say that $u \in \mathrm{dom}(t)$ *belongs* to $\tilde{\sigma}$ if u belongs to the set encoded by it (i.e. if $\tilde{\sigma}(u) = 1$).

Proposition 6.15. *There exists a* $\mathbf{\Sigma}^1_1$ *relation* R_{QSkel} *on* $\mathrm{Tr}_{A_{\mathrm{R}}} \times \mathrm{Tr}_{\{0,1\}^2}$ *such that:*

1. *for every pair* $(t, \tilde{\sigma}) \in R_{\mathrm{QSkel}}$ *we have* $\mathrm{dom}(t) = \mathrm{dom}(\tilde{\sigma})$, $\tilde{\sigma}(\epsilon) = 0$, *and* $\tilde{\sigma}$ *contains (treated as a set of nodes of* t*) exactly one node from each pair of siblings in* t,
2. *for every thin tree* t *there exists a tree* $\tilde{\sigma}$ *such that* $(t, \tilde{\sigma}) \in R_{\mathrm{QSkel}}$,
3. *if* t *is a thin tree and* $(t, \tilde{\sigma}) \in R_{\mathrm{QSkel}}$ *then* $\tilde{\sigma}$ *encodes a skeleton of* t.

A tree $\tilde{\sigma}$ *such that* $(t, \tilde{\sigma}) \in R_{\mathrm{QSkel}}$ *is called a* quasi-skeleton of t.

Note that R_{QSkel} may contain some pairs $(t, \tilde{\sigma})$ with a thick tree t. In that case $\tilde{\sigma}$ encodes some set of nodes of t but not a skeleton.

We define $R_{\mathrm{QSkel}} \subseteq \mathrm{Tr}_{A_{\mathrm{R}}} \times \mathrm{Tr}_{\{0,1\}^2}$ as the set of pairs $(t, \tilde{\sigma})$ such that:

- $\mathrm{dom}(\tilde{\sigma}) = \mathrm{dom}(t)$,
- $\epsilon \notin \tilde{\sigma}$,
- for every pair of siblings in t exactly one of them is in $\tilde{\sigma}$,
- for every internal node u of t such that $ud \in \tilde{\sigma}$ we have

$$(t\restriction_{u\bar{d}}, t\restriction_{ud}) \in R_{\mathrm{Embed}}, \tag{6.12}$$

i.e. the subtree under the sibling of ud *embeds* into the subtree under ud.

Fact 6.63. *Since* R_{Embed} *is analytic and analytic sets are closed under countable intersections, the relation* R_{QSkel} *is also analytic.*

The following two lemmas prove Items 2 and 3 of Proposition 6.15.

Lemma 6.18. *Let* t *be a thin tree. There exists a quasi-skeleton* $\tilde{\sigma}$ *for* t.

Proof. Let t be a thin tree. We show that the canonical skeleton $\sigma(t)$ of t defined in the proof of Proposition 6.12 is a quasi-skeleton of t. Let τ be the spine of t and let u_{L} and u_{R} be two siblings in t. By the inductive construction of $\sigma(t)$ we can assume that at least one of these siblings ud belongs to τ. If $u\bar{d} \notin \tau$ then $\mathrm{rank}(ud, t) > \mathrm{rank}(u\bar{d}, t)$ so (6.12) is satisfied. Now assume that both $ud, u\bar{d}$ belong to τ. In that case we have

$$\mathrm{rank}(ud, t) = \mathrm{rank}(u\bar{d}, t),$$

so (6.12) is also satisfied, no matter which of the siblings belongs to $\sigma(t)$. ∎

Lemma 6.19. *If* t *is a thin tree and* $\tilde{\sigma}$ *is a quasi-skeleton of* t *then* $\tilde{\sigma}$ *(treated as a set of nodes of* t*) is a skeleton of* t.

Proof. Take any infinite branch α of t. We need to show that almost all nodes on α belong to $\tilde{\sigma}$. Assume contrary. Let $u_0 \prec u_1 \prec \ldots \prec \alpha$ be the sequence of nodes on α that do not belong to $\tilde{\sigma}$. By the definition of $\tilde{\sigma}$ for every node u_i the sibling u_i' of u_i satisfies $(t\restriction_{u_i}, t\restriction_{u_i'}) \in R_{\mathrm{Embed}}$. Since t is thin this property implies that

$$\mathrm{rank}(u_i, t) \leqslant \mathrm{rank}(u_i', t).$$

Since ordinal numbers are well-founded, we can assume without loss of generality that all the ranks $\mathrm{rank}(u_i, t)$ are equal some ordinal $\eta < \omega_1$. Since $u_i \prec u_{i+1}'$ so we can also assume that for every i we have $\mathrm{rank}(u_i') = \eta$. Let $t' = t\restriction_{u_0}$ and let τ be the spine of t'. Note that $\mathrm{rank}(t') = \eta$ so by the definition τ contains all the nodes of rank η in t. In particular τ contains all nodes u_i and u_i'. But this is a contradiction, since $u \in B_{\mathrm{CB}}$ so it cannot contain infinitely many branching nodes. ∎

Remark 6.7. Assume that t is a thin tree, $\tilde{\sigma}$ is a quasi-skeleton of t, and $u \in \mathrm{dom}(t)$ is a node of t. The main branch of $\tilde{\sigma}$ from u can be defined in the same way as in the case of skeletons. The only difference is that if $\tilde{\sigma}$ is not a skeleton then not every infinite branch of t is main.

6.3.2 *Proof of Proposition 6.13*

Assume that $L \subseteq \mathrm{Th}_{A_R}$ is a regular language of thin trees, we want to show that $L \in \mathbf{\Pi}^1_1(\mathrm{Tr}_{A_R})$. Let $L' = \mathrm{Tr}_{A_R} \setminus L$ be the complement of L among all ranked trees. L' is a regular language of ranked trees. Let $\mathcal{A}$ be a non-deterministic tree automaton recognizing L'. We will write L' as a sum

$$L' = \left(\mathrm{Tr}_{A_R} \setminus \mathrm{Th}_{A_R}\right) \cup K, \tag{6.13}$$

for some language K that will be defined this way to be analytic and to satisfy the following condition:

$$K \cap \mathrm{Th}_{A_R} = L' \cap \mathrm{Th}_{A_R}.$$

Therefore, Eq. (6.13) will hold and will be an analytic definition of L'.

Let K contain those trees t such that there exists a quasi-skeleton $\tilde{\sigma}$ of t and a run ρ of the automaton $\mathcal{A}$ on t such that for every node $u \in \mathrm{dom}(t)$ the limes inferior of priorities of ρ is even along the main branch of $\tilde{\sigma}$ from u. More formally:

$$\begin{aligned} K = \Big\{ t \in \mathrm{Tr}_{A_R} : \exists_{\tilde{\sigma},\rho}\ & (t,\tilde{\sigma}) \in R_{\mathrm{QSkel}} \text{ and} \\ & \rho \text{ is a run of } \mathcal{A} \text{ on } t \text{ and} \\ & \forall_{u \in \mathrm{dom}(t)}.\ \text{the lim inf of priorities of } \rho \\ & \qquad \text{on the main branch of } \tilde{\sigma} \text{ from } u \text{ is even} \Big\}. \end{aligned}$$

Observe that K is defined by existential quantification over trees $\tilde{\sigma}$ and runs ρ. The inner properties are analytic (the later two are in fact Borel). Therefore, K is analytic. Note that we do not express explicitly that ρ is an accepting run.

Observe that if $t \in L' \cap \mathrm{Th}_{A_R}$ then $t \in K$: there is some quasi-skeleton $\tilde{\sigma}$ for t and there is an accepting run ρ of $\mathcal{A}$. Since ρ is accepting so it is accepting on all main branches of $\tilde{\sigma}$.

What remains is to show that if $t \in K \cap \mathrm{Th}_{A_R}$ then $t \in L'$. Take a thin tree $t \in K$. Assume that $\tilde{\sigma}, \rho$ are a quasi-skeleton and a run given by the definition of K. Since t is a thin tree, $\tilde{\sigma}$ is actually a skeleton of t. We take any infinite branch α of t and show that ρ is accepting along α. By Lemma 6.54 we know that there is a node $u \in \mathrm{dom}(t)$ such that α is the main branch of $\tilde{\sigma}$ from u. Therefore, by the definition of K, the run ρ is accepting on α.

This concludes the proof of Proposition 6.13.

6.4 Characterisation of WMSO-definable Languages

In this section we prove a decidable characterisation of languages of thin trees that are WMSO-definable among all trees. It will be achieved by proving that the following conditions are equivalent.

Proposition 6.16. *Let $L \subseteq \mathrm{Th}_{A_R}$ be a regular language of thin trees over a ranked alphabet $A_R = (A_{R0}, A_{R2})$ and let $\mathcal{B}$ be a non-deterministic automaton recognising L among all trees. The following conditions are equivalent:*

1. *for $M = |Q^{\mathcal{B}}| \cdot |A_{R2}| + 1$ and every $t \in L$ we have $\mathrm{rank}(t) \leqslant M$,*
2. *there exists $M \in \mathbb{N}$ such that every tree $t \in L$ satisfies $\mathrm{rank}(t) \leqslant M$,*
3. *L is WMSO-definable among all trees,*
4. *there exists $N \in \mathbb{N}$ such that $L \in \mathbf{\Sigma}^0_N(\mathrm{Tr}_{A_R})$,*
5. *L is not $\mathbf{\Pi}^1_1(\mathrm{Tr}_{A_R})$-hard.*

Moreover, it is decidable if these conditions hold.

The implications (1) ⇒ (2), (3) ⇒ (4), and (4) ⇒ (5) are trivial — any language definable in WMSO is on a finite level of the Borel hierarchy, thus not $\mathbf{\Pi}^1_1$-hard. The remaining two implications are proved in the following subsections. The decidability follows from Remark 6.8.

A relation between definability in WMSO and boundedness of a certain rank is also exploited in Chap. 4.

6.4.1 *Implication (2) ⇒ (3)*

We need to prove that if for some M every tree $t \in L$ satisfies $\mathrm{rank}(t) \leqslant M$ then L is WMSO-definable among all trees. This will be achieved by an explicit construction (via induction on M) of a WMSO formula defining L among all trees.

In our constructions we use the following additional notion. Assume that $t \in \mathrm{Tr}_{A_R}$ is a tree and $u \preceq w$ are two nodes of t. We say that a node z is *off the path from u to w* if z is not an ancestor of w ($z \npreceq w$) but there exists u' such that $u \preceq u' \prec w$ and z is a child of u'.

The proofs of this section go by induction on M (the bound on the ranks of thin trees). In all of the cases the base step is trivial as there is no thin tree of rank 0.

We start with the following lemma. The constructed formula φ_m will serve as a basis in the following constructions.

Lemma 6.20. *For every $m \in \mathbb{N}$ there exists a WMSO formula φ_m defining among all ranked trees the language of thin trees of rank at most m (denoted $\mathrm{Th}^{\leqslant m}_{A_R}$, see Sect. 6.1.4).*

Proof. The proof goes by induction on m. The base step is trivial.

Assume that the thesis holds for m — we have defined a formula φ_m. Consider a WMSO formula φ_{m+1} that for a given ranked tree $t \in \mathrm{Tr}_{A_R}$ says that:

there exists a finite tree s with $\mathrm{dom}(s) \subseteq \mathrm{dom}(t)$ such that
for every internal node w of t such that $w \notin \mathrm{dom}(s)$
there exists a child wd of w such that
the subtree $t\restriction_{wd}$ has rank at most m (i.e. the formula φ_m holds on $t\restriction_{wd}$).

First assume that φ_{m+1} holds on a given tree t and take s as in the statement. Let $\tau \subseteq \mathrm{dom}(t)$ be the set of nodes $u \in \mathrm{dom}(t)$ such that $\mathrm{rank}(u, t) > m$. Observe that by φ_m if u is a branching node of τ then $u \in \mathrm{dom}(s)$. Therefore $\tau \in B_{\mathrm{CB}}$ and $\mathrm{rank}(t) \leqslant m + 1$.

Now assume that $\mathrm{rank}(t) \leqslant m + 1$. If $\mathrm{rank}(t) < m$ then φ_m is trivially satisfied by any finite tree s. Assume that $\mathrm{rank}(t) = m + 1$ and let $\tau = \mathrm{Dv}^m(\mathrm{dom}(t))$ be the spine of t. Since $\tau \in B_{\mathrm{CB}}$ so τ has finitely many branching nodes. Let us take as s a finite tree with $\mathrm{dom}(s) \subseteq \mathrm{dom}(t)$ and such that s contains all the branching nodes of τ. By the definition of τ, for every internal node w of t that is outside s, at least one of the children of w has rank at most m. ■

The above lemma implies that the set of thin trees of rank at most $m \in \mathbb{N}$ is MSO-definable. Therefore, given a regular language of thin trees it is decidable if it contains a tree of rank greater than a given number m. This gives us the following remark.

Remark 6.8. Condition (1) from Proposition 6.17 is decidable.

The crucial inductive part of the proof of the implication (2) ⇒ (3) is expressed by the following proposition. The rest of this section is devoted to its proof. The implication (2) ⇒ (3) follows when we take as f the automaton homomorphism for an automaton $\mathcal{A}$ recognising L and as m the bound M from Condition (2).

Proposition 6.17. *Let (H, V) be a finite thin algebra over a ranked alphabet A_R. Let $f\colon \mathrm{Th}_{A_R} \to (H, V)$ be the unique homomorphism assigning to thin trees their types. For every type $h \in H$ and number $m \in \mathbb{N}$ there exists a* WMSO *formula φ_m^h that defines those ranked trees $t \in \mathrm{Tr}_{A_R}$ such that $t \in \mathrm{Th}_{A_R}$, $\mathrm{rank}(t) = m$, and the type of t is h with respect to f (i.e. $f(t) = h$).*

The base step for $m = 0$ is trivial. Assume that the thesis of the proposition holds for all types h and all numbers less or equal than m. We show it for $m + 1$.

First we write a formula $\psi_m(u, w)$ expressing that for a given pair of nodes u, w of a given tree t:

$u \preceq w$,
the subtrees $t\restriction_u$ and $t\restriction_w$ have ranks exactly m (we check it using φ_m and $\neg\varphi_{m-1}$),
and for every z that is off the path from u to w
the rank of $t\restriction_z$ is at most $m - 1$ (i.e. φ_{m-1} holds on $t\restriction_z$).

The following lemma expresses the crucial properties of formulae $\psi_m(u, w)$.

Lemma 6.21. *Assume that for a given ranked tree $t \in \mathrm{Tr}_{A_R}$ and a node u of t there are infinitely many nodes w such that $\psi_m(u, w)$. Then $\mathrm{rank}(t\restriction_u) = m$ and the set of nodes of rank equal m below u in t forms a single infinite branch α of t.*

Moreover, $\psi_m(u, w)$ holds for some $w \in \mathrm{dom}(t)$ if and only if $u \preceq w \prec \alpha$.

Proof. Take a ranked tree t and a node $u \in \mathrm{dom}(t)$ as in the statement. Without loss of generality we can assume that $u = \epsilon$, because φ_m talks only about the subtree $t\restriction_u$. Observe that $\mathrm{rank}(t\restriction_\epsilon) = \mathrm{rank}(t) = m$. Let $\tau \subseteq \mathrm{dom}(t)$ be the set of nodes $w \in \mathrm{dom}(t)$ such that $\psi_m(\epsilon, w)$ holds. Observe that if $u \preceq w_1 \preceq w_2 \in \mathrm{dom}(t)$ and $w_2 \in \tau$ then $w_1 \in \tau$. Since there are infinitely many nodes w satisfying $\psi_m(\epsilon, w)$ so τ is infinite. Observe also that τ does not contain any branching node. Therefore τ is a single infinite branch α. Clearly, if w is not a prefix of α then $\mathrm{rank}(w, t) < m$. ■

The above lemma states that the formula $\psi_m(u, w)$ enables us to fix in a WMSO-definable way a particular branch α in our tree such that almost all nodes that are off this branch have ranks smaller than m. What remains is to compute the type of the subtree rooted in the node u from the types of the subtrees that are off α and from α itself. The following formula is an intermediate step in this construction.

Fact 6.64. *For nodes u, w_1, w_2 and a type $s \in V$ there exists a WMSO formula $\gamma^s_m(u, w_1, w_2)$ expressing the following facts:*

- *$u \preceq w_1 \preceq w_2$,*
- *$\psi_m(u, w_2)$ holds (it implies $\psi_m(u, w_1)$),*
- *$f(t\restriction_{[w_1, w_2)}) = s$ — the type of the the context rooted in w_1 with the hole in w_2 is s.*

To achieve the last item of the list, the formula computes the types of the subtrees rooted in the nodes off the path from w_1 to w_2 using the inductive formulae $\varphi^h_{m'}$ for $m' < m$ and $h \in H$. Then the formula executes the multiplication in V on the finite path from w_1 to w_2.

Now we show how to compute a type of a tree with a spine consisting of one infinite branch. The formula is based on a construction from [Tho80] that enables to verify the type of a given ω-word in FO logic using predicates of the form "the type of the infix between the positions w_1 and w_2 is e".

Definition 6.12. *Let u be a node of a tree t and $h \in H$ be a type. Let the formula $\delta^h_m(u)$ express the following facts:*

there are infinitely many nodes w such that $\psi_m(u, w)$ holds,
there exists a pair of context types $s, e \in V$ such that $se^\infty = h$,
there exists a node z_0 such that $\gamma^s_m(u, u, z_0)$ holds (i.e. $f\left(t\restriction_{[u, z_0)}\right) = s$), and
for every node w_1 such that $\psi_m(u, w_1)$
there exists a pair of nodes w_2, w_3 such that

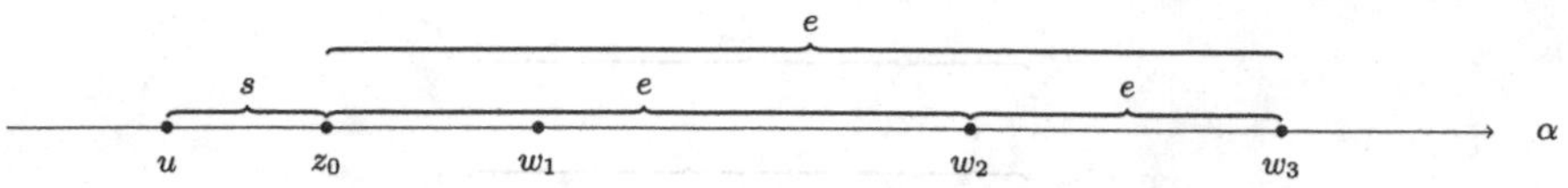

Fig. 6.3 An illustration of properties expressed by the formulae $\delta_m^h(u)$.

$w_1 \prec w_2 \prec w_3$,
$\psi_m(u, w_3)$ holds (it implies $\psi_m(u, w_2)$), and
the formulae $\gamma_m^e(u, z_0, w_2)$, $\gamma_m^e(u, z_0, w_3)$, and $\gamma_m^e(u, w_2, w_3)$ hold
(i.e. the types of the three contexts equal e, see Fig. 6.1).

Lemma 6.22. *Let t be a tree and u be a node of t such that there are infinitely many nodes w satisfying $\psi_m(u, w)$. Then $f(t\restriction_u) = h$ if and only if $\delta_m^h(u)$ holds on t.*

Proof. Again, without loss of generality $u = \epsilon$. First assume that $t \models \delta_m^h(\epsilon)$ for some $h \in H$. Let α be the branch defined by the predicate $\psi_m(\epsilon, w)$ as in Lemma 6.21.

We show that the formula $\gamma_m^h(\epsilon)$ gives rise to a sequence of nodes $z_0 \prec z_1 \prec z_2 \dots$ on α such that for some types s, e satisfying $se^\infty = h$ we have:

$$f\left(t\restriction_{[\epsilon,z_0)}\right) = s, \quad f\left(t\restriction_{[z_i,z_{i+1})}\right) = e. \tag{6.14}$$

Having done so, we conclude that the type of $t = t\restriction_\epsilon$ is h.

Let us fix z_0 as in the definition of $\delta_m^h(\epsilon)$. By the definition we know that $f(t\restriction_{[\epsilon,z_0)}) = s$. We will set w_1 to various nodes along α obtaining nodes w_2, w_3 such that $w_1 \prec w_2 \prec w_3 \prec \alpha$.

Let us start with w_1 equal z_0 and consider w_2, w_3 given by $\delta_m^h(\epsilon)$. Let $z_1 = w_2$ and $z_1' = w_3$. Our inductive invariant is that the types of all three contexts $t\restriction_{[z_0,z_i)}$, $t\restriction_{[z_0,z_i')}$, and $t\restriction_{[z_i,z_i')}$ equal e. For $i = 1$ we get it by the definition of $\delta_m^h(\epsilon)$. Assume that $z_i \prec z_i'$ are defined for some $i > 0$. Let us take $w_1 = z_i'$ and consider w_2, w_3 as in the definition of $\delta_m^h(\epsilon)$. Let us put $z_{i+1} = w_2$ and $z_{i+1}' = w_3$. By the definition, the types of $t\restriction_{[z_0,z_{i+1})}$ and $t\restriction_{[z_0,z_{i+1}')}$ are e. Consider the type of the context $t\restriction_{[z_i,z_{i+1})}$ (see Fig. 6.4):

$$\begin{aligned}
f\left(t\restriction_{[z_i,z_{i+1})}\right) &= f\left(t\restriction_{[z_i,z_i')}\right) \cdot f\left(t\restriction_{[z_i',z_{i+1})}\right) \\
&= e \cdot f\left(t\restriction_{[z_i',z_{i+1})}\right) \\
&= f\left(t\restriction_{[z_0,z_i')}\right) \cdot f\left(t\restriction_{[z_i',z_{i+1})}\right) \\
&= f\left(t\restriction_{[z_0,z_{i+1})}\right) \\
&= e.
\end{aligned}$$

Therefore, the constructed sequence $z_0 \prec z_1 \prec z_2 \prec \dots$ satisfies (6.14).

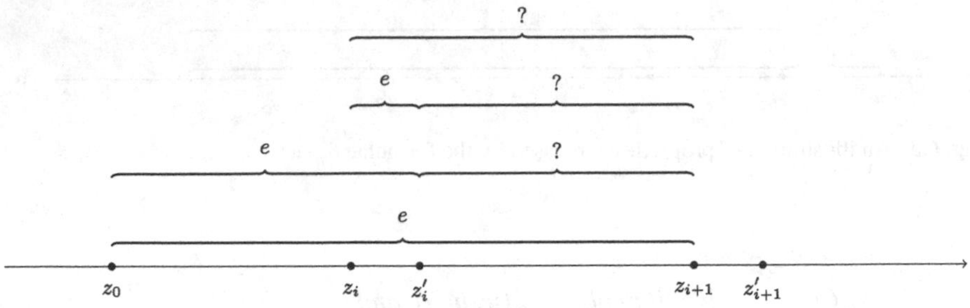

Fig. 6.4 The reasoning used in the proof of Lemma 6.22.

For the other direction take a thin tree t and a branch α of t as in Lemma 6.21. Using Ramsey's theorem (see Theorem 1.1 on page 3) along α, with respect to the function assigning to a pair $u \prec w \prec \alpha$ the type $f(t\restriction_{[u,w)}) \in V$, we find a pair of types s, e and an infinite sequence of nodes $(z_i)_{i\in\mathbb{N}}$ along α satisfying (6.14) and such that $e = e^2$. Since $f(t) = h$, $se^\infty = h$. Therefore, we can satisfy the formula $\delta^h_m(\epsilon)$ using s, e and the successive nodes $(z_i)_{i\in\mathbb{N}}$. ■

We are now ready to construct the formula φ^h_m from Proposition 6.17. It will be obtained by rewriting the formula φ_m from Lemma 6.20 so that it additionally verifies the type of the given ranked tree. In φ^h_m will fix a finite tree s with some leafs $u_1, \ldots, u_n$ of s and a sequence of types $h_1, \ldots, h_n \in H$. We then write $s(h_1, \ldots, h_n)$ for the type obtained by the evaluation of the term represented by s on the given types in the algebra (H, V). Take $m > 0$, $h \in H$ and define φ^h_m that says:

> there exists a finite tree s with $\mathrm{dom}(s) \subseteq \mathrm{dom}(t)$,
> a number of leafs $u_1, \ldots, u_n$ of s, and
> a sequence of types $h_1, \ldots, h_n$ such that
>   the type of $s(h_1, h_2, \ldots, h_n)$ is h and
>   for every leaf $u_i (i = 1, \ldots, n)$
> $\delta^{h_i}_m(u_i)$ holds and
>   there are infinitely many nodes w such that $\psi_m(u_i, w)$ holds.

Lemma 6.23. *A tree $t \in \mathrm{Tr}_{A_R}$ satisfies φ^h_m if and only if* $\mathrm{rank}(t) = m$ *and* $f(t) = h$.

Proof. First assume that $\mathrm{rank}(t) = m$ and $f(t) = h$. Let τ be the spine of t and take s as a finite tree containing all the branching nodes of τ. A leaf u of s is included in the list $u_1, \ldots, u_n$ if $\mathrm{rank}(u, t) = m$. We take as h_i the type $f(t\restriction_{u_i})$. Clearly the type of $s(h_1, \ldots, h_n)$ is the type of t that is h. Also, since $\mathrm{rank}(u_i, t) = m$ for $i = 1, \ldots, n$ so $\psi_m(u_i, w)$ holds for infinitely many w. Lemma 6.22 says that $\delta^{h_i}_m(u_i)$ is satisfied.

Now assume that φ_m^h is satisfied. Again, by Lemma 6.22 we know that for $i = 1, \ldots, n$.

$$f(t\restriction_{u_i}) = h_i.$$

Therefore, $f(t) = s(h_1, \ldots, h_n) = h$. ■

This concludes the proof of Proposition 6.17 and of the implication (2) ⇒ (3).

6.4.2 Implication (5) ⇒ (1)

Now we want to prove that if L is not $\mathbf{\Pi}_1^1(\mathrm{Tr}_{A_R})$-hard then every tree $t \in L$ has rank at most $M = |Q^{\mathcal{B}}| \cdot |A_{R2}| + 1$.

We assume contrary that there exists a thin tree $t \in L$ such that $\mathrm{rank}(t) > M$. Our aim is to show that L is $\mathbf{\Pi}_1^1(\mathrm{Tr}_{A_R})$-hard. The proof consists of two parts: first we find a *pumping scheme* within the tree t and then we construct a continuous reduction f from the set of well-founded ω-trees $WF \subseteq \omega\mathrm{PTr}$ (see Sect. 1.6.3, page 17) into $L \subseteq \mathrm{Tr}_{A_R}$. The idea is that for $\tau \in WF$ the reduction f gives a thin tree in L and if $\tau \notin WF$ then $f(\tau) \notin \mathrm{Th}_{A_R}$, so in particular $f(\tau) \notin L$. Since the set of well-founded ω-trees is $\mathbf{\Pi}_1^1$-complete (see Theorem 1.11 on page 17), it will prove that L is $\mathbf{\Pi}_1^1$-hard.

Let us take $m > 0$ and a ranked tree $t \in \mathrm{Tr}_{A_R}$. A *pumping scheme of depth m in t* (see Fig. 6.5) is a function $P\colon \omega^{\leqslant m} \to \mathrm{dom}(t)$ such that:

- for every $u \in \omega^{\leqslant m}$ the node $P(u)$ is an internal node of t,
- for every $u \prec w \in \omega^{\leqslant m}$ we have $P(u) \prec P(w)$,
- for every $k \leqslant m$ and $u \neq w \in \omega^k$ we have $P(u) \npreceq P(w)$ and $P(w) \npreceq P(u)$,
- for every $k \leqslant m$ and $u, w \in \omega^k$ we have $t(P(u)) = t(P(w))$.

Note that the last condition implies that there exists a function $P_S\colon \{0, 1, \ldots, m\} \to A_{R2}$ assigning to a number $k \leqslant m$ the unique letter $P_S(k) \in A_{R2}$ such that if $u \in \omega^k$ then $t(P(u)) = P_S(k)$. This function is called the *signature of P*.

Lemma 6.24. *If t is a thin tree and* $\mathrm{rank}(t) > m + 1$ *then there exists a pumping scheme of depth m in t.*

Before proving the lemma we extract an observation crucial for the inductive step.

Fact 6.65. *Let t be a thin tree and $(u_i)_{i\in\mathbb{N}}$ be a sequence of nodes of t. Assume that the nodes u_i are pairwise incomparable with respect to the prefix order $\preceq$. If each subtree $t\restriction_{u_i}$ has a pumping scheme of depth m and of a fixed signature P_S (one for all i) then t has a pumping scheme of depth $m + 1$.*

Proof. We just combine the schemes for all the nodes $(u_i)_{i\in\mathbb{N}}$ and put $P(\epsilon) = \epsilon$. ■

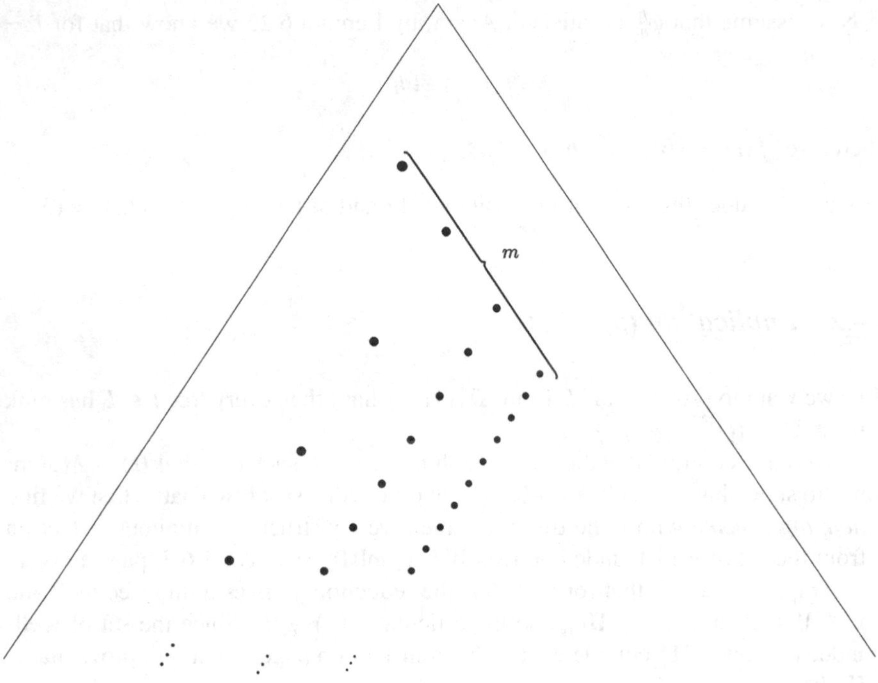

Fig. 6.5 An example of a pumping scheme P_S of depth $m = 3$ in a tree t. The highest dot is the node $P_S(\epsilon)$. Under it we have an anti-chain of nodes $P_S(i)$ for $i \in \mathbb{N}$. Under each node $P_S(i)$ we again have an anti-chain of nodes $P_S(ij)$ for $j \in \mathbb{N}$. The lowest line consists of nodes of the form $P_S(ijk)$ for $k \in \mathbb{N}$.

Proof of Lemma 6.24. The proof is inductive in m. For $m = 0$ we can take as P the function $\epsilon \mapsto \epsilon$ — ϵ is not a leaf of t because t has rank at least 2. Assume that the thesis holds for m. Let $\mathrm{rank}(t) > m + 1$ and let τ be the spine of t. Let $(u_i)_{i\in\mathbb{N}}$ be a sequence of nodes of t that are off τ and have rank at least $m + 1$ in t. This sequence is infinite, otherwise $\mathrm{rank}(t) \leqslant m + 1$. Note that since the nodes $(u_i)_{i\in\mathbb{N}}$ are off τ so they are pairwise incomparable with respect to the prefix order $\preceq$.

By the inductive assumption, for every i there is a pumping scheme P_i of depth m in $t\restriction_{u_i}$. Since there are only finitely many distinct signatures of pumping schemes of depth m so for some infinite subsequence of $(P_i)_{i\in\mathbb{N}}$ all the signatures are equal. By Fact 6.65 we obtain that there exists a pumping scheme of depth $m + 1$ in t. ■

Now we can move to the construction of a continuous reduction f of $WF \subseteq \omega\mathrm{PTr}$ to $L \subseteq \mathrm{Tr}_{A_R}$. Recall that we have fixed a thin tree $t \in L$ such that $\mathrm{rank}(t) > |Q^{\mathcal{B}}| \cdot |A_{R2}| + 1$ for a non-deterministic automaton $\mathcal{B}$ recognising L among all trees. Let $\rho \in \mathrm{Tr}_{(Q^{\mathcal{B}}, Q^{\mathcal{B}})}$ be an accepting run of $\mathcal{B}$ on t. Let $t' = t \otimes \rho$ be the tree over the product alphabet.

Let $m = |Q^{\mathcal{B}}| \cdot |A_{R2}|$. Because $\mathrm{rank}(t') = \mathrm{rank}(t) > m + 1$ so there exists a pumping scheme P of depth m in t'. Let P_S be the signature of P. By the pigeonhole principle there are two numbers $0 \leqslant k < k' \leqslant m$ such that $P_S(k) = P_S(k')$. Let $u = P(0^k)$ — the image by P of the vector of k zeros and let $w_i = P\big(0^k \cdot i \cdot 0^{k'-k-1}\big)$.

Fact 6.66. *By the definition we obtain that:*

- *the nodes w_i are pairwise incomparable w.r.t. the prefix order $\preceq$,*
- *$u \prec w_i$ and $t'(u) = t'(w_i)$ for every $i \in \mathbb{N}$.*

Let w_i' be the word such that $uw_i' = w_i$. This sequence of nodes enables us to cut t' into the following pieces:

- p_I is the thin context obtained from t' by putting the hole in u (i.e. $p_I \stackrel{\text{def}}{=} t'[u \leftarrow \Box]$),
- r_M is the thin tree over the ranked alphabet $(A_{R2} \times Q, A_{R0} \times Q \sqcup \{\Box\})$ obtained from $t'{\restriction}_u$ by putting a leaf labelled by $\Box$ in all the nodes w_i' for $i \in \mathbb{N}$,
- t_F is the subtree $t'{\restriction}_{w_0}$.

Observe that the nodes of r_M labelled by $\Box$ are naturally numbered by natural numbers. Assume that $(t_i)_{i\in\mathbb{N}}$ is a sequence of trees over the alphabet $(A_{R2} \times Q, A_{R0} \times Q)$. Then, $r_M(t_0, t_1, \ldots)$ is the tree obtained by putting, for every i, the root of t_i into the i'th hole of r_M (i.e. into the node w_i' of r_M). Using these notions we can write

$$t' = p_I\big(r_M(t_F, t'{\restriction}_{w_1}, t'{\restriction}_{w_2}, t'{\restriction}_{w_3}, \ldots)\big).$$

We define a function $f_0\colon \omega\mathrm{PTr} \to \mathrm{Tr}_{A_R \times (Q,Q)}$ by co-induction (see Sect. 1.6.5, page 18). Then $f(\tau)$ is defined as $p_I(f_0(\tau))$. If τ is empty then let $f_0(\tau) = t_F$. Otherwise, assume that $(\tau_i)_{i\in\mathbb{N}}$ is the list of subtrees $\tau{\restriction}_{(i)}$. Let

$$f_0(\tau) = r_M\big(f_0(\tau_0),\, f_0(\tau_0),\; f_0(\tau_1),\, f_0(\tau_1),\; \ldots\big).$$

Note that each subtree $f_0(\tau_i)$ is inserted twice into r_M.

Since the root of r_M is its internal node, the function f_0 is continuous — the more is known about τ the bigger fragment of $f_0(\tau)$ can be produced. Therefore, f is also continuous. Observe also that for every τ the result $f(\tau)$ is a product of two trees $f^1(\tau) \otimes f^2(\tau)$ with $f^1(\tau)$ over the ranked alphabet A_R and $f^2(\tau)$ over (Q, Q). Because of Fact 6.66 the tree $f^2(\tau)$ is a run of $\mathcal{B}$ on $f^1(\tau)$. The value of the run $f^2(\tau)$ (i.e. $f^2(\tau)(\epsilon)$) is the same as the value of ρ.

What remains is to prove the following lemma.

Lemma 6.25. *An ω-tree $\tau \in \omega\mathrm{PTr}$ is well-founded (belongs to WF) if and only if $f^1(\tau) \in L$.*

Proof. First assume that $\tau \in WF$. In that case, every branch of $f(\tau)$ from some point on reaches a copy of t_F or stays forever in some copy of p_I or r_M. Thus, the run $f^2(\tau)$ is parity-accepting on every branch of $f^1(\tau)$. So $f^1(\tau) \in L$.

Now take $\tau \notin WF$. Assume that $\alpha \in \omega^\omega$ is an infinite branch of τ. We show how to embed the complete binary tree $\{0, 1\}^*$ into $\mathrm{dom}(f(\tau))$ thus showing that it is not thin. Since $L \subseteq \mathrm{Th}_{A_R}$, $f(\tau) \notin L$.

We take a branch $\beta \in \{0, 1\}^\omega$ and construct a sequence of vertices $z_0 \prec z_1 \prec z_2 \prec, \ldots$ in $f(\tau)$. First we put $z_0 = u$ (the hole of the ranked context p_I). From that moment on we will traverse infinitely many copies of r_M. The invariant is that for every i

$$f(\tau)\restriction_{z_i} = f_0(\tau\restriction_{\alpha\restriction_i}).$$

For $i = 0$ the invariant is satisfied. In a step i we define as z_n the vertex (hole) $w'_{2\cdot\alpha(i)+\beta(i)}$ in the current copy of r, i.e.

$$z_{i+1} = z_i \cdot w'_{2\cdot\alpha(i)+\beta(i)}.$$

Since $\tau\restriction_{\alpha\restriction_i}$ is non-empty for every i, the invariant is satisfied. Let $\pi_\beta \in \{\mathrm{L}, \mathrm{R}\}^\omega$ be the branch of $f(\tau)$ defined by the sequence of vertices $z_0 \prec z_1 \prec \ldots$. Observe that for any $\beta' \neq \beta$ we have $\pi_{\beta'} \neq \pi_\beta$. So indeed the tree $t(\tau)$ is not thin — it contains the complete binary tree as a minor. ■

This concludes the proof of the implication (5) ⇒ (1) and Proposition 6.16.

6.5 Conclusions

This chapter studies descriptive complexity of regular tree languages that contain only thin trees. First of all it is shown that each such language is $\mathbf{\Pi}^1_1$ among all trees. It is a noticeable collapse comparing to general regular tree languages that belong to $\mathbf{\Delta}^1_2$.

The second part of the chapter is devoted to studying when a regular language containing only thin trees can be defined in WMSO. It turns out that this problem relates the following three notions:

- definability in WMSO,
- topological complexity (i.e. $\mathbf{\Pi}^1_1$-complete sets),
- certain ranks of thin trees.

These links show that Conjecture 2.2 from page 32 is true in the case of regular languages containing only thin trees. Additionally, a pumping argument (see Lemma 6.24) is presented that shows that one of the conditions equivalent to WMSO-definability is decidable.

The results of this chapter do not solve the problem of definability in WMSO among thin trees. This is stated as the following conjecture.

Conjecture 6.6. It is decidable if a given regular language L of thin trees is WMSO-definable among thin trees, i.e. if there exists a WMSO formula φ such that

$$L \;=\; \{t \in \mathrm{Th}_{A_{\mathrm{R}}} \;:\; t \;\models\; \varphi\} = \mathrm{L}(\varphi) \cap \mathrm{Th}_{A_{\mathrm{R}}}.$$

Since the language of all thin trees is WMSO-definable among thin trees (by the formula $\top$) so the method of ranks does not seem to be useful in this case.

This chapter is based on [BIS13].

Chapter 7
Recognition by Thin Algebras

In both cases of finite words and ω-words the class of regular languages can be equivalently defined as the class of languages recognisable by homomorphisms to appropriate finite algebras (monoids and ω-semigroups respectively, see Sect. 1.5, page 10). This algebraic approach to recognition turned out to be fruitful by entailing many effective characterizations [Sch65, Sim75, BW08]. However, there is no satisfactory algebraic approach to infinite trees, nor even a canonical way to represent a given regular tree language. Proposed algebras (see [BI09, Blu11]) either have no finite representation or yield no new effective characterisations.

This chapter can be seen as an attempt to use thin algebra defined in Chap. 6 to recognise languages of general ranked infinite trees (i.e. not necessarily thin). As observed in Sect. 6.2 (see page 102), the pair $(\mathrm{Tr}_{A_R}, \mathrm{Con}_{A_R})$ of all ranked trees and all ranked contexts over a ranked alphabet A_R has a natural structure of a thin algebra with a subalgebra $(\mathrm{Th}_{A_R}, \mathrm{ThCon}_{A_R})$ consisting of all thin trees and all thin contexts. It can be shown [Idz12] that $(\mathrm{Th}_{A_R}, \mathrm{ThCon}_{A_R})$ is free (formally initial) in the class of thin algebras over A_R. The problem is that the thin algebra $(\mathrm{Tr}_{A_R}, \mathrm{Con}_{A_R})$ is richer than $(\mathrm{Th}_{A_R}, \mathrm{ThCon}_{A_R})$; in particular, for some finite thin algebras S over the ranked alphabet A_R there may be many homomorphisms

$$f\colon (\mathrm{Tr}_{A_R}, \mathrm{Con}_{A_R}) \to S.$$

The notion of *prophetic thin algebras*, introduced in this chapter, can be seen as a natural constraint guaranteeing that there is at most one homomorphism f as above. A natural problem arises what is the class of languages that can be recognised by homomorphisms to finite prophetic thin algebras. Example 7.3 presented in this chapter shows that not every regular tree language is recognised in this way. The following theorem constitutes a characterisation of the languages recognisable by finite prophetic thin algebras.

Theorem 7.5. *A language of infinite trees L is recognised by a homomorphism into a finite prophetic thin algebra if and only if L is bi-unambiguous, i.e. both L and the complement L^{c} can be recognised by unambiguous automata.*

M. Skrzypczak, *Set Theoretic Methods in Automata Theory*, LNCS 9802
DOI: 10.1007/978-3-662-52947-8_7

Blumensath in [Blu11, Blu13] undertook the task of designing an algebraic framework for infinite trees that would allow to recognise precisely the regular tree languages. The relations between prophetic thin algebras and the concept of path-continuity of Blumensath are discussed in Sect. 7.1.

It turns out that bi-unambiguous languages and prophetic thin algebras are closely related to Conjecture 2.1 from page 31 saying that there is no MSO-definable choice function in the class of thin trees. These relations are studied in Chap. 8, see Theorem 8.74 on page 143). In Sect. 7.4 we prove that if Conjecture 2.1 holds then the class of bi-unambiguous languages is decidable among all regular tree languages (see Theorem 7.70). The consequences of Conjecture 2.1 regarding prophetic thin algebras are studied in Sect. 7.3. For instance, Conjecture 2.1 implies that the class of finite prophetic thin algebras is a pseudo-variety.

The chapter is organized as follows. In Sect. 7.1 we introduce prophetic thin algebras. Section 7.2 is devoted to a proof of Theorem 7.5. In Sects. 7.3 and 7.4 we study consequences of Conjecture 2.1. Finally, in Sect. 7.5 we conclude.

7.1 Prophetic Thin Algebras

In this section we introduce the notion of *prophetic thin algebras*. The aim of this definition is to guarantee that if S is a prophetic thin algebra over a ranked alphabet A_{R} then there is at most one homomorphism

$$f\colon (\mathrm{Tr}_{A_{\mathrm{R}}}, \mathrm{Con}_{A_{\mathrm{R}}}) \to S,$$

similarly as in Fact 1.5 on page 12 in the case of ω-semigroups. Example 7.3 below shows that for general (non-prophetic) thin algebras there may be more than one such homomorphism.

Let $S = (H, V)$ be a thin algebra over a ranked alphabet $A_{\mathrm{R}} = (A_{\mathrm{R}2}, A_{\mathrm{R}0})$ and let $t \in \mathrm{Tr}_{A_{\mathrm{R}}}$ be a ranked tree. Observe that every homomorphism $f\colon (\mathrm{Tr}_{A_{\mathrm{R}}}, \mathrm{Con}_{A_{\mathrm{R}}}) \to S$ induces a natural labelling τ_f of t by elements in H:

$$\tau_f(u) \stackrel{\text{def}}{=} f\big(t\restriction_u\big) \text{ for } u \in \mathrm{dom}(t).$$

The labelling τ_f is called the *marking induced by* f *on* t. Intuitively, it declares in advance the f-type of all the subtrees of t.

The axioms of thin algebra and the fact that f is a homomorphism imply that τ_f satisfies many *consistency constraints*. The following two definitions formalise these *consistency constraints* by introducing a notion of a *consistent marking*. The definition reflects the axioms of thin algebra in such a way to guarantee Lemma 7.26.

The first definition says that a labelling τ is supposed to be consistent with respect to the *local* operations of thin algebra: $\mathrm{Node}(a, d, h)$ and $\mathrm{Leaf}(b)$.

Definition 7.13. *Let (H, V) be a thin algebra over a ranked alphabet A_R and let $t \in \mathrm{Tr}_{A_R}$. A labelling $\tau \in \mathrm{Tr}_{(H,H)}$ of t is a* marking of t by types in H *if:*

– *for every internal node u of t we have*

$$\tau(u) = t(u)\big(\tau(u\mathrm{L}), \tau(u\mathrm{R})\big) \quad (i.e.\ \tau(u) = \mathrm{Node}(t(u), \mathrm{R}, \tau(u\mathrm{L})) \cdot \tau(u\mathrm{R})),$$

– *for every leaf u of t we have*

$$\tau(u) = t(u)() \quad (i.e.\ \tau(u) = \mathrm{Leaf}(t(u))).$$

The second definition reflects the infinite product operation $\prod$, it can be seen as a counterpart of Fact 6.57 from page 103.

Definition 7.14. *Fix a thin algebra (H, V) over a ranked alphabet A_R. Let $t \in \mathrm{Tr}_{A_R}$ be a ranked tree, τ be a marking of t by types in H, and α be an infinite branch of t. Assume that $\alpha = d_0 d_1 \ldots$ and let $u_0 \prec u_1 \prec \ldots$ be the sequence of vertices of t along α. Let us put $a_i = t(u_i)$ (the i'th letter of t along α), and $h_i = \tau(u_i \bar{d}_i)$ (the value of τ in the i'th node that is off α).*

The sequence of types of contexts $\mathrm{Node}(a_i, d_i, h_i) \in V$ *for $i = 0, 1, \ldots$ is called the* decomposition of τ along α. *We say that τ is* consistent on α *if for every $i \in \mathbb{N}$ we have*

$$\tau(u_i) = \prod_{j=i,i+1,\ldots} \mathrm{Node}(a_j, d_j, h_j). \tag{7.1}$$

A marking τ is consistent *if it is* consistent on α *for every infinite branch α of t.*

Remark 7.9. By the definition of a marking and axiom (6.6) of thin algebra, it is enough to require (7.1) for infinitely many $i \in \mathbb{N}$.

Lemma 7.26. *The marking τ_f induced by a homomorphism f on a tree t is a consistent marking.*

Proof. It follows directly from the axioms of thin algebra, see also Fact 6.57 on page 103. ■

Intuitively, a marking is consistent if the operations of thin algebra are not enough to prove its inconsistency.

The following example shows that some thin algebras S admit more than one homomorphism from $(\mathrm{Tr}_{A_R}, \mathrm{Con}_{A_R})$ into S. In particular, the analogue of Fact 1.5 from page 12 does not hold here.

Example 7.3. Fix the ranked alphabet $A_b = (\{n\}, \{b\})$. Let $L_b \subseteq \mathrm{Tr}_{A_b}$ contain exactly these trees which have at least one leaf. The following homomorphism recognises L_b: $H_{L_b} = \{h_a, h_b\}$, $V_{L_b} = \{s_a, s_b\}$, and $f_{L_b}(t) = h_b$ (resp. $f_{L_b}(p) = s_b$) if and only if the tree t (resp. the context p) contains any leaf (not counting the hole of p).

Let t_n be the complete binary tree equal everywhere n. Observe that t_n does not belong to L_b and the marking $\tau_{f_{L_b}}(t_n)$ induced by f_{L_b} on t_n equals h_a in every vertex.

Consider another marking τ' of t_n that equals h_b everywhere. Note that τ' is consistent — along every infinite branch of t it looks like a marking induced by f_{L_b} (on a different tree). Therefore, t has two consistent markings.

Going further, one can construct a homomorphism $f'\colon (\mathrm{Tr}_{A_b}, \mathrm{Con}_{A_b}) \to (H_{L_b}, V_{L_b})$ that assigns h_b to the tree t_n. Therefore, there are two distinct homomorphisms from $(\mathrm{Tr}_{A_b}, \mathrm{Con}_{A_b})$ to (H_{L_b}, V_{L_b}).

Recall that the language L_b used above is known to be ambiguous, see [NW96]. Using the notions of Sect. 7.4.1, one can check that (H_{L_b}, V_{L_b}) is a pseudo-syntactic thin algebra of L_b.

Now we can define prophetic thin algebras as those that admit at most one consistent marking.

Definition 7.15. *We say that a thin algebra (H, V) over a ranked alphabet A_{R} is* prophetic *iff for every ranked tree $t \in \mathrm{Tr}_{A_{\mathrm{R}}}$ there exists at most one consistent marking of t by types in H.*

Blumensath [Blu11, Blu13] has proposed recently an algebraic framework for infinite trees. His *path-continuous ω-hyperclones* recognise precisely the class of regular languages of infinite trees. The construction has some disadvantages though. One of the disadvantages of the construction is that the use of an ideal (see [Blu13, Definition 2.7]) together with existential quantification over its elements (the supremum taken in the definition of $\pi(a^{\square})$) is an algebraic translation of runs of the automata. A more precise formulation of this objection is that path-continuous ω-hyperclones are not closed under homomorphic images.

There is some inherent difficulty when designing a way to recognise regular languages of infinite trees. The source of the problem seems to be that there is no reasonable way of decomposing an infinite tree in such a way that the types of the parts can be computed separately. Both known solutions: non-deterministic automata of Rabin and path-continuous ω-hyperclones of Blumensath involve an essential existential quantification that corresponds to guessing some kind of a witness. The case of prophetic thin algebras is different: it is enough to verify the types path-wise (using the standard Ramsey's theorem) and already path-wise consistency guarantees global consistency (there is no way to *cheat*). The cost one has to pay is that prophetic thin algebras do not recognise all regular tree languages. Therefore, the results of this chapter can be seen as an indication where the difficulty lays.

The concepts of prophetic thin algebras and path-continuous ω-hyperclones were defined independently.

Note that if $f\colon (\mathrm{Tr}_{A_{\mathrm{R}}}, \mathrm{Con}_{A_{\mathrm{R}}}) \to S$ is a homomorphism and S is prophetic then, for every ranked tree $t \in \mathrm{Tr}_{A_{\mathrm{R}}}$, the only consistent marking of t is the marking τ_f induced by f. In particular, we obtain the following remark.

Remark 7.10. If S is prophetic then there is at most one homomorphism of the form

$$f\colon (\mathrm{Tr}_{A_{\mathrm{R}}}, \mathrm{Con}_{A_{\mathrm{R}}}) \to S.$$

Since the property that a given finite thin algebra is prophetic can be expressed in MSO on the complete binary tree, we obtain the following fact.

Fact 7.67. *It is decidable whether a given finite thin algebra (H, V) is prophetic.*

7.2 Bi-unambiguous Languages

In this section we show that the languages recognised by finite prophetic thin algebras are precisely the bi-unambiguous languages.

Theorem 7.5. *A language of infinite trees L is recognised by a homomorphism into a finite prophetic thin algebra if and only if L is bi-unambiguous, i.e. both L and the complement L^c can be recognised by unambiguous automata.*

In this section we implicitly assume that the automata are *pruned*: every state q of an automaton is productive and reachable: there exists an accepting run ρ of $\mathcal{A}$ on some tree t and a node $u \in \mathrm{dom}(\rho)$ such that $\rho(u) = q$. Every non-deterministic automaton recognising non-empty language can be pruned by removing some states. The result recognises the same language and this removal does not influence unambiguity.

The proof of Theorem 7.5 is split into the following three subsections.

7.2.1 *Prophetic Thin Algebras Recognise only Bi-unambiguous Languages*

The "only if" part of Theorem 7.5 (i.e. that every language recognised by a finite prophetic thin algebra is bi-unambiguous) is expressed by the following lemma.

Lemma 7.27. *Let $f\colon (\mathrm{Tr}_{A_R}, \mathrm{Con}_{A_R}) \to (H, V)$ be a homomorphism into a finite prophetic thin algebra (H, V) and $h_0 \in H$. The language $L_{h_0} = f^{-1}(h_0)$ is unambiguous.*

The construction used in the following proof is motivated by *algebraic automata* proposed by Bilkowski in [Bil11].

Proof. The desired automaton $\mathcal{C}$ is built as a product of two automata $\mathcal{A}$ and $\mathcal{D}$. The automaton $\mathcal{D}$ is deterministic and computes the priorities of states of $\mathcal{C}$. First we describe the automaton $\mathcal{A}$. Let $A_R = (A_{R2}, A_{R0})$, $Q_0 = H \times A_{R0}$, $Q_2 = H \times A_{R2} \times H$, and $Q^{\mathcal{A}} = Q_0 \sqcup Q_2$. Let us define $J\colon Q \to H$ as $J(h, b) = h$ and $J(h_L, a, h_R) = a(h_L, h_R)$. $J(q)$ is called *the value of a state $q \in Q$*. Let $I^{\mathcal{A}} = \{q \in Q^{\mathcal{A}} : J(q) = h_0\}$. Now $\delta_0^{\mathcal{A}}$ consists of all pairs $((h, b), b)$ such that $b() = h$ and $\delta_2^{\mathcal{A}}$ consists of all pairs $((h_L, a, h_R), a, q_L, q_R)$ such that $J(q_L) = h_L$ and $J(q_R) = h_R$.

Let $t \in \mathrm{Tr}_{A_{\mathrm{R}}}$ be any ranked tree. It is easy to verify that there is a 1-1 correspondence between runs ρ of $\mathcal{A}$ on t and markings τ_ρ by types in H. A state $(h_{\mathrm{L}}, a, h_{\mathrm{R}})$ in a node $u \in \mathrm{dom}(t)$ denotes that $t(u) = a$ and the marking τ_ρ equals h_{L} and h_{R} in $u\mathrm{L}$, $u\mathrm{R}$ respectively. What remains is to verify that the marking τ_ρ is consistent. Let $\alpha = d_0 d_1 \ldots$ be an infinite branch of t and let $q_0, q_1, \ldots$ be the sequence of states of ρ on α. Since every state q_i contains types of both subtrees under $\alpha\restriction_i$, basing on $q_0, q_1, \ldots$ we can define the decomposition $s_0, s_1, \ldots$ of τ_ρ along α (see Definition 7.14). Now, the condition expressed by (7.1) is ω-regular (see Fact 6.55 on page 103). Therefore, there exists a deterministic parity automaton $\mathcal{D}$ on ω-words that reads a sequence of directions $\alpha = (d_i)_{i \in \mathbb{N}}$ and states $(q_i)_{i \in \mathbb{N}}$ and verifies that the marking encoded by $(q_i)_{i \in \mathbb{N}}$ is consistent on the branch α.

Let $\mathcal{C}$ guess a run of $\mathcal{A}$ on a given tree and then run $\mathcal{D}$ independently on all the branches of t. Let the priorities of $\mathcal{C}$ equal the priorities of $\mathcal{D}$. By the construction, every parity-accepting run ρ of $\mathcal{C}$ encodes a consistent marking τ_ρ of t. And vice versa: every consistent marking can be encoded into a parity-accepting run.

Since the algebra (H, V) is prophetic, there is at most one accepting run of $\mathcal{C}$ on every tree. Therefore, $\mathcal{C}$ is unambiguous. $t \in L_{h_0}$ if and only if there exists a consistent marking of t with the value h_0, what is equivalent to the existence of an accepting run of $\mathcal{C}$ on t. So $\mathrm{L}(\mathcal{C}) = L_{h_0}$. ∎

7.2.2 *Markings by the Automaton Algebra for an Unambiguous Automaton*

Before proving the "if" part of Theorem 7.5 we first study some properties of consistent markings by the automaton algebra $S_{\mathcal{A}}$ (see Sect. 6.2.1, page 104) for an unambiguous automaton $\mathcal{A}$.

Just to recall results of Sect. 6.2.1: for every non-deterministic tree automaton $\mathcal{A}$ one can effectively construct a finite thin algebra $S_{\mathcal{A}} = (H_{\mathcal{A}}, V_{\mathcal{A}})$ that recognises $\mathrm{L}(\mathcal{A})$ (by the homomorphism $f_{\mathcal{A}}$); additionally, the elements of $H_{\mathcal{A}}$ are sets of states of $\mathcal{A}$, see (6.9), page 104.

The aim of this section is the following proposition. Intuitively it says that a consistent marking may cheat but only in one direction — it may underestimate the real $f_{\mathcal{A}}$-type of a given subtree.

Proposition 7.18. *Let $t \in \mathrm{Tr}_{A_{\mathrm{R}}}$ be a ranked tree and $(H_{\mathcal{A}}, V_{\mathcal{A}})$ be the automaton algebra for an unambiguous automaton $\mathcal{A}$. Assume that τ is a consistent marking of t by elements of $H_{\mathcal{A}}$. Then, for every node u of the tree t we have $\tau(u) \subseteq f_{\mathcal{A}}(t\restriction_u)$.*

We begin with an analysis of the operations of the automaton algebras, see (6.11a) on page 105 for an explicit definition of these operations.

Lemma 7.28. *Let $\mathcal{A}$ be a non-deterministic tree automaton over a ranked alphabet A_{R} and $t \in \mathrm{Tr}_{A_{\mathrm{R}}}$ be a ranked tree. Let $S_{\mathcal{A}} = (H_{\mathcal{A}}, V_{\mathcal{A}})$ be the automaton algebra for*

$\mathcal{A}$ and assume that τ is a consistent marking of t by types in $H_{\mathcal{A}}$. Let $\alpha = d_0 d_1 \ldots$ be an infinite branch of t.

A state $q \in Q^{\mathcal{A}}$ belongs to $\tau(\epsilon)$ if and only if there exists a sequence $(\delta_i)_{i\in\mathbb{N}}$ of transitions of $\mathcal{A}$ with $\delta_i = (q^i, a^i, q^i_{\mathrm{L}}, q^i_{\mathrm{R}})$ and $q^0 = q$ that encodes a parity-accepting run of $\mathcal{A}$ on α:

- *the sequence of states $(q^i)_{i\in\mathbb{N}}$ satisfies the parity condition,*
- *for every i, the state $q^i_{d_i}$ equals q^{i+1} — the transitions agree with each other,*
- *for every i and $d \in \{\mathrm{L}, \mathrm{R}\}$ the state q^i_d belongs to $\tau(d_0 \cdots d_{i-1} \cdot d)$ — the states used in the transitions belong to the respective sets $\tau(u)$ for $u \prec \alpha$ as well as for u that is off α.*

Proof. First take a state $q \in \tau(\epsilon)$. Let $(s_i)_{i\in\mathbb{N}}$ be the decomposition of τ along α as in Definition 7.14. Since $(V_{\mathcal{A}}, \cdot)$ is a semigroup, we can apply Ramsey's Theorem (Theorem 1.7 on page 13) to obtain a linked pair $(s, e) \in V^2_{\mathcal{A}}$ and a sequence of numbers $0 < n_0 < n_1 < \ldots$ such that

$$s_0 \cdot \ldots \cdot s_{n_0} = s \text{ and for every } i \geqslant 0 \text{ we have } s_{n_i+1} \cdot \ldots \cdot s_{n_{i+1}} = e. \tag{7.2}$$

Since $q \in \tau(\epsilon) = s \cdot e^{\infty}$ so by (6.11a) and (6.11d) (see page 105) it is witnessed by:

- an element $(q, j, q') \in s$,
- an element $(q', j', q') \in e$ with j' even (we use the fact that e is an idempotent).

Using (6.11a), (6.11e), and (7.2) we find a sequence of transitions as in the statement.

Now assume that there exists a sequence $(\delta_i)_{i\in\mathbb{N}}$ of transitions as in the statement, we want to show that $q \in \tau(\epsilon)$. As before, let $(s_i)_{i\in\mathbb{N}}$ be the decomposition of τ along α. We will construct a Ramsey decomposition of α with respect to both sequences $(s_i)_{i\in\mathbb{N}}$ and $(\delta_i)_{i\in\mathbb{N}}$ at the same time. For $i < j$ let

$$\alpha(i, j) = \big(s_i \cdot \ldots \cdot s_{j-1},\ (q^i, \min_{i \leqslant k < j} \Omega^{\mathcal{A}}(q^k))\big).$$

Since the set of values of α is finite[1], we can find a Ramsey decomposition with respect to α (see Theorem 1.1 on page 3): a sequence of numbers $0 < n_0 < n_1 < \ldots$ such that (7.2) is satisfied and for some fixed j' and every $i \geqslant 0$ we have:

$$q^{n_i} = q^{n_{i+1}}, \qquad \min_{n_i \leqslant k < n_{i+1}} \Omega^{\mathcal{A}}(q^k) = j'. \tag{7.3}$$

Since the run encoded by $(\delta_i)_{i\in\mathbb{N}}$ is parity-accepting so j' is even. Therefore, by (6.11a) and (7.2) we know that:

[1] It is possible to define a structure of semigroup on $\mathrm{rg}(\alpha)$ but Theorem 1.1 works for any function α.

- $(q, j, q^{n_0}) \in s$ for some j,
- $(q^{n_0}, j', q^{n_0}) \in e$.

It implies that $q \in s \cdot e^{\infty} = \tau(\epsilon)$ by (6.11d). ∎

Now, we will be interested in finding runs of an automaton $\mathcal{A}$ on a ranked tree t that are *contained* in a marking τ of t by types in $H_{\mathcal{A}}$: for every $u \in \mathrm{dom}(t)$ we require that $\rho(u) \in \tau(u)$.

Lemma 7.29. *Let $\mathcal{A}$ be a non-deterministic tree automaton over a ranked alphabet A_{R}, $t \in \mathrm{Tr}_{A_{\mathrm{R}}}$ be a ranked tree, and τ be a consistent marking of t by types in $H_{\mathcal{A}}$. Let $q \in Q^{\mathcal{A}}$ be a state of $\mathcal{A}$. The following conditions are equivalent:*

- *$q \in \tau(\epsilon)$*
- *There exists a run (possibly not parity-accepting) ρ of $\mathcal{A}$ on t with the value q, that is contained in τ. Additionally, for every infinite branch α of t there exists a run ρ_α of $\mathcal{A}$ on t with the value q, that is contained in τ, such that ρ_α satisfies the parity condition on α.*

Proof. First assume that $q \in \tau(\epsilon)$. We inductively show that there exists a run of $\mathcal{A}$ on t satisfying $\rho(u) \in \tau(u)$. Assume that $t = a(t_{\mathrm{L}}, t_{\mathrm{R}})$ for a pair of ranked trees $t_{\mathrm{L}}, t_{\mathrm{R}}$. Let $h = \tau(u)$, $h_{\mathrm{L}} = \tau(u\mathrm{L})$, and $h_{\mathrm{R}} = \tau(u\mathrm{R})$. By (6.11e) and (6.11b) there exists a transition $(q, a, q_{\mathrm{L}}, q_{\mathrm{R}}) \in \delta_2^{\mathcal{A}}$ such that $q_{\mathrm{L}} \in h_{\mathrm{L}}$ and $q_{\mathrm{R}} \in h_{\mathrm{R}}$. Therefore, we can proceed inductively in $u\mathrm{L}$ and $u\mathrm{R}$ in states q_{L} and q_{R} respectively. Note that by (6.11f) if u is a leaf of t and $q \in \tau(u)$ then $(q, t(u)) \in \delta_0^{\mathcal{A}}$, so the constructed run agrees with the transitions over leafs.

Now take an infinite branch α of t. Using the above observation, it is enough to construct a run ρ along α that satisfies $\rho(u) \in \tau(u)$ for every u that is off α — it will extend to a run on the subtree $t{\restriction}_u$. The existence of a parity-accepting run along α follows from Lemma 7.28.

Now assume that the second bullet of the statement is satisfied. We want to show that $q \in \tau(\epsilon)$. If the tree t is finite then $q \in \tau(\epsilon)$ by induction on the height of t. Otherwise, there exists an infinite branch α of t and similarly as above, any run ρ_α that is parity-accepting on α is a witness that $q \in h$. ∎

Before we prove Proposition 7.18 let us observe the following *local* property of unambiguous automata (it is slightly related to Lemma 3.2 on page 39).

Lemma 7.30. *Let $\mathcal{A}$ be an unambiguous automaton and let $f_{\mathcal{A}}\colon (\mathrm{Tr}_{A_{\mathrm{R}}}, \mathrm{Con}_{A_{\mathrm{R}}}) \to S_{\mathcal{A}}$ be the automaton morphism for $\mathcal{A}$. Let $h = a(h_{\mathrm{L}}, h_{\mathrm{R}})$ for a triple of types $h, h_{\mathrm{L}}, h_{\mathrm{R}} \in H_{\mathcal{A}}$ and a letter $a \in A_{\mathrm{R}2}$. Then for every $q \in h$ there exists exactly one transition of the form $(q, a, q_{\mathrm{L}}, q_{\mathrm{R}}) \in \delta_2^{\mathcal{A}}$ such that $q_{\mathrm{L}} \in h_{\mathrm{L}}$ and $q_{\mathrm{R}} \in h_{\mathrm{R}}$.*

Proof. At least one such a transition exists by (6.11e) and (6.11b). Assume that there are two transitions as in the statement.

Let p be a context that has an accepting run ρ with the value q in the hole — we use the fact that the automaton $\mathcal{A}$ is pruned (every state appears in some accepting run).

Let t_L, t_R be trees of $f_{\mathcal{A}}$-types respectively h_L, h_R. In that case the tree $p \cdot a(t_L, t_R)$ has two different accepting runs: both these runs equal ρ on p, then use two distinct transitions in the hole of p, and extend to parity-accepting runs on t_L, t_R by the fact that h_L, h_R are $f_{\mathcal{A}}$-types of t_L, t_R respectively (see (6.9) on page 104). ∎

Finally we can conclude with the proof of Proposition 7.18, saying that for an unambiguous automaton and a consistent marking τ of t by types in $H_{\mathcal{A}}$ we have $\tau(u) \subseteq f_{\mathcal{A}}(t\restriction_u)$, for every $u \in \mathrm{dom}(t)$.

Proof of Proposition 7.18 Without loss of generality we can assume that $u = \epsilon$. Let us take any state $q \in \tau(\epsilon)$, we want to show that $q \in f_{\mathcal{A}}(t)$. Let us take the run ρ constructed inductively in Lemma 7.30 for the state q (i.e. ρ is contained in τ and $\rho(\epsilon) = q$). What remains is to show that ρ is parity-accepting.

Take any infinite branch α of t. By Lemma 7.29 there exists a run ρ_α on t that is contained in τ and satisfies the parity condition on α. But Lemma 7.30 shows inductively that for every $u \prec \alpha$ we have $\rho(u) = \rho_\alpha(u)$. So, since ρ_α satisfies the parity condition on α, ρ also satisfies it on α. Therefore, $q \in f_{\mathcal{A}}(t)$. ∎

7.2.3 Every Bi-unambiguous Language is Recognised by a Prophetic Thin Algebra

Now we prove the "if" part of Theorem 7.5: if a language $L \subseteq \mathrm{Tr}_{A_R}$ is bi-unambiguous then there exists a finite prophetic thin algebra S and a homomorphism $f\colon (\mathrm{Tr}_{A_R}, \mathrm{Con}_{A_R}) \to S$ such that f recognises L.

The algebra S is the product of the automaton algebras (see Sect. 6.2.1, page 104) for the two unambiguous automata recognising L and the complement L^c. Proposition 7.18 together with a combinatorial observation in Lemma 7.31 will imply that S is prophetic.

Let $\mathcal{A}$, $\mathcal{B}$ be two unambiguous automata such that $\mathrm{L}(\mathcal{A}) = L$ and $\mathrm{L}(\mathcal{B}) = \mathrm{Tr}_{A_R} \setminus L$. Let $f_{\mathcal{A}}$, $S_{\mathcal{A}}$ and $f_{\mathcal{B}}$, $S_{\mathcal{B}}$ be the respective automaton morphisms. Consider the surjective homomorphism $f_U\colon (\mathrm{Tr}_{A_R}, \mathrm{Con}_{A_R}) \to (H_U, V_U)$ obtained as the product of the above algebras:

- $f_U(t) = (f_{\mathcal{A}}(t), f_{\mathcal{B}}(t))$,
- $f_U(p) = (f_{\mathcal{A}}(p), f_{\mathcal{B}}(p))$,
- $H_U = f_U(\mathrm{Tr}_{A_R}) \subseteq H_{\mathcal{A}} \times H_{\mathcal{B}}$, and
- $V_U = f_U(\mathrm{Con}_{A_R}) \subseteq V_{\mathcal{A}} \times V_{\mathcal{B}}$.

The following lemma states that there is a trade-off between types in $H_{\mathcal{A}}$ and $H_{\mathcal{B}}$.

Lemma 7.31. *The set H_U is an anti-chain with respect to the coordinate-wise inclusion order.*

Proof. Assume contrary, by the symmetry between h and h', that:

- there are $h = (h_{\mathcal{A}}, h_{\mathcal{B}}), h' = (h'_{\mathcal{A}}, h'_{\mathcal{B}}) \in H_U$,
- $h_{\mathcal{A}} \subseteq h'_{\mathcal{A}}$ and $h_{\mathcal{B}} \subseteq h'_{\mathcal{B}}$,
- there exists a state $q' \in h'_{\mathcal{A}}$ but $q' \notin h_{\mathcal{A}}$ (the symmetry is used here).

Let t, t' be ranked trees such that $f_U(t) = h$ and $f_U(t') = h'$ and let p be a ranked context with an accepting run ρ' of $\mathcal{A}$ that has the value q' in the hole of p. Note that by the definition $p \cdot t' \in \mathrm{L}(\mathcal{A})$ — the run ρ' can be extended to t'.

Consider two cases:

1. $p \cdot t \in \mathrm{L}(\mathcal{A})$. Let ρ be the accepting run of $\mathcal{A}$ that witnesses that. Let q be the value of ρ in the hole of p. Then $q \in h_{\mathcal{A}} \subseteq h'_{\mathcal{A}}$. It means that we have two distinct accepting runs of $\mathcal{A}$ on $p \cdot t'$: the first one equals ρ on p and then extends to t' by the assumption that $q \in h'_{\mathcal{A}}$ and the second one equals ρ' on p and then extends to t' by the assumption that $q' \in h'_{\mathcal{A}}$. A contradiction.
2. $p \cdot t \in \mathrm{L}(\mathcal{B})$. Let ρ be the accepting run of $\mathcal{B}$ that witnesses that. Let q be the value of ρ in the hole of p. Then $q \in h_{\mathcal{B}} \subseteq h'_{\mathcal{B}}$. So we can construct an accepting run of $\mathcal{B}$ on $p \cdot t'$ by using ρ on p and extending it to t'. So $p \cdot t' \in \mathrm{L}(\mathcal{B})$ — a contradiction, since we assumed that the languages $\mathrm{L}(\mathcal{A})$ and $\mathrm{L}(\mathcal{B})$ are disjoint.

■

Lemma 7.32. *Let* $f_U \colon (\mathrm{Tr}_{A_R}, \mathrm{Con}_{A_R}) \to (H_U, V_U)$ *be the homomorphism constructed above for a pair of unambiguous automata* $\mathcal{A}$, $\mathcal{B}$. *If* τ *is a consistent marking of a given ranked tree* t *by types in* H_U *then it is equal to the marking* τ_{f_U} *induced by* f_U *on* t.

Proof. Take any vertex $u \in \mathrm{dom}(t)$. By the definition $\tau(u) \in H_U$. By Proposition 7.18 we have $\tau(u) \subseteq f_U(t\restriction_u) = \tau_{f_U}(u)$ coordinate-wise. Using Lemma 7.31 we obtain that $\tau(u) = \tau_{f_U}(u)$. ■

The following fact concludes the proof of Theorem 7.5.

Fact 7.68. *The homomorphism* f_U *defined above is surjective and recognises* $\mathrm{L}(\mathcal{A})$, *the algebra* (H_U, V_U) *is prophetic.*

Proof. f_U is surjective by the definition; it recognises L because $f_{\mathcal{A}}$ recognises L; Lemma 7.32 implies that (H_U, V_U) is prophetic. ■

7.3 Consequences of Conjecture 2.1

In this section we study properties of the class of prophetic thin algebras under the assumption of Conjecture 2.1 from page 31 (stating that there is no MSO-definable choice function on thin trees). It turns out that this conjecture implies that finite prophetic thin algebras form a pseudo-variety (see [BS81] for an introduction to universal algebra and [Ban83] for pseudo-varieties of finite algebras) and have unique homomorphisms from $(\mathrm{Tr}_{A_R}, \mathrm{Con}_{A_R})$. Roughly speaking it means that prophetic thin algebras and bi-unambiguous languages are as well-behaved as ω-semigroups and ω-regular languages.

To emphasise that the presented results use Conjecture 2.1, we explicitly put it as an assumption in brackets. The results of this section depend highly on consequences of Conjecture 2.1 proved in Chap. 8.

Proposition 7.19 (Conjecture 2.1). *Let (H, V) be a finite prophetic thin algebra over a ranked alphabet A_R. There exists a unique homomorphism $f\colon (\mathrm{Tr}_{A_R}, \mathrm{Con}_{A_R}) \to (H, V)$.*

Proof. The existence of at most one such homomorphism was observed in Sect. 7.1. By Theorem 8.74 proved in Chap. 8 (see page 143) and the fact that (H, V) is prophetic, every tree $t \in \mathrm{Tr}_{A_R}$ has exactly one consistent marking τ_t by types in H. Let us define $f(t) = \tau_t(\epsilon)$. The condition of consistency of a marking implies that f is a homomorphism. ■

Proposition 7.20 (Conjecture 2.1). *Let $g\colon S \to S'$ be a surjective homomorphism between two finite thin algebras. If S is prophetic then S' is also prophetic.*

Proof. First fix the homomorphism $f\colon (\mathrm{Tr}_{A_R}, \mathrm{Con}_{A_R}) \to S = (H, V)$ given by Proposition 7.19. Note that $g \circ f\colon (\mathrm{Tr}_{A_R}, \mathrm{Con}_{A_R}) \to (H, V)$ is a homomorphism. Assume that S' is not prophetic, so there exists a ranked tree t with two consistent markings σ, σ' by types of S'. Without loss of generality we can assume that σ is the marking induced by $g \circ f$ and $\sigma'(\epsilon) \neq \sigma(\epsilon)$. Let τ be the marking by types in S induced by f on t. Observe that pointwise $g(\tau) = \sigma$. By Proposition 8.22 proved in Chap. 8 (see page 145) there exists a consistent marking τ' of t such that pointwise $g(\tau') = \sigma'$. Therefore, τ and τ' are two distinct consistent markings of t by types in H — a contradiction. ■

Theorem 7.69 (Conjecture 2.1). *The class of finite prophetic thin algebras over a fixed ranked alphabet A_R is a pseudo-variety: it is closed under homomorphic images, subalgebras, and finite direct products.*

Proof. The closure under subalgebras and finite direct products follows directly from the definition. Proposition 7.20 implies that (under the assumption of Conjecture 2.1), a homomorphic image of a finite prophetic thin algebra is also prophetic. ■

7.4 Decidable Characterisation of the Bi-unambiguous Languages

In this section we prove that, assuming Conjecture 2.1, the class of bi-unambiguous languages of complete binary trees is decidable among all regular tree languages, as expressed by the following decision problem.

Problem 7.3 (Characterisation of bi-unambiguous languages).

– **Input** A non-deterministic parity tree automaton $\mathcal{A}$.
– **Output** "yes" if the language $\mathrm{L}(\mathcal{A})$ is bi-unambiguous.

The proposed effective procedure P deciding this problem always terminates and is sound, only the completeness of the procedure depends on Conjecture 2.1. Additionally, Bilkowski proved (see [BS13, Item 3 of Theorem 5] that the procedure P is complete if the given language is deterministic.

Theorem 7.70. *Assuming Conjecture 2.1, the decision problem if a given regular tree language is bi-unambiguous (i.e. Problem 7.3) is decidable.*

The proof of this theorem relies on a construction of a *pseudo-syntactic thin algebra* of a given regular language of complete trees L. The construction of this algebra is effective and Conjecture 2.1 implies that if there is any prophetic thin algebra recognising L then the pseudo-syntactic one is also prophetic. Since $(\mathrm{Tr}_{A_\mathrm{R}}, \mathrm{ThCon}_{A_\mathrm{R}})$ is not free in the class of thin algebras over the ranked alphabet A_R, some special care has to be taken when defining the pseudo-syntactic thin algebra.

7.4.1 Pseudo-syntactic Morphisms

Intuitively, the pseudo-syntactic algebra can be seen as a minimal algebra recognising a given language. Chapter 3 of [Idz12] presents a generic way of constructing syntactic algebras for languages. However, the constructions presented there work if a given language is a subset of the free algebra. Example 7.3 implies that $(\mathrm{Tr}_{A_\mathrm{R}}, \mathrm{Con}_{A_\mathrm{R}})$ is not free in the class of thin algebras. Therefore, the notion of syntactic morphism for a given language has to be adopted to the case of non-thin trees.

We start by recalling the classical notions of free algebras and syntactic morphisms in the setting of thin algebras. Since thin algebras already contain alphabets, we use the term *free algebra* having in mind the empty set of generators (i.e. a thin algebra over a ranked alphabet A_R is free if it is initial in the category of thin algebras over A_R, see the following definition).

In this section we work with ranked alphabets, a language of complete trees $L \subseteq \mathrm{Tr}_A$ can be seen as a language over the ranked alphabet $A_\mathrm{R} = (A, \emptyset)$.

Definition 7.16. *A thin algebra S over a ranked alphabet A_R is* free *if for every thin algebra S' over A_R there exists a unique homomorphism $f\colon S \to S'$.*

Let $F = (H_F, V_F)$ be a free thin algebra over a ranked alphabet A_R and $L \subseteq H_F$. A homomorphism $f_L\colon F \to S_L = (H_L, V_L)$ is the syntactic morphism of L *if:*

1. *f_L is surjective,*
2. *f_L recognises L (i.e. $L = f_L^{-1}(X)$ for some $X \subseteq H_L$),*
3. *for every surjective homomorphism $f\colon F \to S'$ that recognises L there exists a unique homomorphism $g\colon S' \to S_L$ such that*

$$g \circ f = f_L.$$

Observe that up to an isomorphism the free thin algebra over a given ranked alphabet is unique. The following fact summarizes the relations between thin trees and thin algebras, see [Idz12, Lemma 22, Lemma 23, Theorem 54].

Fact 7.71. *The thin algebra* $(\mathrm{Th}_{A_R}, \mathrm{ThCon}_{A_R})$ *is a free thin algebra over* A_R. *For every language* $L \subseteq \mathrm{Th}_{A_R}$ *there exists a syntactic morphism of* L. *If* L *is regular then the syntactic algebra of* L *(denoted* S_L*) is finite and can be effectively computed basing on any representation of* L.

Sketch of a proof. Let $F = (\mathrm{Th}_{A_R}, \mathrm{ThCon}_{A_R})$. The uniqueness of a homomorphism $f\colon F \to S'$ can be proved by induction on the rank of a thin tree. Therefore, F is a free thin algebra over A_R.

To construct a syntactic morphism it is enough to divide the free thin algebra $(\mathrm{Th}_{A_R}, \mathrm{ThCon}_{A_R})$ by the syntactic congruence $\sim_L$ (see [Idz12, Lemma 19]). Since there exists a finite thin algebra recognising a given regular tree language L (namely the automaton algebra from Sect. 6.2.1, page 104), S_L is finite.

To effectively compute S_L one can use the Moore's algorithm, see [Idz12, Lemma 23] ■

The following definition formalizes the notion of a pseudo-syntactic thin algebra. The conditions are much weaker than in the case of syntactic algebras, however they are strong enough to serve for the purpose of our effective characterisation.

Definition 7.17. *Let* $L \subseteq \mathrm{Tr}_{A_R}$ *be a regular tree language. We say that a finite thin algebra* S_L *is a* pseudo-syntactic algebra of L *if* S_L *recognises* L *and for every finite thin algebra* S' *recognising* L *there exists a subalgebra* $S'' \subseteq S'$ *and a surjective homomorphism* $f\colon S'' \to S_L$.

If we required the homomorphisms under consideration to satisfy additional constraints of *compositionality*, we could obtain a more rigid notion of syntactic algebra for a language $L \subseteq \mathrm{Tr}_{A_R}$. However, it is not needed in this chapter, so we use the weaker (and much simpler) notion of pseudo-syntactic algebra.

The aim of this section is to prove the following proposition. By taking $A_{R0} = \emptyset$ we reduce the statement to the case of languages of complete binary trees $L \subseteq \mathrm{Tr}_A$ for $A = A_{R2}$.

Proposition 7.21. *For every regular tree language* $L \subseteq \mathrm{Tr}_{A_R}$ *one can effectively construct a pseudo-syntactic thin algebra of* L.

Let $\mathcal{A}$ be a non-deterministic tree automaton recognising a regular tree language L. Let $S_{\mathcal{A}} = (H_{\mathcal{A}}, V_{\mathcal{A}})$ be the automaton algebra and $f_{\mathcal{A}}$ be the automaton morphism of $\mathcal{A}$, see Sect. 6.2.1, page 104. By the definition $f_{\mathcal{A}}$ is surjective. Consider the ranked alphabet $A_R \sqcup H_{\mathcal{A}} \stackrel{\text{def}}{=} (A_{R2}, A_{R0} \sqcup H_{\mathcal{A}})$. As we have already seen, $S_{\mathcal{A}}$ can be seen as a thin algebra over $A_R \sqcup H_{\mathcal{A}}$.

Let $F = \left(\mathrm{Th}_{A_R \sqcup H_{\mathcal{A}}}, \mathrm{ThCon}_{A_R \sqcup H_{\mathcal{A}}}\right)$ be the free thin algebra over $A_R \sqcup H_{\mathcal{A}}$. Our aim is to define a homomorphism

$$\iota \colon F \to (\mathrm{Tr}_{A_R}, \mathrm{Con}_{A_R}). \tag{7.4}$$

For every type $h \in H_{\mathcal{A}}$ let us fix a tree $t_h \in \mathrm{Tr}_{A_R}$ such that $f_{\mathcal{A}}(t_h) = h$. Now let $\iota(t)$ be the tree obtained by putting t_h in every leaf $u \in \mathrm{dom}(t)$ such that $t(u) = h \in H_{\mathcal{A}}$. $\iota(p)$ is defined in the same way for thin contexts p. Since the substitution is done only in the leafs, the function ι defined this way is a homomorphism[2].

Now let $f = f_{\mathcal{A}} \circ \iota$ and $L' = \iota^{-1}(L) \subseteq \mathrm{Th}_{A_R \sqcup H_{\mathcal{A}}}$. Observe that $f \colon F \to S_{\mathcal{A}}$ is a surjective homomorphism that recognises L'.

Since F is free, we can apply Fact 7.71 to the homomorphism f to effectively compute the syntactic thin algebra S_L of L'.

We will show that S_L is a pseudo-syntactic algebra of L. Consider any thin algebra S' that recognises L using a homomorphism f_2. Let $f' = f_2 \circ \iota$. Let $S'' \subseteq S'$ be the image of F under f' — S'' is a subalgebra of S'. Clearly, $f' \colon F \to S''$ is a surjective homomorphism recognising L'. By the universal property of S_L we know that there exists a unique surjective homomorphism $g \colon S'' \to S_L$.

This concludes the proof of Proposition 7.21.

7.4.2 Decidable Characterisation

Now we can prove Theorem 7.70 stating that assuming Conjecture 2.1 it is decidable if a given regular tree language is bi-unambiguous. The crucial technical part of the proof is based on Theorem 8.74 from Chap. 8 on page 143.

Consider the following decision procedure P:

1. Input a non-deterministic automaton $\mathcal{A}$ recognising a regular tree language L.
2. Compute a pseudo-syntactic thin algebra S_L of L.
3. Answer "yes" if S_L is prophetic, otherwise answer "no".

Observe that by Proposition 7.21 and Fact 7.67 all the operations performed by P are effective. Observe also that by Proposition 7.21 and Theorem 7.5, if the answer of P is "yes" then L is bi-unambiguous (the algebra S_L is a witness). What remains is to prove the following lemma.

Lemma 7.33. *Assuming Conjecture 2.1, if L is bi-unambiguous then every pseudo-syntactic thin algebra of L is prophetic.*

Proof. Since L is bi-unambiguous, by Theorem 7.5 there exists a surjective homomorphism $f \colon (\mathrm{Tr}_{A_R}, \mathrm{Con}_{A_R}) \to (H, V)$ that recognises L and such that (H, V) is a finite prophetic thin algebra. Since S_L is a pseudo-syntactic thin algebra of L so there exists a subalgebra (H', V') of (H, V) and a surjective homomorphism $g \colon (H', V') \to S_L$. By the definition of prophetic thin algebras we know that (H', V') is prophetic. By Proposition 7.20 we obtain that S_L is also prophetic. ■

This concludes the proof of Proposition 7.70.

[2] We treat F as a thin algebra over A_R when we say that ι is a homomorphism.

7.5 Conclusions

In this chapter we study which regular tree languages can be recognised by thin algebras. It turns out that bi-unambiguous languages of complete binary trees and regular languages of thin trees are strongly related. The main result of this chapter provides an algebraic framework for the class of bi-unambiguous languages using thin algebras. As a side effect of these considerations a new conjecture about MSO-definability of choice functions was posed (Conjecture 2.1).

If Conjecture 2.1 holds then the bi-unambiguous languages form a well-behaved class of regular tree languages: not only it would be decidable if a given language is bi-unambiguous but also prophetic thin algebras would provide a good algebraic framework for studying these languages. Therefore, proving Conjecture 2.1 would open the following line of research:

- prove Conjecture 2.2 for bi-unambiguous languages: if L is bi-unambiguous and Borel then L is WMSO-definable,
- provide an effective (or even equational) characterisation of bi-unambiguous languages that are WMSO-definable,
- provide equational characterisations of bi-unambiguous languages in certain classes of the Borel hierarchy (similarly to the characterisation from [BP12] of regular tree languages that belong to $\mathcal{BC}(\Sigma_1^0)$),
- study the Wadge hierarchy of bi-unambiguous languages,
- and more…

The idea to study relations between bi-unambiguous languages and thin trees was given by Bilkowski [Bil11]. In particular, he posed the following conjecture. Recall that for a pair of partial trees t, t' by $t \subseteq t'$ we mean that $\mathrm{dom}(t) \subseteq \mathrm{dom}(t')$ and for every vertex $u \in \mathrm{dom}(t)$ we have $t(u) = t'(u)$.

Conjecture 7.7 (Bilkowski [Bil11]). A regular tree language $L \subseteq \mathrm{Tr}_A$ is bi-unambiguous if and only if every tree $t \in \mathrm{Tr}_A$ has a "thin core": there exists a partial tree $\bar{t} \in \mathrm{PTr}_A$ such that $\bar{t}$ has countably many branches, $\bar{t} \subseteq t$ and for every complete tree $t' \in \mathrm{Tr}_A$ such that $\bar{t} \subseteq t'$ we have

$$t \in L \iff t' \in L.$$

In other words, every tree has a "thin core" that guarantees whether t belongs to L or not.

This conjecture remains open, even its relations with Conjecture 2.1 are still unclear.

This chapter is based on [BS13].

Chapter 8
Uniformization on Thin Trees

As the axiom of choice implies, for every relation $R \subseteq X \times Y$ there exists a graph of a total function $f : \pi_X(R) \to Y$ that is contained in R (such a graph is called a *uniformization of* R). A natural question asks in which cases such a function f is *definable*. A particular instance of this problem is, when R is an MSO-definable set of pairs of trees and we ask about MSO-definable f. This question is known as *Rabin's uniformization question*. The negative answer to this question was given by Gurevich and Shelah [GS83] (see [CL07] for a simplified proof). They proved that there is no MSO formula $\psi(x, X)$ that *chooses* from every non-empty subset X of the complete binary tree a unique element x of X. This result is known as undefinability of a choice function on the complete binary tree. On the other hand, the formula saying that x is the $\leq$-minimal element of X is a choice formula on ω-words. In [Sie75, LS98, Rab07] it is proved that any MSO-definable relation on ω-words admits an MSO-definable uniformization.

In this chapter we study the following conjecture about a uniformizability on thin trees, the statement here is a bit more formal than the one in Introduction.

Conjecture 8.1. *There is no* MSO*-definable choice function on thin trees — there is no formula* $\psi(x, X)$ *such that for every thin tree* t *and every non-empty* $X \subseteq \mathrm{dom}(t)$*, the formula* $\psi(x, X)$ *is satisfied for a unique* $x \in X$.

This conjecture is a strengthening of the result by Gurevich and Shelah [GS83] as the class of admissible sets X is smaller (they have to be contained in thin trees). Unfortunately, the author was unable to prove that Conjecture 8.1 holds. This chapter presents a study of Conjecture 8.1 and some related uniformization problems.

As observed by Niwiński and Walukiewicz [NW96] (cf. [CLNW10]), the non-existence of an MSO-definable choice function implies that the language $L_b = \{t \in \mathrm{Tr}_{\{a,b\}} : \exists_{u \in \mathrm{dom}(t)}\ t(u) = b\}$ is ambiguous (there is no unambiguous automaton recognising L_b). To the author's best knowledge, all the known examples of ambiguous tree languages are derived from the language L_b. Also, the choice formula and its variants remain the only known MSO-definable relations on trees that do not have any MSO-definable uniformization. In this chapter a new technique of proving non-uniformizability is introduced that allows to prove that:

M. Skrzypczak, *Set Theoretic Methods in Automata Theory*, LNCS 9802
DOI: 10.1007/978-3-662-52947-8_8

- there is no MSO-definable uniformization of the relation saying that σ is a skeleton of a tree t: there is no MSO formula that defines, for every thin tree t, a unique skeleton σ of t (we treat σ as a set of vertices of t),
- the language of all thin trees is ambiguous among all trees.

Liefsches and Shelah studied uniformization problems on trees in [LS98]. In particular, it is proved there that on thin trees every MSO-definable relation has an MSO-definable uniformization if we allow additional monadic parameters (that are adjusted appropriately to a given tree). The crucial difference here is that we do not allow any additional parameters.

The following theorem summarizes results of this chapter.

Theorem 8.6. *Conjecture 8.1 is equivalent to the fact that every finite thin algebra admits some consistent marking on every infinite tree.*

The relation $\varphi(\sigma, t)$ stating that t is a thin tree and σ is a skeleton of t does not admit any MSO*-definable uniformization of σ.*

The language of all thin trees is ambiguous (i.e. it is not recognised by any unambiguous automaton).

The chapter is organised as follows. Section 8.2 presents a technical construction of a transducer that is useful in the remaining sections. In Sect. 8.3 we prove some statements that are equivalent to Conjecture 8.1, in particular we show that Conjecture 8.1 is strongly related to prophetic thin algebras studied in Chap. 7. Then, in Sect. 8.4 we prove the above non-uniformizability results. In Sect. 8.5 we conclude.

8.1 Basic Notions

We will work with trees over ranked alphabets, as introduced in Sect. 6.1, page 95. The main interest of this chapter will be on uniformizations, as expressed by the following definition.

Definition 8.18. *Let $\varphi(X, \boldsymbol{P})$ be a formula of* MSO *on trees over a ranked alphabet with monadic variables X and $\boldsymbol{P} = P_1, \dots, P_n$. We say that $\psi(X, \boldsymbol{P})$ is a* uniformization of *$\varphi(X, \boldsymbol{P})$ if the following conditions are satisfied for every ranked tree t, values of $\boldsymbol{P}$, and sets $X_1, X_2 \subseteq \mathrm{dom}(t)$:*

$$
\begin{aligned}
(\exists_X \, \psi(X, \boldsymbol{P})) &\iff (\exists_X \, \varphi(X, \boldsymbol{P})) \\
\psi(X_1, \boldsymbol{P}) &\implies \varphi(X_1, \boldsymbol{P}) \\
(\psi(X_1, \boldsymbol{P}) \wedge \psi(X_2, \boldsymbol{P})) &\implies X_1 = X_2
\end{aligned}
$$

That is, whenever it is possible to pick some X satisfying $\varphi(X, \boldsymbol{P})$ then $\psi(X, \boldsymbol{P})$ chooses exactly one such X. To simplify the notation, we always assume that the first variable of a formula is the one that should be uniformized, we also allow $\boldsymbol{P}$ to be empty and some of the variables $X, \boldsymbol{P}$ to be first-order variables.

The following two formulae will be of our main interest (both conditions are MSO-definable by Remark 6.6 from page 102):

$$\mathrm{CHOICE}(x, X) \stackrel{\text{def}}{=} \text{“the given tree } t \text{ is thin and } x \in X\text{”},$$

$$\mathrm{LEAF} - \mathrm{CHOICE}(x) \stackrel{\text{def}}{=} \text{“the given tree } t \text{ is thin and } x \text{ is a leaf of } t\text{”}. \qquad (8.1)$$

By the definition, Conjecture 8.1 is equivalent to the fact that the formula CHOICE (x, X) does not have MSO-definable uniformization. We will see in Theorem 8.74 that it is also equivalent to LEAF $-$ CHOICE(x) not having such uniformization.

Recall that for two ranked alphabets A_R and M, we define the product $A_R \times M$ as $(A_{R2} \times M_2, A_{R0} \times M_0)$. Through this chapter we will sometimes treat a language $L \subseteq \mathrm{Tr}_{A_R \times M}$ as a relation $L \subseteq \mathrm{Tr}_{A_R} \times \mathrm{Tr}_M$. We say that L is *uniformized* if for every $t_A \in \mathrm{Tr}_{A_R}$ there is at most one $t_M \in \mathrm{Tr}_M$ with $\mathrm{dom}(t_A) = \mathrm{dom}(t_M)$ such that (t_A, t_M) (formally $t_A \otimes t_M$) belongs to L.

Example 8.4. If $\mathcal{A}$ is an unambiguous tree automaton over a ranked alphabet A_R then the following set of trees over the ranked alphabet $A_R \times (Q^{\mathcal{A}}, Q^{\mathcal{A}})$ is a uniformized relation:

$$\{t \otimes \rho :\ \rho \text{ is an accepting run of } \mathcal{A} \text{ on } t\}.$$

8.2 Transducer for a Uniformized Relation

In this section we introduce a technical construction that will be used in the subsequent sections of this chapter.

Assume that we are given a regular tree language of ranked trees $L_M \subseteq \mathrm{Tr}_{A_R \times M}$ that is uniformized as a relation in $\mathrm{Tr}_{A_R} \times \mathrm{Tr}_M$. It turns out that it is possible to construct a deterministic transducer that maps a given tree $t_A \in \mathrm{Tr}_{A_R}$ into the unique tree $t_A \otimes t_M \in L_M$. The idea is to equip the transducer with an additional knowledge about the *types* of the subtrees of t_A. It will be achieved by presenting a marking of t induced by a homomorphism into a fixed thin algebra (see Section 6.2, page 102). The way this additional information for the transducer is presented is rather arbitrary, we use here thin algebras because of the applications to thin trees.

The crucial property is that the constructed transducer will be deterministic so it will allow us to modify a given input tree t_A into t'_A and reason about the resulting tree t'_M (see Fact 8.72).

Let $A_R = (A_{R2}, A_{R0})$ and $M = (M_2, M_0)$ be a pair of ranked alphabets. A *transducer from* A_R *to* M is a deterministic device $\mathcal{T} = \langle Q^{\mathcal{T}}, q_I^{\mathcal{T}}, \delta^{\mathcal{T}}\rangle$ such that:

1. $Q^{\mathcal{T}}$ is a finite set of *states*,
2. $q_I^{\mathcal{T}} \in Q^{\mathcal{T}}$ is an *initial state*,
3. $\delta^{\mathcal{T}}$ is a pair of functions $\delta_2^{\mathcal{T}}, \delta_0^{\mathcal{T}}$,
4. the function $\delta_2^{\mathcal{T}}$ of the type

$$\delta_2^{\mathcal{T}}\colon Q^{\mathcal{T}} \times (A_{\mathrm{R}2} \cup A_{\mathrm{R}0}) \times A_{\mathrm{R}2} \times (A_{\mathrm{R}2} \cup A_{\mathrm{R}0}) \to Q^{\mathcal{T}} \times M_2 \times Q^{\mathcal{T}}$$

determines transitions in internal nodes,

5. $\delta_0^{\mathcal{T}} : Q^{\mathcal{T}} \times A_{\mathrm{R}0} \to M_0$ determines transitions in leafs.

Note that a transition in an internal node w takes three letters as the input, it will be the letters in: $w\mathrm{L}$, w, and $w\mathrm{R}$. Note also that the transducer does not have any *acceptance condition*, its run on a tree is always successful.

For every tree $t \in \mathrm{Tr}_{A_\mathrm{R}}$ a transducer $\mathcal{T}$ defines inductively a labelling $\mathcal{T}(t)$ of t by letters in M defined inductively as follows. We start in $w = \epsilon$ in the state $q_I^{\mathcal{T}}$. Assume that the transducer reached a vertex $w \in \mathrm{dom}(t)$ in a state q. If w is a leaf then we put $\mathcal{T}(t)(w) = \delta_0^{\mathcal{T}}(q, t(w))$. Otherwise, let $a_\mathrm{L}, a, a_\mathrm{R}$ be the letters of t in $w\mathrm{L}, w, w\mathrm{R}$ respectively. Then let $\delta_2^{\mathcal{T}}(q, a_\mathrm{L}, a, a_\mathrm{R}) = (q_\mathrm{L}, m, q_\mathrm{R})$, put $\mathcal{T}(t)(w) = m$, and continue in $w\mathrm{L}, w\mathrm{R}$ in the states $q_\mathrm{L}, q_\mathrm{R}$ respectively.

Fact 8.72. *The value $\mathcal{T}(t)(w)$ in a vertex $w \in \mathrm{dom}(t)$ depends on the letters of t in vertices of the form u, $u\mathrm{L}$, $u\mathrm{R}$ for $u \prec w$. That is, if t, t' agree on all the vertices u, $u\mathrm{L}$, $u\mathrm{R}$ for $u \prec w$ then $\mathcal{T}(t)(w) = \mathcal{T}(t')(w)$.*

Theorem 8.73. *Let A_R and M be two ranked alphabets. Assume that $L_M \subseteq Tr_{A_\mathrm{R}\times M}$ is a regular tree language, $L_A \subseteq Tr_{A_\mathrm{R}}$ is the projection of L_M onto the ranked alphabet A_R, and*

$$\forall_{t_A \in L_A}\ \exists!_{t_M \in Tr_M}\ t_A \otimes t_M \in L_M \quad \text{(i.e. the relation } L_M \text{ is uniformized).}$$

Then, there exist:

- *a homomorphism $f\colon (Tr_{A_\mathrm{R}}, Con_{A_\mathrm{R}}) \to S$ into a finite thin algebra S (see Sect. 6.2),*
- *a deterministic finite state transducer $\mathcal{T}$ that reads the marking $\tau_f(t_A)$ induced by f on a given tree t_A and outputs the labelling t_M such that $t_A \otimes t_M \in L_M$, whenever such t_M exists:*

$$\forall_{t_A \in L_A} \left[t_A \otimes \mathcal{T}\big(t_A \otimes \tau_f(t_A)\big)\right] \in L_M.$$

Before proving the theorem, consider the following continuation of Example 8.4.

Example 8.5. Let $\mathcal{A}$ be an unambiguous tree automaton over a ranked alphabet A_R. Let $L_A = L(\mathcal{A})$ and L_M contain trees $t \otimes \rho$ where ρ is an accepting run of $\mathcal{A}$ on $t \in \mathrm{Tr}_{A_\mathrm{R}}$. Then, the above theorem states that there exists a transducer that reads the marking induced by some homomorphism f on a given tree $t \in L(\mathcal{A})$ and produces the unique accepting run of $\mathcal{A}$ on t (whenever exists).

A simple proof of Theorem 8.73 can be given using the composition method (see [She75]). This proof was suggested by Bojańczyk as a simplification of an earlier proof given by the author.

Since we are focused on automata, we only sketch the proof based on the composition method here and give a longer self-contained proof below. Assume that there is an MSO formula defining a language L_M that has quantifier depth n. Let $|M| = k$ and let $f\colon (\mathrm{Tr}_{A_\mathrm{R}}, \mathrm{Con}_{A_\mathrm{R}}) \to (H, V)$ be a homomorphism that recognises all the $(n+k+1)$-types of MSO over A_R. In a vertex w of a given ranked tree t the transducer $\mathcal{T}$ can store in its memory the $(n+m+1)$-type of the currently read context $t[w \leftarrow \square]$. Then, given the $(n+k+1)$-types of both subtrees under w, it can compute the $(n+k)$-type of the tree $t[w \leftarrow x]$ with the current vertex w denoted by an additional variable x. The $(n+k)$-type of $t[w \leftarrow x]$ is enough to ask about the truth value of the following formulae (for every $a \in M_2$):

there exists a labelling $t_M \in L_M$ of $t[w \leftarrow x]$ such that $t_M(x) = a$.

If there is any such labelling t_M then the above formula is true for exactly one letter $a \in M_2$. The transducer $\mathcal{T}$ outputs this letter in w and proceeds in $w\mathrm{L}$, $w\mathrm{R}$ updating the type of the context respectively.

The rest of this section is devoted to an automata-based proof of Theorem 8.73.

Let $\mathcal{A}$ be some non-deterministic tree automaton recognising the language L_M. Note that $\mathcal{A}$ itself may not be unambiguous. Consider an automaton denoted $\widehat{\mathcal{A}}$ that is a projection of the automaton $\mathcal{A}$ from the ranked alphabet $A_\mathrm{R} \times M$ to A_R: the working alphabet of $\widehat{\mathcal{A}}$ is A_R, transitions are transitions of $\mathcal{A}$ with the component M of each letter removed, the rest is unchanged. Note that $L(\widehat{\mathcal{A}}) = L_A$.

We will use the notion of ranked contexts from Sect. 6.1.1 (see page 95) with one extension: we allow a context to have the hole $\square$ in the root. The notion of a run of an automaton on a context is unchanged (e.g. if ρ is a run on $t[\epsilon \leftarrow \square]$ then ρ consists of one state).

By the definition, every transition of $\widehat{\mathcal{A}}$ comes from a transition of $\mathcal{A}$. In particular, every run ρ of $\widehat{\mathcal{A}}$ on a tree t_A corresponds to (at least one) labelling of $\mathrm{dom}(t_A)$ by letters in M. Similarly, a run of $\widehat{\mathcal{A}}$ on a context p_A induces an M-labelled context p_M with the same domain and the same hole as p_A. We call these labellings the *M-labellings consistent with* ρ. A letter of such a labelling is called *the M-letter of* ρ.

For technical reasons we assume that there is some fixed linear order on the sets M_2, M_0 that enables to pick minimal elements from non-empty sets of letters.

Let $f_{\widehat{\mathcal{A}}}$ be the automaton morphism into the automaton algebra $(H_{\widehat{\mathcal{A}}}, V_{\widehat{\mathcal{A}}})$ for $\widehat{\mathcal{A}}$ (see Sect. 6.2.1, page 104). Let $t_A \in \mathrm{Tr}_{A_\mathrm{R}}$ be a tree and let $\tau(t_A)$ be the marking induced by the automaton morphism $f_{\widehat{\mathcal{A}}}$ on t_A. We will encode $\tau(t_A)$ as a tree over the ranked alphabet $G = (H_\mathcal{A}, H_\mathcal{A})$.

The construction goes as follows. The input ranked alphabet is $A_\mathrm{R} \times G$. The set of states $Q^\mathcal{T}$ of $\mathcal{T}$ is $\mathsf{P}(Q^\mathcal{A})$. The initial state $q_I^\mathcal{T}$ is the singleton $\{q_I^\mathcal{A}\}$.

We start by stating an invariant that will be satisfied by the constructed transducer $\mathcal{T}$: if $\mathcal{T}$ is in a vertex w of a tree t_A and it have assigned letters $m_u \in M_2$ to all the vertices $u \prec w$ then the state S_w of $\mathcal{T}$ in w satisfies:

$$S_w = \big\{q \in Q^\mathcal{A} : \exists_\rho\ \rho \text{ is an accepting run of } \widehat{\mathcal{A}} \text{ on } t_A[w \leftarrow \square] \\ \text{and the } M\text{-letters of } \rho \text{ in the vertices } u \prec w \text{ are } m_u\big\}. \quad (8.2)$$

We will show that the invariant can be preserved. Let us fix a moment during the computation of $\mathcal{T}$: we are in a vertex $w \in \mathrm{dom}(t_A)$.

If w is a leaf of t_A then we use the following transition over leafs: given a state S_w and a letter $b \in A_{\mathrm{R}0}$ output a minimal element m_0 of the set

$$P_w \stackrel{\text{def}}{=} \{m_0 : \exists_{(q,(b,m_0))\in\delta_0^{\mathcal{A}}}\} \subseteq M_0,$$

or some fixed m_0 if the set is empty.

Now assume that w is an internal node of t_A. Assume that we have already assigned letters $m_u \in M_2$ to all the nodes $u \prec w$. The marking $\tau(t_A)$ gives us sets $Q_{w\mathrm{L}}, Q_{w\mathrm{R}} \subseteq Q^{\mathcal{A}}$ in nodes $w\mathrm{L}$, $w\mathrm{R}$ respectively (i.e. $Q_{wd} = f_{\widehat{\mathcal{A}}}(t\restriction_{wd})$). The current state of $\mathcal{T}$ is a set of states $S_w \subseteq Q$.

Consider the following set of letters:

$$P_w = \left\{m_2 \in M_2 : \exists_{(q,(t_A(w),m_2),q_{\mathrm{L}},q_{\mathrm{R}})\in\delta_2^{\mathcal{A}}}\ q \in S_w \wedge q_{\mathrm{L}} \in Q_{w\mathrm{L}} \wedge q_{\mathrm{R}} \in Q_{w\mathrm{R}}\right\} \subseteq M_2.$$

If $P_w = \emptyset$ then we output some fixed letter $m_2 \in M_2$. In that case, the state of $\mathcal{T}$ will always stay $\emptyset$ and the invariant will be satisfied — there will be no accepting run of $\mathcal{A}$ on the currently read context. We will show that during the run of $\mathcal{T}$ on any tree $t_A \in L_A$ the sets P_w are non-empty and have at most one element each.

If $P_w \neq \emptyset$ let $\mathcal{T}$ output the minimal element $m_w \in P_w$ and proceed in the vertices wd for $d = \mathrm{L}, \mathrm{R}$ in the state

$$S_{wd} \stackrel{\text{def}}{=} \left\{q_d : \exists_{(q,(t_A(w),m_w),q_{\mathrm{L}},q_{\mathrm{R}})\in\delta_2^{\mathcal{A}}}\ q \in S_w \wedge q_{\bar{d}} \in Q_{w\bar{d}}\right\}.$$

Clearly the invariant (8.2) is satisfied. This finishes the definition of $\mathcal{T}$ — the transitions described above can be easily encoded in the functions $\delta_2^{\mathcal{T}}$, $\delta_0^{\mathcal{T}}$ of appropriate types.

Lemma 8.34. *During the run of $\mathcal{T}$ on any tree $t_A \in Tr_{A_{\mathrm{R}}}$ in every vertex $w \in \mathrm{dom}(t)$ the set P_w contains at most one letter.*

Proof. First assume that w is a leaf of t_A. For a contradiction assume that there are two distinct letters $m_0, m_0' \in P_w$ and let (q, m_0), (q', m_0') be the respective transitions. Using the invariant (8.2) we can find two accepting runs ρ, ρ' of $\mathcal{A}$ on $t_A[w \leftarrow \square]$ with values q and q' in the hole w respectively. Let p_M, p_M' be some M-labellings consistent with ρ and ρ'. Let $t_M = p_M(m_0())$ be the tree obtained by putting the single-node tree $m_0()$ into the hole of p_M (similarly $t_M' = p_M'(m_0'()))$. Clearly $t_M \neq t_M'$ and the runs ρ, ρ' can be extended to accepting runs on $t_A \otimes t_M$ and $t_A \otimes t_M'$ using the above transitions. This gives us two distinct labellings of the tree t_A, both in the language L_M.

Now assume that w is an internal node of t_A, this case is similar to the above one but more technical. Let $t_A(w) = a$ and assume contrary that there are two distinct letters $m_2, m_2' \in P_w$. Consider the respective transitions $(q, (a, m_2), q_{\mathrm{L}}, q_{\mathrm{R}})$ and $(q, (a, m_2'), q_{\mathrm{L}}', q_{\mathrm{R}}')$.

Since $q, q' \in S_w$ so by (8.2) there are two accepting runs ρ, ρ' of $\widehat{\mathcal{A}}$ on $t_A[w \leftarrow \square]$ that assign letters m_u to $u \prec w$ and have values q, q' respectively in the hole w. Let p_M, p'_M be some M-labellings of consistent with the runs ρ, ρ' respectively.

For $d \in \{\mathrm{L}, \mathrm{R}\}$ let $t_d, t'_d \in \mathrm{Tr}_M$ be trees and ρ_d, ρ'_d be parity-accepting runs of $\mathcal{A}$ that witness that $q_d, q'_d \in Q_{wd}$, i.e. ρ_d is a parity-accepting run of $\mathcal{A}$ on $t_A\restriction_{wd} \otimes t_d$ with value q_d, similarly for t'_d, ρ'_d, and q'_d.

Consider now two trees over the ranked alphabet $A_\mathrm{R} \times M \times Q^{\mathcal{A}}$:

$$t = \big(t_A[w \leftarrow \square] \otimes p_M \otimes \rho\big) \cdot (a, m_2, q)\Big(t_A\restriction_{w\mathrm{L}} \otimes t_\mathrm{L} \otimes \rho_\mathrm{L},\ \ t_A\restriction_{w\mathrm{R}} \otimes t_\mathrm{R} \otimes \rho_\mathrm{R}\Big),$$

$$t' = \big(t_A[w \leftarrow \square] \otimes p'_M \otimes \rho'\big) \cdot (a, m'_2, q')\Big(t_A\restriction_{w\mathrm{L}} \otimes t'_\mathrm{L} \otimes \rho'_\mathrm{L},\ \ t_A\restriction_{w\mathrm{R}} \otimes t'_\mathrm{R} \otimes \rho'_\mathrm{R}\Big).$$

Note that:

- both t, t' equal t_A on the A_R'th coordinate,
- they differ in the vertex w on the M'th coordinate,
- the Q'th coordinate of t, t' denotes an accepting run of $\mathcal{A}$ on the $A_\mathrm{R} \times M$ coordinates.

Therefore, we have a contradiction: t_A has two different labellings t_M, t'_M such that $(t_A, t_M) \in L_M$ and $(t_A, t'_M) \in L_M$. ∎

Now take any tree $t_A \in L_A$ and consider the result $t_R = \mathcal{T}(t_A \otimes \tau(t_A))$. Let t_M be the unique labelling of t_A such that $(t_A, t_M) \in L_M$. Let ρ be an accepting run of $\mathcal{A}$ on $t_A \otimes t_M$. We show inductively that $t_R = t_M$ what finishes the proof. Let w be a node of t_A and assume that for all $u \prec w$ we have $t_R(u) = t_M(u)$. Let $(q, (a, m_2), q_\mathrm{L}, q_\mathrm{R})$ be the transition used by ρ in w. By the definition of P_w this transition is a witness that $m_2 \in P_w$. Therefore, P_w is non-empty and $t_R(w) = m_2 = t_M(w)$ by Lemma 8.34.

This concludes the construction of the transducer and the proof of Theorem 8.73.

8.3 Choice Hypothesis

In this section we study equivalent formulations of Conjecture 8.1, as expressed by the following theorem. The formulations bind Conjecture 8.1 with consistent markings as defined in Definition 7.14 on page 123 in Sect. 7.1. The implications of this theorem regarding prophetic thin algebras are discussed in Sect. 7.3. (see page 130).

Theorem 8.74. *The following conditions are equivalent:*

1. *There is no uniformization of* CHOICE(x, X) *(i.e. Conjecture 8.1 holds).*
2. *There is no uniformization of* LEAF − CHOICE(x) *(see* (8.1)*).*
3. *For every finite thin algebra (H, V) over a ranked alphabet $A_\mathrm{R} = (A_{\mathrm{R}2}, A_{\mathrm{R}0})$ and every ranked tree $t \in Tr_{A_\mathrm{R}}$ there exists a consistent marking of t by types in H.*

4. *For every finite thin algebra* (H, V) *over the ranked alphabet* $A_b = (\{n\}, \{b\})$ *there exists a consistent marking of the unique complete binary tree* $t_n \in Tr_{A_b}$ *by types in* H.

The proof of the above theorem is split over the following sections. Clearly (3) implies (4).

8.3.1 *Equivalence* (1) ⇔ (2)

We start by observing that LEAF − CHOICE(x) and CHOICE(x, X) is essentially the same uniformization problem. However, LEAF − CHOICE(x) turns out to be much easier to work with. First observe that if $\psi(x, X)$ is a uniformization of CHOICE(x, X) then

$$\widehat{\psi}(x) \stackrel{\text{def}}{=} \psi\big(x, \{y : y \text{ is a leaf }\}\big)$$

uniformizes LEAF − CHOICE(x). What remains is to show the following lemma.

Lemma 8.35. *If* LEAF − CHOICE(x) *has a uniformization then* CHOICE(x, X) *also has one.*

Proof. We show how to MSO-interpret any set X contained in a thin tree as a set of leafs of another thin tree.

Take non-empty a set $X \subseteq \mathrm{dom}(t)$ for a thin tree t. Without loss of generality we can assume that X is prefix-free (i.e. there are no $u, w \in X$ with $u \prec w$), otherwise we can start by restricting to $\prec$-minimal elements of X. Now consider the upward closure $\bar{X}$ of X defined as

$$\bar{X} = \{u \in \mathrm{dom}(t) : \exists_{w \in X}\ u \preceq w\}.$$

We say that a vertex $u \in \bar{X}$ is X*-branching* if $u\mathrm{L}, u\mathrm{R} \in \bar{X}$. Similarly, a vertex $u \in \bar{X}$ is a X*-leaf* if $u\mathrm{L}, u\mathrm{R} \notin \bar{X}$ (equivalently if $u \in X$). Let us consider the set $Y \subseteq \bar{X}$ that contains all the X-branching vertices of $\bar{X}$ and all the X-leaf vertices of $\bar{X}$. Note that Y is MSO-definable from X. Additionally, Y with the prefix and lexicographic orders (treated as a relational structure) is isomorphic to the set of vertices of some thin tree t'. The leafs of t' correspond to the elements of X. Therefore, we can use an uniformization of LEAF − CHOICE(x) to choose a unique leaf of t' by interpreting this formula on $(Y, \preceq, \leq_{\mathrm{lex}})$. Therefore, a uniformization of LEAF − CHOICE(x) gives a uniformization of CHOICE(x, X). ■

8.3.2 Implication (2) ⇒ (3)

Now we prove that non-existence of a uniformization of LEAF − CHOICE(x) implies that every finite thin algebra labels every ranked tree. It is achieved by proving a stronger statement, namely Proposition 8.22. It is designed in such a way to imply other consequences of Conjecture 8.1 from Sect. 7.1, page 122.

Proposition 8.22. *Assume that Conjecture 8.1 holds and that $f\colon (H, V) \to (H', V')$ is a surjective homomorphism between two finite thin algebras over a ranked alphabet A_{R}. Let $t \in Tr_{A_{\mathrm{R}}}$ be a ranked tree and τ' be a consistent marking of t by H'. Then there exists a consistent marking τ of t by H such that*

$$\forall_{u\in\mathrm{dom}(t)}\ f(\tau(u)) = \tau'(u). \tag{8.3}$$

The rest of this section is devoted to a proof of this proposition. The implication (2) ⇒ (3) follows from it by taking as (H', V') the singleton thin algebra $(\{h_0\}, \{v_0\})$ and the unique homomorphism $f\colon (H, V) \to (H', V')$—then the constant marking by h_0 is always a consistent marking and its *preimage* given by Proposition 8.22 is a consistent marking of a given tree, therefore (3) of Theorem 8.74 is satisfied.

We start the proof with the following lemma that can be seen as a reformulation of Fact 6.58 from page 104 in the language of consistent markings.

Lemma 8.36. *If $t \in Tr_{A_{\mathrm{R}}}$ is a thin tree and (H, V) is a thin algebra over a ranked alphabet A_{R} then there exists exactly one consistent marking of t. In particular, all the homomorphisms $f\colon (Tr_{A_{\mathrm{R}}}, Con_{A_{\mathrm{R}}}) \to (H, V)$ must agree on thin trees.*

Proof. The proof is inductive on the rank of a given thin tree t, see Sect. 6.1.4, page 98. Assume that for all thin trees of rank smaller than η the thesis holds. Assume that $\mathrm{rank}(t) = \eta$ and let τ_S be the spine of t (i.e. τ_S is the set of nodes in t of rank precisely η). For every node u that is off τ_S there is a unique consistent marking of $t{\restriction}_u$ by induction hypothesis. Since τ_S is a thin tree of rank 1, it consists of finitely many infinite branches. The values of the marking on these branches are uniquely determined by (7.1) from page 123. Finally, the conditions of the marking determine the values of the marking in the finitely many branching nodes of τ_S. ■

Now we move to the proof of Proposition 8.22. Assume the contrary. Since all the properties are MSO-definable, by Rabin's theorem (Theorem 1.17 on page 20) we can find a regular ranked tree with a marking $t_0 \otimes \tau' \in \mathrm{Tr}_{A_{\mathrm{R}}\times(H',H')}$ such that there is no consistent marking τ of t_0 by H that satisfies (8.3). Let G be a finite graph such that:

- the edges of G are labelled by $\{\mathrm{L}, \mathrm{R}\}$,
- there are functions $\widehat{t_0}\colon G \to A_{\mathrm{R}2} \cup A_{\mathrm{R}0}$ and $\widehat{\tau'}\colon G \to H'$ labelling nodes of G by A_{R} and H',
- the unfolding of G from a vertex $g_0 \in G$ gives (via $\widehat{t_0}, \widehat{\tau'}$) $t_0 \otimes \tau'$.

We denote by $\widehat{u} \in G$ the vertex of G that corresponds to a vertex $u \in \mathrm{dom}(t_0)$. If g is a non-leaf vertex of G and $d \in \{\mathrm{L}, \mathrm{R}\}$ then by $g \cdot d$ we denote the unique d-successor of g.

Consider the following perfect information finite arena game $\mathcal{G}$ with players $\exists$ and $\forall$. The arena of $\mathcal{G}$ is

$$\{(h, g) \in H \times G : f(h) = \widehat{\tau'}(g)\} \cup \{\epsilon\}.$$

The initial position is ϵ. $\exists$ can move from ϵ to one of the positions $(h_0, g_0) \in \mathcal{G}$ for $h_0 \in H$. After such a move, a sequence of *rounds* is played. Assume that an j'th round starts in a position (h_j, g_j). If g_j is a leaf of t_0 then the game ends. Otherwise let $a = \widehat{t_0}(g_j)$ and:

- first $\exists$ gives a pair of types $h_{j,\mathrm{L}}, h_{j,\mathrm{R}} \in H$ such that

$$a(h_{j,\mathrm{L}}, h_{j,\mathrm{R}}) = h_j \ \wedge\ f(h_{j,\mathrm{L}}) = \widehat{\tau'}(g_j \cdot \mathrm{L}) \ \wedge\ f(h_{j,\mathrm{R}}) = \widehat{\tau'}(g_j \cdot \mathrm{R}),$$

- then $\forall$ picks a direction $d_j \in \{\mathrm{L}, \mathrm{R}\}$ and the game proceeds in the position $(h_{j+1}, g_{j+1}) \stackrel{\text{def}}{=} (h_{j,d}, g_j \cdot d)$.

If a play reaches a position (h_j, g_j) such that g_j is a leaf of G then $\exists$ wins if and only if $\mathrm{Leaf}(\widehat{t_0}(g_j)) = h_j$ (i.e. h_j is the type of the root-only tree labelled by $\widehat{t_0}(g_j)$). Assume that a play π is infinite and let α be the sequence of directions $d_0, d_1, \ldots$ played by $\forall$. π is winning for $\exists$ if the marking defined by the played types $h_{j,\mathrm{L}}$,$h_{j,\mathrm{R}}$ along the path α they followed in t_0 is consistent (see (7.1), page 123); formally if for every $i \in N$ we have

$$h_i = \prod_{j=i,i+1,\ldots} \mathrm{Node}\big(\widehat{t_0}(g_j), d_j, h_{j,\bar{d}_j}\big). \tag{8.4}$$

Fact 8.75. *Winning strategies for $\exists$ in $\mathcal{G}$ are in* $1-1$ *correspondence with consistent markings τ of t_0 that satisfy* (8.3).

Proof. Every strategy induces a function $\tau : \mathrm{dom}(t_0) \to H$ and if it is winning then τ is a consistent marking. By the definition of the arena, such a marking satisfies (8.3).

Similarly, every consistent marking τ as in the statement induces a strategy: first play $\tau(\epsilon)$, then inductively ensure that after obtaining directions $u = d_0, d_1, \ldots, d_{j-1}$ from $\forall$ the reached position (h_j, g_j) satisfies $h_j = \tau(u)$. When asked for a pair of types play $(\tau(u\mathrm{L}), \tau(u\mathrm{R}))$. If a leaf is reached then we know that $\exists$ wins because τ is a marking. Otherwise, an infinite path is followed and since τ is consistent so (8.4) is satisfied. ■

Note that the winning condition of $\mathcal{G}$ is ω-regular, so the game is determined. Since we assumed that there is no appropriate consistent marking, $\forall$ has a finite-memory strategy in $\mathcal{G}$. Let us fix such a strategy $\sigma_\forall$ with a memory structure M.

Plan for the rest of the proof. Now, our plan is to take a thin tree $t \in \mathrm{Th}$ and interpret it as a subset $\bar{t}$ of $\mathrm{dom}(t_0)$. Then, using Fact 8.36, we can compute the unique marking $\bar{\tau}$ of $\bar{t}$ by types in H in such a way that the image of $\bar{\tau}$ by f equals τ' pointwise. Finally, we run the strategy $\sigma_\forall$ against $\bar{\tau}$ what results in a path α in $\bar{t}$. By the definition of the game $\mathcal{G}$ the path α has to reach a vertex corresponding to a leaf of t, otherwise the play would be winning for $\exists$ what contradicts the assumption that $\sigma_\forall$ is winning.

Let $T \subseteq \mathrm{dom}(t_0)$ be the set of vertices $u \in \mathrm{dom}(t_0)$ such that the tree $t_0\restriction_u$ is not thin. Clearly T is prefix-closed. By Fact 8.36 we know that T is non-empty — otherwise t_0 would be thin and both H, H' would have exactly one consistent marking of t_0 and (8.3) would be satisfied by these markings.

Let $W \subseteq T$ be the set of branching vertices in T. By the definition of T, for every vertex $u \in T$ there exists $u' \in W$ such that $u \preceq u'$ — otherwise $T\restriction_u$ is a single infinite branch and therefore $t_0\restriction_u$ is thin.

Since both sets T and W are defined basing only on the subtree of t under a given node, in fact T and W correspond to unfoldings of subsets $\widehat{T}$ and $\widehat{W}$ of G.

Let $\iota\colon \{\mathrm{L}, \mathrm{R}\}^* \to W$ be the unique bijection that preserves the prefix and the lexicographical order.

Let us fix some type $P(h') \in H$ for every $h' \in H'$ in such a way that $f(P(h')) = h'$ — it is possible by the fact that f is surjective. We can assume that the types $P(h')$ are fixed in our construction since there are only finitely many $h' \in H'$.

Let $A_\mathrm{R} \sqcup H = (A_{\mathrm{R}2}, A_{\mathrm{R}0} \sqcup H)$ be the extension of the ranked alphabet by types in H. Note that we can treat the algebra (H, V) as an algebra over the ranked alphabet $A_\mathrm{R} \sqcup H$ by putting $\mathrm{Leaf}(h) = h$.

Now we take a thin tree $t \in \mathrm{Th}$. We will try to choose a leaf of t in a way MSO-definable on t. The following fact expresses an important consequence of the definition of ι and the fact that G is a finite graph.

Fact 8.76. *The labelling t_G of the given thin tree t by vertices of G such that $u \in \mathrm{dom}(t)$ is labelled by $\widehat{\iota(u)} \in \widehat{W} \subseteq G$ is MSO-definable on t.*

Additionally, for every $ud \in \mathrm{dom}(t)$ the path between $\iota(u)$ and $\iota(ud)$ in t_0 is of length at most $|G|$. We can define in MSO on t for a given node ud what is the sequence of vertices of G on the corresponding path from $\widehat{\iota(u)}$ to $\widehat{\iota(ud)}$.

Proof. By the definition of T and W we know that for every vertex $g \in G$ such that $g \in \widehat{T} \setminus \widehat{W}$ there exists a unique finite path π_g that starts in g and contains only vertices in $\widehat{T} \setminus \widehat{W}$ until it reaches a vertex $\mathrm{next}(g) \in \widehat{W}$. It implies that for every node $z \in T \setminus W$ there is a unique $\preceq$-minimal node $\mathrm{next}(z)$ such that $z \preceq \mathrm{next}(z) \in W$ and

$$\widehat{\mathrm{next}(z)} = \mathrm{next}(\widehat{z}).$$

In particular, by the definition of ι, for every $u \in \{\mathrm{L}, \mathrm{R}\}^*$ and $d \in \{\mathrm{L}, \mathrm{R}\}$ we have

$$\iota(ud) = \mathrm{next}(\iota(u) \cdot d). \tag{8.5}$$

Therefore, we can construct the desired labelling of t by vertices g and paths π_g by inductively following the function $g \mapsto \mathrm{next}(g \cdot d)$ in G. ■

Let us construct a thin tree $\bar{t}$ over the ranked alphabet $A_\mathrm{R} \sqcup H$ such that $\mathrm{dom}(\bar{t}) \subseteq \mathrm{dom}(t_0)$. First let

$$I \stackrel{\text{def}}{=} \{w \in \mathrm{dom}(t_0) : \exists_{u \in \mathrm{dom}(t)}\ w \preceq \iota(u)\}. \tag{8.6}$$

Now, for $u \in \mathrm{dom}(t)$:

- if $u \preceq \iota(u')$ for some internal node $u' \in \mathrm{dom}(t)$ then $u \in \mathrm{dom}(\bar{t})$ and $\bar{t}(u) = t_0(u)$,
- if $u = \iota(u')$ for some leaf u' of t then $u \in \mathrm{dom}(\bar{t})$ and $\bar{t}(u) = P(\tau'(u))$,
- if $u \notin T$ but the maximal prefix u' of u that belongs to T satisfies $u' \in I$ then $u \in \mathrm{dom}(\bar{t})$ and $\bar{t}(u) = t_0(u)$,
- otherwise $u \notin \mathrm{dom}(\bar{t})$.

Note that $\bar{t}$ is thin because t is thin and all the subtrees $t_0\restriction_u$ for $u \notin T$ are thin. Intuitively, $\mathrm{dom}(\bar{t})$ consists of the set I and all the thin subtrees of t_0 of the form $t_0\restriction_u$ such that the sibling of u is in I.

By Fact 8.36 there is a unique consistent marking $\bar{\tau}$ of $\bar{t}$ by types in H.

Fact 8.77. *For every $u \in \mathrm{dom}(\bar{t})$ we have $f(\bar{\tau}(u)) = \tau'(u)$.*

Proof. If u is a leaf of $\bar{t}$ and $\bar{t}(u) \in H$ then by the definition $\bar{t}(u) = P(\tau'(u))$ so

$$f(\bar{\tau}(u)) = f(\bar{t}(u)) = \tau'(u).$$

Therefore, since $\bar{t}$ is thin and f is a homomorphism, we obtain that for every $u \in \mathrm{dom}(\bar{t})$ we have $f(\bar{\tau}(u)) = \tau'(u)$. ■

The following lemma shows that $\bar{\tau}$ can be encoded on the thin tree t.

Lemma 8.37. *The labelling (denoted $\tau\restriction_W$) of the nodes u of t by the types $\bar{\tau}(\iota(u)) \in H$ is* MSO*-definable on t.*

Proof. Take any pair of nodes u, u' in t such that u' is a child of u. By Fact 8.76 we can assume that we have an access to the vertices of G $\widehat{\iota(u)}$ and $\widehat{\iota(u')}$ as well as to paths π between them in G. We will define an element $s_{u,u'} \in V \sqcup \{1\}$ called *context type between u and u'* representing what happens in t_0 on the path from $w = \iota(u)$ to $w' = \iota(u')$.

Assume that $w' = wd_0d_1 \ldots d_n$. Take any $i \in \{1, \ldots, n\}$ and consider the node $z = wd_0 \cdots d_{i-1}\bar{d}_i$ (i.e. a node that is off the path from wd_0 to w' in t_0). Since there are no elements of W on the path from w to w' (except the end-points), we know that $wd_0 \cdots d_{i-1} \notin W$ so $z \notin T$ (i.e. the subtree of t_0 under z is thin).

Lemma 8.36 implies that there is a unique consistent marking of the subtree $t_0\restriction_z$ by types in H. As observed before, this marking must satisfy (8.3). The value h_i of this marking in z depends only on the subtree, so we can assume that this value is fixed together with the finite graph G.

Now, the context type $s_{u,u'}$ between u and u' is the multiplication of the types of contexts along the path $wd_0, \ldots, w'$:

$$s_{u,u'} \stackrel{\text{def}}{=} \prod_{i=1,\ldots,n} \text{Node}\big(t_0(wd_0 \cdots d_{i-1}), d_i, h_i\big).$$

Therefore, we have shown an extension of Fact 8.76 stating that we have an access in MSO on t to the types of the contexts between every pair u, u' with u' a child of u in t.

Now we can guess a labelling of t by types in H and verify that it encodes a consistent marking on t_0 (via ι, as in the statement) by additionally multiplying all the contexts by the context types between each parent and child (we assume that $s \cdot 1 = 1 \cdot s = s$). Since $\bar{\tau}$ is unique, the guessed labelling must equal $\tau\restriction_W$ as in the statement. ■

Now we consider the sequence of directions $\pi \in \{\text{L}, \text{R}\}^{\leqslant\omega}$ played by $\forall$ according to $\sigma_\forall$ when $\exists$ is playing $\bar{\tau}$ (see Fact 8.75). If the play reaches a vertex $u \in \text{dom}(\bar{t})$ such that $u = \iota(u')$ for a leaf u' of t then the play stops and the sequence π is finite — $\exists$ is unable to produce successive types.

Consider the following cases:

- π reached a leaf u of t_0. In this case $\exists$ wins π since the marking $\bar{\tau}$ is consistent. Contradiction to the fact that $\sigma_\forall$ is a winning strategy of $\forall$.
- π is an infinite play. In this case the marking given by $\exists$ is consistent along π since it comes from a consistent marking $\bar{\tau}$. So again $\exists$ wins the play and we have a contradiction.
- π reached a vertex $w \in \text{dom}(t_0)$ such that $w = \iota(u)$ for a leaf u of t. In this case we call u *the selected leaf of* t.

Therefore, the only possible case is that a leaf u of t was selected. What remains is to observe the following fact.

Fact 8.78. *The play π can be simulated in* MSO *on t. In particular we can define in* MSO *on t the unique selected leaf u.*

Proof. Since the strategy $\sigma_\forall$ as well as the arena of the game $\mathcal{G}$ are finite, it is enough to show how to simulate the strategy of $\exists$ that corresponds to $\bar{\tau}$. Therefore, $\exists$ should be aware what is the currently played sequence of directions $u \in \{\text{L}, \text{R}\}^*$ to be able to play the types $\bar{\rho}(u\text{L})$ and $\bar{\rho}(u\text{R})$ (see Fact 8.75). By the above case study, we know that the play has to reach a node $w \in \text{dom}(t_0)$ such that $w = \iota(u)$ for a leaf u of t. In particular, the play will always stay in the set I as defined in (8.6).

Observe that every element $w \in I$ either belongs to W (and can be represented by $\iota^{-1}(w)$) or has a unique decomposition $w = udz$ with maximal $u \in W$. In the latter case $w \prec \text{next}(ud)$ and in particular $|z| < |G|$ (z must correspond to a prefix of the path from w to $\text{next}(w)$, see Fact 8.76). Therefore, for a given $u \in \text{dom}(t)$ there is finitely many possible $w \in I$ with the decomposition as above. Additionally, Lemma 8.37 implies that knowing the decomposition $w = udz$ we can compute what are the values of $\bar{\tau}$ in $w\text{L}$ and $w\text{R}$. ■

Using this fact we can write a formula $\psi(x)$ that inputs a thin tree t, performs all the above constructions on t, and checks whether x is the selected leaf of t. This formula is a uniformization of LEAF – CHOICE(x); therefore, by Lemma 8.35 it contradicts Conjecture 8.1 and finishes the proof of Proposition 8.22.

8.3.3 Implication (4) ⇒ (2)

We need to prove that if every thin algebra over the ranked alphabet $A_b = (\{n\}, \{b\})$ has a consistent marking of the complete binary tree $t_n \in \mathrm{Tr}_{A_b}$ then there is no uniformization of LEAF – CHOICE(x).

Assume for the contradiction that $\psi(x)$ is a formula uniformizing $LEAF - CHOICE(x)$: for every thin tree $t \in \mathrm{Tr}_{A_b}$ there exists exactly one vertex $u \in \mathrm{dom}(t)$ such that $t \models \psi(u)$ and this vertex is a leaf of t. We want to show that there exists a thin algebra (H, V) such that there is no consistent marking of the complete binary tree t_n by types in H.

Let $M = (\{\mathtt{L}, \mathtt{R}, \star\}, \{b\})$ and let L_M be the language of trees over the ranked alphabet $A_b \times M$ that contains a pair $t_A \otimes t_M$ if the following are satisfied:

1. t_A is a thin tree,
2. all leafs of t_M are labelled by b,
3. let w be the leaf of t_A selected by ψ (i.e. $t_A \models \psi(w)$),
4. $t_M(u) = \star$ for all internal nodes $u \in \mathrm{dom}(t)$ except those that $u \prec w$,
5. for $u \prec w$ we have $t_M(u) = d$ where $d \in \{\mathtt{L}, \mathtt{R}\}$ is the direction such that $ud \preceq w$.

Note that L_M is a regular tree language and the relation L_M is uniformized:

$$\forall_{t_A \in \mathrm{Th}_{A_b}} \exists!_{t_M \in \mathrm{Tr}_M}\ t_A \otimes t_M \in L_M.$$

Using Theorem 8.73 there exists a transducer $\mathcal{T}$ that reads t_A and $\tau_f(t_A)$ for a homomorphism $f\colon (\mathrm{Tr}_A, \mathrm{Con}_A) \to (H, V)$ into a finite thin algebra (H, V) and outputs the only labelling t_M of t_A such that $t_A \otimes t_M \in L_M$ (if such a labelling exists). By the definition of L_M we have the following fact.

Fact 8.79. *For every thin tree t_A the path indicated by letters $\{\mathtt{L}, \mathtt{R}\}$ in $\mathcal{T}(t_A \otimes \tau_f(t_A))$ leads to a leaf u of t_A. Moreover, $t_A \models \psi(u)$.*

Let (H', V') be the subalgebra of (H, V) that is the image of $(\mathrm{Th}_{A_b}, \mathrm{ThCon}_{A_b})$ under f.

For the purpose of contradiction assume that τ is a consistent marking of the complete binary tree t_n by the types of H' — it may not be the marking of t_n induced by f since possibly $H' \subsetneq H$. Let $\alpha \in \{\mathtt{L}, \mathtt{R}\}^{\leq\omega}$ be the sequence of directions output by $\mathcal{T}$ when run on $t_n \otimes \tau$.

First assume that α is an infinite branch of t_n. Consider a tree t' that results in plugging a thin tree of type $\tau(u)$ under u for every vertex u that is off α. Note that t'

is thin and $\tau_f(t')$ equals τ for every $u \prec \alpha$ and for every u that is off α. Therefore, the run of $\mathcal{T}$ on $t' \otimes \tau_f(t')$ is the same as on $t \otimes \tau$ for every $u \prec \alpha$ (see Fact 8.72). So $\mathcal{T}$ labels an infinite branch of t' by letters $\{\text{L}, \text{R}\}$, a contradiction with Fact 8.79.

If α is finite then the same argument holds (since t_n is complete, α cannot reach a leaf of t_n) — we can change the subtrees along α and the two subtrees under αL, αR obtaining a thin tree on which the sequence of letters $\{\text{L}, \text{R}\}$ does not reach any leaf.

This concludes the proof of the last implication of Theorem 8.74.

8.4 Negative Results

In this section we show two non-uniformizability results. Both rely on the transducers described in Sect. 8.2 and a construction of a consistent marking of a thick tree presented in Sect. 8.4.2. The construction is based on Green's relations (see [Gre51]) that provide an insight into the structure of finite semigroups.

8.4.1 Green's Relations

We start by recalling definitions and standard facts about these relations. The definitions follow [PP04, Annex A]. Let M be a finite semigroup. Let M^1 be defined as M if M is a monoid and as $M \sqcup \{1\}$ with $1 \cdot m = m$ for $m \in M^1$ in the other case. Clearly M^1 is a monoid and M is a sub-semigroup of M^1.

If $s \in M$ then by $s \cdot M^1$ we denote the set $\{s \cdot m \,:\, m \in M^1\}$ or equivalently $\{s\} \cup \{s \cdot m \,:\, m \in M\}$. $M^1 \cdot s$ is defined symmetrically and $M^1 s M^1$ is obtained by taking $\{m \cdot s \cdot m' \,:\, m, m' \in M^1\}$.

Let s, s' be two elements of M. We say that

$$s \leq_{\mathcal{R}} s' \quad \text{if} \quad s \cdot M^1 \subseteq s' \cdot M^1,$$
$$s \leq_{\mathcal{L}} s' \quad \text{if} \quad M^1 \cdot s \subseteq M^1 \cdot s',$$
$$s \leq_{\mathcal{J}} s' \quad \text{if} \quad M^1 \cdot s \cdot M^1 \subseteq M^1 \cdot s' \cdot M^1.$$

Let $T \in \{\mathcal{R}, \mathcal{L}, \mathcal{J}\}$. We say that s and s' are T-*comparable* if $s \leq_T s'$ or $s' \leq_T s$. We say that s and s' are T-*equivalent* (denoted $s \sim_T s'$) if $s \leq_T s'$ and $s' \leq_T s$. We additionally say that s and s' are $\mathcal{H}$-*equivalent* if they are $\mathcal{R}$- and $\mathcal{L}$-equivalent. For $T \in \{\mathcal{R}, \mathcal{L}, \mathcal{J}, \mathcal{H}\}$ the equivalence classes of the T-equivalence are called T-classes of M.

The following results summarize properties of these relations that will be used here.

Theorem 8.80. *Let M be a finite semigroup.*

1. *If $s \sim_{\mathcal{H}} s'$ then $s \sim_{\mathcal{J}} s'$.*
2. *For $T \in \{\mathcal{R}, \mathcal{L}\}$ if $s \sim_{\mathcal{J}} s'$ and $s \leq_T s'$ then $s \sim_T s'$.*

3. *There exists a $\leq_{\mathcal{J}}$-minimal $\mathcal{J}$-class of M.*
4. *The minimal $\mathcal{J}$-class of M contains an idempotent.*

Proposition 8.23 (Proposition 2.4 in Annex A of [PP04]). *If an $\mathcal{H}$-class $G \subseteq M$ of a semigroup M contains an idempotent then the product $\cdot$ of any two elements of G belongs to G and $(G, \cdot)$ is a group*[1].

Remark 8.11. If G is an $\mathcal{H}$-class of a semigroup M that contains an idempotent e and e' is an idempotent in G then $e = e'$.

Proof. Assume that m_1 is the unit of the group G and let e be an idempotent in G. Let e^{-1} be the inverse of e in the group G (i.e. $e \cdot e^{-1} = m_1$). Then

$$m_1 = e \cdot e^{-1} = e \cdot e \cdot e^{-1} = e \cdot m_1 = e. \qquad \blacksquare$$

8.4.2 A Marking of a Thick Tree

As proved in Theorem 8.74, Conjecture 8.1 is equivalent to the fact that every finite thin algebra has a consistent marking on every tree (see Item (3) of the theorem). Unfortunately, the author was unable to prove this fact. On the other hand, by Lemma 8.36, every finite thin algebra has a consistent marking on every thin tree. The following proposition can be seen as an intermediate result: every finite thin algebra has a consistent marking on *some* non-thin (i.e. thick) tree. The construction of this thick tree is motivated by a result of Bojańczyk [Boj10a, Theorem 4.1] stating that, in the context of finite trees, every preclone contains a certain "idempotent sub-preclone".

Proposition 8.24. *For every finite thin algebra (H, V) over a ranked alphabet $A_{\mathrm{R}} = (A_{\mathrm{R}2}, A_{\mathrm{R}0})$ with $A_{\mathrm{R}0} \neq \emptyset$ there exists a thick tree $t \in \mathrm{Tr}_{A_{\mathrm{R}}}$ and a consistent marking τ of t by types in H.*

We assume that $A_{\mathrm{R}0} \neq \emptyset$ because otherwise all ranked trees over $A_{\mathrm{R}0}$ have $\{\mathtt{L}, \mathtt{R}\}^*$ as the domain so $\mathrm{Tr}_{A_{\mathrm{R}}}$ contains only complete trees.

During the proof we extensively use facts about Green's relations (see Sect. 8.4.1). Note that by Axiom 6.1 of thin algebra (see Sect. 6.2, page 102), the set V with the operation $\cdot$ is a semigroup.

By Fact 6.58 from page 104 we know that there is a unique homomorphism f from $(\mathrm{Th}_{A_{\mathrm{R}}}, \mathrm{ThCon}_{A_{\mathrm{R}}})$ into (H, V). First we can assume that (H, V) contains only types that are represented as f-types of thin trees and thin contexts (we use the fact that $A_{\mathrm{R}0}$ is non-empty and we restrict ourselves to the subalgebra generated by $\{b() : b \in A_{\mathrm{R}0}\}$). Let e be an idempotent in the lowest $\mathcal{J}$-class of V. Let G be the

[1]More formally, one can pick an element m_1 of G and define an operation $m \mapsto m^{-1}$ on G such that $(G, m_1, \cdot, .^{-1})$ is a group.

$\mathcal{H}$-class of e (i.e. the intersection of the $\mathcal{L}$- and $\mathcal{R}$-class of e). By Proposition 8.23 we know that G is a group because it contains an idempotent.

It turns out that e acts as a certain *attractor*, as expressed by the following lemma.

Lemma 8.38. *For every $s \in V$ we have $(ese)^\infty = e^\infty$.*

Proof. Note that ese is $\mathcal{R}$- and $\mathcal{L}$-comparable with e. Since e is in the lowest $\mathcal{J}$-class of V so $ese \sim_{\mathcal{J}} e$ and therefore ese is $\mathcal{H}$-equivalent with e, hence $ese \in G$. Therefore, since e is the only idempotent of G (see Remark 8.11) so the idempotent power of ese is e (i.e. $(ese)^\sharp = e$) and we have $(ese)^\infty = \left((ese)^\sharp\right)^\infty = e^\infty$. ■

Now we move to the construction of a thick tree t. Let p_1 be a thin context of f-type e. Let $a \in A_{\mathrm{R}2}$ be any letter. We define the following tree p_2 over the ranked alphabet $A_{\mathrm{R}} \sqcup \{\square\}$ (it can be seen as a context with two holes):

$$p_2 \stackrel{\text{def}}{=} p_1 \cdot a\,(p_1 \cdot \square, p_1 \cdot \square)\,.$$

Let u_{L}, u_{R} be the positions of the two holes put explicitly in the above definition. Let us consider the tree $\bar{t}$ that is obtained from p_2 by putting trees p_1^∞ instead of u_{L}, u_{R}. This tree is thin, let τ be the unique consistent marking of $\bar{t}$. Note that $\tau(u_{\mathrm{L}}) = \tau(u_{\mathrm{R}}) = e^\infty$.

Let $s_{\mathrm{L}} = a(\square, e^\infty)$ and $s_{\mathrm{R}} = a(e^\infty, \square)$. Note that

$$\tau(\epsilon) = e \cdot s_{\mathrm{L}} \cdot e \cdot e^\infty = (es_{\mathrm{L}}e) \cdot (es_{\mathrm{L}}e)^\infty = (es_{\mathrm{L}}e)^\infty = e^\infty.$$

Let $t_T \otimes \tau_T$ be the tree obtained from $p_2 \otimes \left(\tau[u_{\mathrm{L}} \leftarrow \square, u_{\mathrm{R}} \leftarrow \square]\right)$ by looping vertices u_{L}, u_{R} back to the root of p_2 (see Fig. 8.1). Since $\tau_T(u_{\mathrm{L}}) = \tau_T(u_{\mathrm{R}}) = \tau_T(\epsilon) = e^\infty$, τ_T is a marking of t_T. The constructed tree t_T is thick but it is not complete — many subtrees of t_T are thin and contain leafs.

Consider any infinite branch α of t_T. If α does not pass through infinitely many copies of the root of p_2 then α is from some point on contained in one copy of p_2. In that case α is from some point on consistent (by the consistency of τ). Consider the opposite case and observe that

$$\alpha = u_{d_0} \cdot u_{d_1} \cdot \ldots,$$

for a sequence of directions $d_0, d_1, \ldots$ It is enough to show that the value $\tau_T(\epsilon) = e^\infty$ is consistent with the product $\prod$ of contexts along α (see Remark 7.9 on page 123). We can group the decomposition of α in t_T in the following way:

$$(es_{d_0}e) \cdot (es_{d_1}e) \cdot (es_{d_2}e) \cdot \ldots$$

Let $\bar{s} \cdot \bar{e}^\infty$ be a Ramsey decomposition of the above infinite product. In that case $\bar{s} = exe$ and $\bar{e} = eye$ for some $x, y \in V$. Therefore,

$$\bar{s} \cdot \bar{e}^\infty = (exe) \cdot (eye)^\infty = (exe) \cdot (exe)^\infty = (exe)^\infty = e^\infty.$$

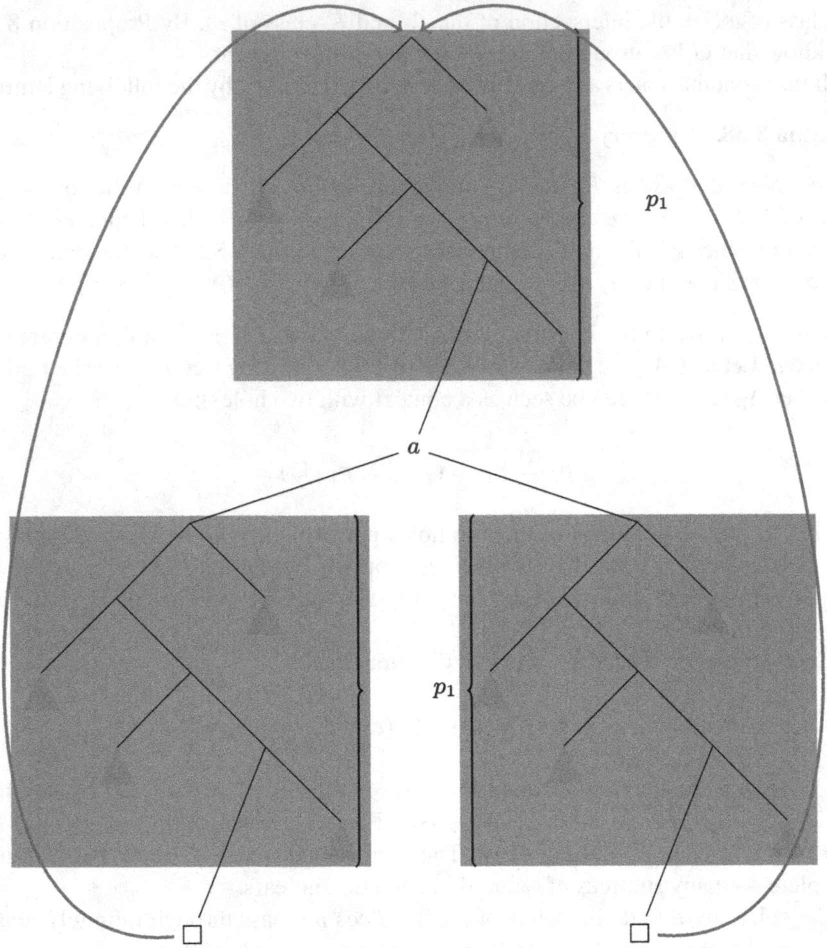

Fig. 8.1 The looping of the two holes of p_2 to obtain a thick tree. The gray subtrees are thin. The second coordinate (i.e. $\tau[u_L \leftarrow \square, u_R \leftarrow \square]$) is skipped for the sake of simplicity.

This proves that τ is consistent and therefore the proof of Proposition 8.24 is finished.

8.4.3 Non-uniformizability of Skeletons

We identify here a set $\sigma \subseteq \mathrm{dom}(t)$ with its characteristic function $\sigma \in \mathrm{Tr}_{(\{0,1\},\{0,1\})}$. By $\mathrm{SKEL}(\sigma)$ we denote the MSO formula expressing that σ is a skeleton of a given tree t.

Theorem 8.81. *There is no* MSO *formula uniformizing* SKEL(σ).

Proof. Assume contrary that $\psi(\sigma)$ uniformizes SKEL(σ). Consider a transducer $\mathcal{T}$ that, given a thin tree t_A and the marking $\tau_f(t_A)$ constructs the labelling $t_S \in \mathrm{Tr}_{\{0,1\}^2}$ that encodes a skeleton of t_A satisfying $\psi(t_S)$.

Assume that $\mathcal{T}$ uses a homomorphism f into a finite thin algebra (H, V) and let (H', V') be the subalgebra that is the image of $(\mathrm{Th}_{A_R}, \mathrm{ThCon}_{A_R})$. Let $t \otimes \tau$ be a thick tree with a consistent marking by types in H' given by Proposition 8.24. Consider the result $t_S = \mathcal{T}(t \otimes \tau)$. By Proposition 6.12 from page 101 t_S does not encode a skeleton of t.

First assume that there exists an infinite branch α of t such that infinitely many vertices $u \prec \alpha$ does not belong to t_S. Let t' be the tree obtained by putting a thin tree of type $\tau(w)$ under vertex w for every w that is off α. Note that t' is thin. Let τ' be the only consistent marking of t'. Let $t'_S = \mathcal{T}(t' \otimes \tau')$. By the definition, if $u \prec \alpha$ or u is off α then $\tau'(u) = \tau(u)$. By Fact 8.72 for every $u \prec \alpha$ we have $t'_S(u) = t_S(u)$, so t'_S also does not encode a skeleton of t'. A contradiction.

Now assume that t_S does not satisfy the local constraint of skeletons in some vertices u, u' (i.e. $u = u' = \epsilon \in t_S$ or u, u' are siblings and it is not true that exactly one of them belongs to t_S). The proof of this case is essentially the same — it is enough to substitute finitely many subtrees along the paths leading to u, u' and the subtrees under u, u'. ■

8.4.4 *Ambiguity of Thin Trees*

Theorem 8.82. *The language* $\mathrm{Th}_{A_b} \subset \mathrm{Tr}_{A_b}$ *of thin trees over the ranked alphabet* $A_b = (\{n\}, \{b\})$ *is ambiguous (i.e. it is not recognised by any unambiguous automaton).*

We use the ranked alphabet A_b for simplicity, the same construction works for any ranked alphabet A_R with $A_{R0} \neq \emptyset$.

Proof. The proof follows the same lines as the proof of Theorem 8.81. We assume that $\mathcal{A}$ is an unambiguous automaton recognising Th_{A_b}. We define L_M as the language of trees $t \otimes \rho$ where t is a ranked tree and ρ is an accepting run of $\mathcal{A}$ on t (as in Example 8.4). The relation defined by L_M is uniformized so there exists a transducer $\mathcal{T}$ and a homomorphism f such that given a thin tree t_A and the marking $\tau_f(t_A)$ it constructs the unique accepting run $\rho = \mathcal{T}(t \otimes \tau)$ of $\mathcal{A}$ on t_A.

We consider a thick tree with a respective marking $t \otimes \tau$ given by Proposition 8.24 and construct the labelling $\rho = \mathcal{T}(t \otimes \tau)$ of t by states $Q^{\mathcal{A}}$. Since $t \notin \mathrm{Th}_{A_b}$ so ρ is not an accepting run. The rest of the proof is the same as in Theorem 8.81: either ρ violates local constraints or is not parity-accepting along some infinite branch of t. In both cases we can define a thin tree t' such that the run constructed by $\mathcal{T}$ on $t' \otimes \tau_f(t')$ is also not accepting. ■

8.5 Conclusions

This chapter is devoted mainly to Conjecture 8.1 stating that there is no MSO-definable choice function on thin trees. These statement is somehow non-constructive: there is no MSO-formula that defines a choice function. The results of this chapter provide an equivalent statement that has a more constructive form: in order to prove Conjecture 8.1 it is enough to find, for every thin algebra S, a consistent marking of the complete binary tree by elements of S.

Although the author was unable to find a construction of such a marking, a weaker construction of a consistent marking of a thick tree is provided. Already this weaker construction turns out to be enough to obtain two new non-uniformizability examples:

- an essentially new MSO-definable relation that does not admit any MSO-definable uniformization,
- an essentially new example of an ambiguous language.

To the author's best knowledge, all the examples existing before were based on [GS83]:

- it was proved by Gurevich and Shelah [GS83] (see also [CL07]) that the relation $x \in X$ does not admit any MSO-definable uniformization,
- basing on this observation, Niwiński and Walukiewicz [NW96] (cf. [CLNW10]) proved that the language "exists a node labelled by a" is ambiguous.

It seems that proving Conjecture 8.1 is a hard task that requires a better understanding of the relations between regular tree languages and conditions that can be verified pathwise.

This chapter is based on [BS13].

Part III

Extensions of Regular Languages

Chapter 9
Descriptive Complexity of MSO+U

MSO logic is quite expressive, in particular it covers most of other logics used for specifying properties of computer systems. However, MSO is not able to express quantitative properties of structures. A natural example of such a quantitative property is "the delays between a request and the successive answer are uniformly bounded". Bojańczyk in [Boj04] introduced an additional quantifier U, called the *unbounding quantifier*, that allows to express such properties. A formula $\mathsf{U}X.\varphi(X)$ holds if $\varphi(X)$ is satisfied for arbitrarily large finite sets X. Formally, $\mathsf{U}X.\varphi(X)$ is equivalent to:

$$\bigwedge_{n\in\mathbb{N}} \exists X.\ (\varphi(X)\ \wedge\ n < |X| < \infty)\,.$$

The following language is an example of a language of ω-words that is definable in the extended logic MSO+U but is not ω-regular

$$\mathsf{U}X.\,(\forall x{\in}X.\ P_a(x) \wedge \forall x{<}y{<}z.\,(x{\in}X \wedge z{\in}X) \Leftrightarrow y{\in}X)\,,$$

i.e. the language of those ω-words that contain arbitrarily long blocks of consecutive letters a.

One of the crucial open problems about the U quantifier is decidability: is the MSO+U theory of the ω-chain or the complete binary tree decidable? The decidability was proved for various fragments of the MSO+U logic [BC06, Boj11, Boj10b, BT12] but the problem for MSO+U remained open for over 10 years.

In the following two chapters we approach the problem of decidability of MSO+U via descriptive set-theoretical methods. First, in this chapter we prove the following theorem.

Theorem 9.7. *There exists an alphabet A such that for every $i > 0$ there exists an* MSO+U *formula φ_i such that the language $\mathrm{L}(\varphi_i) \subseteq A^\omega$ of ω-words satisfying φ_i is $\boldsymbol{\Sigma}^1_i$-complete.*

The following theorem exploits the above result to show that there is no *simple* automata model for MSO+U on ω-words.

M. Skrzypczak, *Set Theoretic Methods in Automata Theory*, LNCS 9802
DOI: 10.1007/978-3-662-52947-8_9

Theorem 9.83 (Hummel S. [HS12]). *There is no model of alternating nor non-deterministic automata on ω-words with countably many states and projective acceptance condition that captures* MSO+U.

Sketch of a proof. If $\mathcal{A}$ is an alternating automaton with countably many states Q and acceptance condition $W \subseteq Q^\omega$ then the language of $\mathcal{A}$ can be written as

$$\mathrm{L}(\mathcal{A}) = \Big\{\alpha \in A^\omega : \exists_{\sigma_\exists} \text{ — a strategy of } \exists \text{ in } \mathbf{G}(\mathcal{A}, \alpha)$$
$$\forall_\pi \text{ — play consistent with } \sigma_\exists \text{ in } \mathbf{G}(\mathcal{A}, \alpha)$$
$$\pi \text{ satisfies the winning condition } W\Big\}.$$

Therefore, if $W \in \mathbf{\Sigma}^1_n$ for some n then the above formula implies that $\mathrm{L}(\mathcal{A}) \in \mathbf{\Sigma}^1_{n+2}$. But Theorem 9.7 shows that for every n there are MSO+U-definable languages of ω-words that do not belong to $\mathbf{\Sigma}^1_{n+2}$. ■

This result shows that standard technique of proving decidability of variants of MSO by translating into appropriate automata (see e.g. [BT09, Boj11]) is not enough in the case of MSO+U. Chapter 10 further builds on the topological complexity of MSO+U to prove that in a certain sense the MSO+U theory of the complete binary tree is undecidable. The decidability of MSO+U on ω-words is still open.

To prove Theorem 9.7 we first construct an appropriate sequence of languages IF^i of *multi-branching* trees such that the language IF^i is $\mathbf{\Sigma}^1_i$-hard. Then we show how to inductively encode such multi-branching trees into ω-words. These encodings are the technical heart of the proof — their aim is to present a given multi-branching tree in a way understandable for an MSO+U formula. Finally, we construct a sequence of MSO+U formulae φ_i that, given an encoding of a multi-branching tree t, can verify if $t \in \mathrm{IF}^i$. The formula cannot check if a given ω-word encodes any multi-branching tree at all but this is not needed for our needs.

The chapter is organised as follows. In Sect. 9.1 we introduce the concept of multi-branching trees and languages IF^i. Then, in Sect. 9.2 we define the alphabets of ω-words we use and the formulae φ_i. Section 9.3 introduces inductively reductions r_i that encode multi-branching trees into ω-words. It is shown there that r_i is continuous and satisfies an additional technical property of *sequentiality*. In Sect. 9.4 we prove that the functions r_i reduce IF^i to $\mathrm{L}(\varphi_i)$. Finally, in Sect. 9.5 we show upper bounds on topological complexity of the languages $\mathrm{L}(\varphi_i)$ what concludes the proof of Theorem 9.7. In Sect. 9.6 we conclude.

9.1 Basic Notions

Let us recall from Sect. 1.1 (see page 1) that:

- $\omega\mathrm{Tr}_X$ is the family of total functions $\tau : \omega^* \to X$,
- $\omega\mathrm{PTr}$ is the family of prefix-closed subsets of ω^*.

In this chapter we use the so-called *multi-branching* trees. Let $i > 0$. An (i-dimensional) multi-branching tree is a prefix-closed subset of $(\omega^i)^*$. The set of all such trees is denoted $\omega\mathrm{PTr}^i$. Clearly $\omega\mathrm{PTr}^i$ is a Polish space and $\omega\mathrm{PTr}^1 = \omega\mathrm{PTr}$.

Let us fix an order $\sqsubseteq$ of type ω on ω^*, such that $\omega^* = \{v_0, v_1, \ldots\}$. Additionally assume that for all $n \in \mathbb{N}$ we have $|v_n| \leqslant n$. There are infinitely many vertices of length 1 so it is possible.

Definition 9.19. *Consider $i > 0$, a multi-branching tree $\tau \in \omega\mathrm{PTr}^{i+1}$, and a finite word or ω-word $\alpha \in \omega^{\leqslant\omega}$. We define the section $\tau\restriction_\alpha \in \omega\mathrm{PTr}^i$ of the multi-branching tree τ as follows*

$$t\restriction_\alpha = \{u \in (\omega^i)^* : \ |u| \leqslant |\alpha| \wedge (\alpha\restriction_{|u|} \otimes u) \in t\},$$

where

$$(\alpha_0, \alpha_1, \alpha_2, \ldots) \otimes (u_0, u_1, u_2, \ldots) = (\alpha_0 \cdot u_0, \alpha_1 \cdot u_1, \alpha_2 \cdot u_2, \ldots).$$

The dots in the above definition can stand for a finite or an infinite sequence.

Figure 9.1 presents the first two levels of a multi-branching tree t on ω^2 i.e. $t \in \omega\mathrm{PTr}^2$. The children of the root are arranged into a two-dimensional grid. Given a sequence $\alpha \in \omega^{\leqslant\omega}$ the section $t\restriction_\alpha \in \omega\mathrm{PTr}^1$ is defined as the one-dimensional multi-branching tree obtained by selecting particular rows from the grids of children on every level. The position of the selected row is defined by the successive values of α. For example the children of the root in $t\restriction_\alpha$ come from the α_0'th row of the presented grid.

Observe that if u is a finite word, $t\restriction_u$ is a finite-depth tree — its depth is bounded by $|u|$.

For an ω-tree $t \in \omega\mathrm{Tr}_X$ and an ω-word $\alpha \in \omega^\omega$, let

$$t(\alpha) = \big(t(\alpha\restriction_0), t(\alpha\restriction_1), \ldots\big) \in X^\omega.$$

Fig. 9.1 A 2-dimensional multi-branching tree.

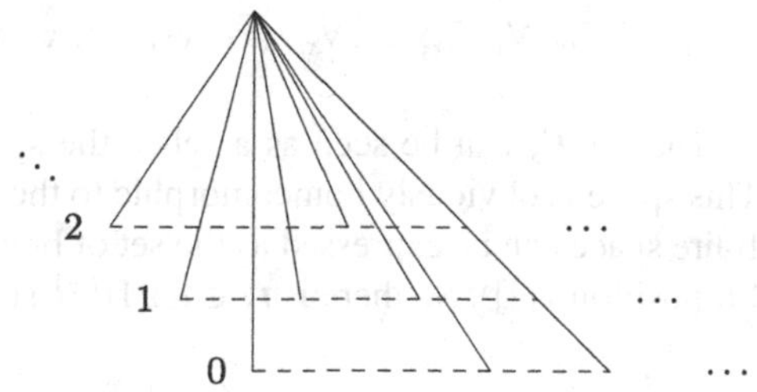

9.1.1 Languages IF^i

To prove that the languages defined in this chapter are $\mathbf{\Sigma}^1_i$-hard we will construct continuous reductions from languages $\mathrm{IF}^i \subseteq \omega\mathrm{PTr}^i$ defined below.

Let IF^1 be the set of all trees $t \in \omega\mathrm{PTr}^1$ that contain an infinite branch (i.e. $\mathrm{IF}^1 =$ IF, see Sect. 1.6.3, page 17).

Take $i > 0$. Let IF^{i+1} be the set of all multi-branching trees $t \in \omega\mathrm{PTr}^{i+1}$ such that there exists an ω-word $\alpha \in \omega^\omega$ such that

$$t\restriction_\alpha \notin \mathrm{IF}^i.$$

Fact 9.84. *For each $i \geqslant 1$ the set IF^i is a $\mathbf{\Sigma}^1_i$-complete subset of $\omega\mathrm{PTr}^i$.*

This fact follows easily from unravelling the definition of an $\mathbf{\Sigma}^1_i$ set. For the sake of completeness we give here a formal proof of this fact.

Proof. First we prove the upper-bound. By the definition, IF^1 is the set of ill-founded trees IF that is known to be $\mathbf{\Sigma}^1_1$-complete (see Sect. 1.6.4, page 17).

We proceed by induction. Assume that $\mathrm{IF}^i \in \mathbf{\Sigma}^1_i$. Let

$$P_i = \left\{(\alpha, t) \in \omega^\omega \times \omega\mathrm{PTr}^{i+1} \,:\, t\restriction_\alpha \notin \mathrm{IF}^i\right\} \in \mathbf{\Pi}^1_i.$$

Note that IF^{i+1} is the projection of P_i, so it is in $\mathbf{\Sigma}^1_{i+1}$.

Let us prove that each $\mathbf{\Sigma}^1_i$ set in ω^ω continuously reduces to IF^i.

As we know (see e.g. [Kec95, Exercise 14.3]), each analytic ($\mathbf{\Sigma}^1_1$) set in a space X is a projection of a closed set in $\omega^\omega \times X$. Recall that, by the definition, each $\mathbf{\Sigma}^1_{i+1}$ set is a projection of some $\mathbf{\Pi}^1_i$ set. Therefore, each $\mathbf{\Sigma}^1_i$ set in ω^ω is of the form[1]:

$$S = \left\{x \,:\, \exists_{x_1\in\omega^\omega} \neg\exists_{x_2\in\omega^\omega} \neg\exists_{x_3\in\omega^\omega} \ldots \neg\exists_{x_i\in\omega^\omega}\ (x_1, x_2, \ldots, x_i, x) \in F_S\right\},$$

for some closed set $F_S \in (\omega^\omega)^{i+1}$. The formula unravels to:

$$\exists_{x_1} \forall_{x_2} \exists_{x_3} \ldots \exists_{x_i}\ (x_1, x_2, \ldots, x_i, x) \in F_S \quad \text{if } i \text{ is odd, and to:}$$
$$\exists_{x_1} \forall_{x_2} \exists_{x_3} \ldots \forall_{x_i}\ (x_1, x_2, \ldots, x_i, x) \notin F_S \quad \text{if } i \text{ is even.}$$

The set F_S can be seen as a set in the space $\left(\omega^{i+1}\right)^\omega$, by simple transposition. This space is obviously homeomorphic to the Baire space ω^ω. Each closed set in the Baire space can be expressed as the set of branches of some ω-tree (see e.g. [Kec95, Proposition 2.4]). So there is $t_S \in \omega\mathrm{PTr}^{i+1}$ such that:

$$F_S = \left\{(x_1 \otimes x_2 \otimes \cdots \otimes x_{i+1}) \in \left(\omega^{i+1}\right)^\omega \,:\, \forall_{n\in\mathbb{N}}\ \left(x_1\restriction_n \otimes x_2\restriction_n \otimes \cdots \otimes x_{i+1}\restriction_n\right) \in t_S\right\} \tag{9.1}$$

[1] Formally, for $i = 1$ the formula takes the form $S = \{x : \exists x_1 \in \omega^\omega.\ (x_1, x) \in F_S\}$.

To simplify the notation, for a prefix-closed set $t \subseteq X^*$, by $[t] \subseteq X^\omega$ we denote the set of infinite branches of t. Using this notation, the above equation can be formulated as

$$F_S = [t_S].$$

We will use the multi-branching tree t_S to define the needed reduction. Let $f\colon \omega^\omega \to \omega\mathrm{PTr}^i$ be defined as follows:

$$f(x) = \left\{(v_1 \otimes v_2 \otimes \cdots \otimes v_i) \in \left(\omega^i\right)^k :\ \left(v_1 \otimes v_2 \otimes \cdots \otimes v_i \otimes x{\restriction_k}\right) \in t_S, k \in \mathbb{N}\right\}.$$

To determine whether a vertex at some level k belongs to $f(x)$ we only need to know the first k numbers in the sequence x, so the function is continuous. To prove that this is a reduction of S to IF^i we need:

$$f(x) \in \mathrm{IF}^i \iff x \in S \tag{9.2}$$

Now we will take a closer look at the sets IF^i. Observe that:

$$\mathrm{IF}^i = \left\{t :\ \exists_{x_1} \forall_{x_2} \exists_{x_3} \ldots \exists_{x_i}\ (x_1 \otimes x_2 \otimes \cdots \otimes x_i) \in [t]\right\} \text{ if } i \text{ is odd, and:}$$
$$\mathrm{IF}^i = \left\{t :\ \exists_{x_1} \forall_{x_2} \exists_{x_3} \ldots \forall_{x_i}\ (x_1 \otimes x_2 \otimes \cdots \otimes x_i) \notin [t]\right\} \text{ if } i \text{ is even.}$$

So the quantifier structure is the same as in case of the above representation of S. Therefore, to obtain (9.2), it suffices to show that for any fixed $x_1, x_2, \ldots, x_i$:

$$(x_1 \otimes x_2 \otimes \cdots \otimes x_i) \in [f(x)] \iff (x_1, x_2, \ldots, x_i, x) \in F_S.$$

By (9.1) it is equivalent to:

$$(x_1 \otimes x_2 \otimes \cdots \otimes x_i) \in [f(x)] \iff (x_1 \otimes x_2 \otimes \cdots \otimes x_i \times x) \in [t_S].$$

But the latter follows immediately from the definition of f. ∎

9.2 Languages H_i

In this section we inductively construct a sequence of languages $(H_i)_{i\in\mathbb{N}}$. We will later show that for each $i \in \mathbb{N}$ the language H_i is MSO+U-definable and $\boldsymbol{\Sigma}^1_i$-hard. Additionally, in Sect. 9.5 we observe that $H_i \in \boldsymbol{\Sigma}^1_i$.

Let us fix a finite alphabet $B_0 = \{a, |_0, b\}$ and define inductively $B_i = B_{i-1} \sqcup \{[_{i-1}, |_i,]_{i-1}\}$ (i.e. B_0 contains 3 letters and B_i contains $3(i+1)$ letters).

The reductions used in the rest of the proof work on the space $(B_i^+)^\omega$. Since we want to build MSO+U formulae over finite alphabets, we need use one additional encoding which is simply a kind of concatenation. For $i \geqslant 0$ consider $j_i\colon (B_i^+)^\omega \to$

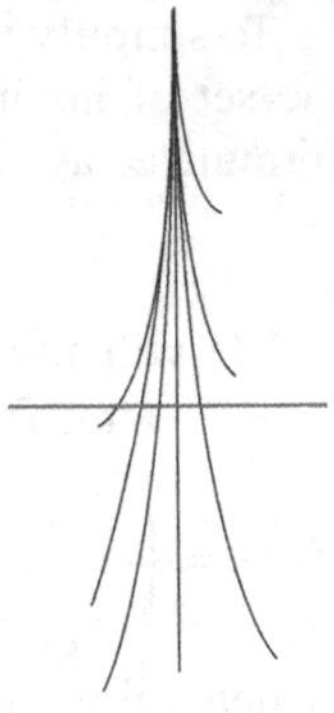

Fig. 9.2 An illustration of the narrow property — any section of finite depth contains only finitely many prefixes of branches in A.

B^{ω}_{i+1} defined as follows

$$j_i(w_0, w_1, \ldots) = [_i w_0]_i \cdot [_i w_1]_i \cdot \ldots$$

Clearly functions j_i defined above are continuous and $1-1$.

For a node $u = (u_1, u_2, \ldots, u_m) \in \omega^*$ of an ω-tree, we will call the word $a^{u_1} b a^{u_2} b \ldots b a^{u_m} b$ the *address* of u in the ω-tree.

Let an *i-block* be a word of the form $[_i w |_i w']_i$ where $w \in (a^* b)^*$ and $w' \in (B_i \setminus \{|_i\})^+$. We will call the word w the *address* of this i-block (since it will be interpreted as an address of a node in an ω-tree) and the word w' the *body* of this i-block.

We will call a set A of addresses of nodes:

deep if the number of letters b in elements of A is unbounded,

narrow if for any set P of some prefixes of elements of A such that the number of letters b in elements of P is bounded, the lengths of sequences a^* in elements of P are bounded.

The following fact provides a way of using the above properties.

Fact 9.85. *An ω-tree $t \subseteq \omega^*$ has an infinite branch if and only if there is a narrow and deep set A of addresses of some nodes in t.*

Proof. First assume that t has an infinite branch $\alpha \in \omega^\omega$. Take as A the set of addresses of vertices in $\{\alpha\restriction_n : n \in \omega\}$. Of course such A is deep. We show that A is narrow. Consider any set P of prefixes of addresses in A, such that the number of letters b in elements of P is bounded by some number $k \in \omega$. In that case, lengths of sequences a^* in P are bounded by $\max_{n \leqslant k} \alpha_n$: in each element of A the sequence a^* before the n'th letter b has length α_{n-1}.

Now take a narrow and deep set A of addresses of some nodes of t. We identify elements of A with those nodes, i.e. $A \subseteq t$. Consider as T the closure of A under prefixes, i.e.:

$$T = \{u \in \omega^* : \exists_{u' \in A}\, u \preceq u'\}.$$

Then T is an infinite tree, because A is deep. Additionally, at each level $k \in \omega$, there are only finitely many vertices in $T \cap \omega^k$, by narrowness of A. So T is a finitely branching ω-tree. Therefore, by König's lemma (see Lemma 1.1, page 4), T contains an infinite branch α. But $T \subseteq t$, so α is also an infinite branch of t. ■

Now we can define the MSO+U formulae defining our languages. Observe that both properties of *deepness* and *narrowness* of a set of addresses can be expressed in MSO+U. It is because in those definitions we only use regular properties and properties like *the number of letters b is unbounded* or *the lengths of sequences a^* are bounded.*

It is easy to see that we can express in MSO that a given ω-word $\alpha \in (B_{i+1})^\omega$ is of the form $b_0 \cdot b_1 \cdot \ldots$ such that each b_n is an i-block. We implicitly assume that all formulae φ_i express it.

Let φ_0 additionally express that a given ω-word is not of the form

$$\left([_0\ (a^*b)^*\ |_0\ a\ ]_0\right)^\omega,$$

i.e. there is at least one 0-block with body different than a.

For $i > 0$, let φ_i express the following property:

There exists a set G containing only whole i-blocks such that:

1. the set of addresses of the i-blocks of G is deep,
2. the set of addresses of the i-blocks of G is narrow,
3. the bodies of the i-blocks of G, when concatenated, form an ω-word that satisfies $\neg\varphi_{i-1}$.

Take $i \geqslant 0$. Since $\mathrm{L}(\varphi_i) \subseteq B_{i+1}^\omega$, we can define

$$H_i = j_i^{-1}\left(\mathrm{L}(\varphi_i)\right) \subseteq (B_i^+)^\omega.$$

Languages H_i defined above are (up to the j_i operator) MSO+U definable.

We will use the following important property of the languages H_i.

Definition 9.20. *A language $L \subseteq X^\omega$ is* monotone *if for any $\alpha, \beta \in X^\omega$*

$$\{\alpha_n : n \in \mathbb{N}\} \subseteq \{\beta_n : n \in \mathbb{N}\} \implies (\alpha \in L \Rightarrow \beta \in L)\,.$$

Note, that belonging to a monotone language depends only on the set of letters occurring in an ω-word, namely we have the following fact.

Fact 9.86. *If $L \subseteq X^\omega$ is a monotone language then for any $\alpha, \beta \in X^\omega$ the following holds*

$$\{\alpha_n : n \in \mathbb{N}\} = \{\beta_n : n \in \mathbb{N}\} \implies (\alpha \in L \Leftrightarrow \beta \in L)\,.$$

The following lemma says that, if we restrict to well-formatted words, the languages H_i are monotone. Since all our formulae implicitly assume that the ω-words are sequences of i-blocks, we can restrict ourselves to the well-formatted pre-images of such ω-words under j_i.

Lemma 9.39. *Let $i \geqslant 0$ and X be the set of words $u \in B_i^+$ such that $[_i u]_i$ is an i-block. Then the languages $H_i \cap X^\omega \subseteq X^\omega$ are monotone.*

Proof. For $i = 0$ it is obvious. For $i > 0$ the formula φ_i expresses that there exists a set of i-blocks such that this set satisfies some additional property. Moreover, it does not matter in what order the i-blocks appear. ∎

9.3 Functions c_i, d_i, and r_i

Now we will show how to continuously reduce the languages of multi-branching trees IF^i to H_i. For technical reasons we will use the following intermediate languages.

Definition 9.21. *For $L \subseteq X^\omega$ let* $\mathrm{EPath}(L) \subseteq \omega\mathrm{Tr}_X$ *be a set of such labelled ω-trees t that there exists an ω-word $\alpha \in \omega^\omega$ such that*

$$t(\alpha) \in L.$$

In other words $\mathrm{EPath}(L)$ *is the set of ω-trees that contain an infinite branch such that labels on this branch form an ω-word in L.*

The languages $\mathrm{EPath}(L)$ were used originally by Szczepan Hummel to prove certain lower bounds on the topological complexity of MSO+U-definable languages of ω-trees.

The construction will be inductive, it will start with $i = 1$ and in each step the picture looks as follows:

$$\begin{array}{ccccc} \omega\mathrm{PTr}^i & \xrightarrow{c_i} & \omega\mathrm{Tr}_{B_{i-1}^+} & \xrightarrow{d_i} & (B_i^+)^\omega \\ \cup\!\mid & & \cup\!\mid & & \cup\!\mid \\ \mathrm{IF}^i & & \mathrm{EPath}\left(H_{i-1}^{\mathrm{c}}\right) & & H_i \end{array}$$

The construction will ensure (see Sect. 9.4) that $d_i^{-1}(H_i) = \mathrm{EPath}\left(H_{i-1}^{\mathrm{c}}\right)$ and $c_i^{-1}\left(\mathrm{EPath}\left(H_{i-1}^{\mathrm{c}}\right)\right) = \mathrm{IF}^i$. Therefore, r_i defined as $d_i \circ c_i$ will reduce IF^i to H_i. We will use the function r_{i-1} to construct a reduction c_i of IF^i to the language $\mathrm{EPath}\left(H_{i-1}^{\mathrm{c}}\right)$ of ω-trees that have a branch labelled with an ω-word $\alpha \notin H_{i-1}$. Then we again encode such labelled ω-trees in ω-words.

Recall our inductively defined alphabets $B_0 = \{a, |_0, b\}$, $B_i = B_{i-1} \sqcup \{[_{i-1}, |_i,]_{i-1}\}$.

First we define $c_1 : \omega\mathrm{PTr}^1 \to \omega\mathrm{Tr}_{B_0^+}$. Take a multi-branching tree $t \in \omega\mathrm{PTr}^1$ and a vertex $v = (u_1, u_2, \ldots, u_m) \in \omega^*$. Put

$$c_1(t)(v) \stackrel{\mathrm{def}}{=} \begin{cases} a^{u_1} b a^{u_2} b \ldots b a^{u_m} b\, |_0\, a & \text{if } v \in t, \\ a^{u_1} b a^{u_2} b \ldots b a^{u_m} b\, |_0\, b & \text{if } v \notin t. \end{cases}$$

That is, $c_1(t)(v)$ consists of the address of v and an additional bit indicating whether $v \in t$.

For $i > 1$ take a multi-branching tree $t \in \omega\mathrm{PTr}^i$ and a vertex $v \in \omega^*$. Let

$$c_i(t)(v) = \left(r_{i-1}(t\restriction_v)\right)_{|v|} \in B^+_{i-1},$$

that is, we apply the reduction r_{i-1} to the section of t along v (such a section is an $(i-1)$-dimensional multi-branching tree) and then we take the first $|v|$ words from the result.

Now we define the function d_i. We encode a tree $t \in \omega\mathrm{Tr}_{B^+_{i-1}}$ into a word $d_i(t) \in (B^+_i)^\omega$ in the following way: let v_n be the n'th vertex with respect to the order $\sqsubseteq$. Let $v_n = (u_1, u_2, \ldots, u_m)$ and let $w_0, w_1, \ldots, w_m \in B^+_{i-1}$ be the list of labels of t on the path from the root to v_n. Then

$$d_i(t)_n \overset{\text{def}}{=} a^{u_1}ba^{u_2}b\ldots ba^{u_m}b \,|_i\, [_{i-1}w_0]_{i-1} \cdot [_{i-1}w_1]_{i-1} \cdot \ldots \cdot [_{i-1}w_m]_{i-1} \in B^+_i.$$

Intuitively $d_i(t)_n$ encodes the vertex v_n in t. Such an encoding consists of two parts: the part before $|_i$ is the address of v_n in the multi-branching tree, while the part after $|_i$ is intended to store labels of t on the path from the root to v_n as $(i-1)$-blocks. The fact that we store not only the label but also the address of the given vertex in this coding will be crucial for the following parts of the construction.

Lemma 9.40. *Functions c_i, d_i defined above are continuous.*

Proof. For d_i it holds by the definition. The continuity of c_i can be proved by induction together with the continuity of r_i, since they cyclically depend on each other. The function r_{i+1} is continuous as a composition of continuous functions, likewise c_i at each coordinate v is a composition of continuous operations: $\_\restriction_v$, r_{i-1}, $\_{}_{|v|}$. ■

The following lemma states that the functions r_i are in some sense *sequential.*

Lemma 9.41. *For any $i > 0$ and any $m \in \mathbb{N}$ if $t_1, t_2 \in \omega\mathrm{PTr}^i$ agree on all $v \in (\omega^i)^*$ such that $|v| \leqslant m$ then*

$$r_i(t_1)_m = r_i(t_2)_m.$$

Proof. Recall that $r_i(t) = d_i(c_i(t))$. First observe that for a given ω-tree $t' \in \omega\mathrm{Tr}_X$, by the definition of d_i, the value $d_i(t')_m$ depends only on v_m and the labels of t' on the path from the root to v_m.

Now use an induction on i and consider the labels of $c_i(t_1)$ and $c_i(t_2)$ on the path from the root to v_m. For $i = 1$ they depend only on t_1, t_2 up to the depth of $|v_m|$, and $|v_m| \leqslant m$, thanks to our assumption about the order $\sqsubseteq$.

Take $i > 1$ and a vertex $v \preceq v_m$ (where $\preceq$ denotes the prefix order). By the definition $c_i(t)(v) = r_{i-1}(t\restriction_v)_{|v|}$. So, by the inductive assumption, this value also depends only on t at the depth of at most $|v| \leqslant |v_m| \leqslant m$. ■

From the above lemma we conclude that the labels on each branch $\alpha \in \omega^\omega$ in $c_i(t)$ code the multi-branching tree $t\restriction_\alpha$. Formally:

Lemma 9.42. *For $i > 1$, a given multi-branching tree $t \in \omega\mathrm{PTr}^i$ and an infinite branch $\alpha \in \omega^\omega$ we have:*

$$c_i(t)(\alpha) = r_{i-1}(t\restriction_\alpha) \in \left(B_{i-1}^+\right)^\omega .$$

Proof. Take any $m \in \mathbb{N}$ and consider $v = \alpha\restriction_m \in \omega^m$. By the definition

$$(c_i(t)(\alpha))_m = c_i(t)(\alpha\restriction_m) = \left(r_{i-1}(t\restriction_v)\right)_m .$$

Since $t\restriction_v$ and $t\restriction_\alpha$ agree on all vertices up to the depth m, by Lemma 9.41, we have

$$\left(r_{i-1}(t\restriction_v)\right)_m = \left(r_{i-1}(t\restriction_\alpha)\right)_m . \qquad \blacksquare$$

9.4 Reductions

In this section we show that r_i is a reduction of IF^i to H_i. We do it in two steps.

Lemma 9.43. *For $i > 0$ the function $d_i : \omega\mathrm{Tr}_{B_{i-1}^+} \to (B_i^+)^\omega$ is a reduction of $\mathrm{EPath}\left(H_{i-1}^{\mathrm{c}}\right)$ to H_i.*

Proof. We have to prove that for any $t \in \omega\mathrm{Tr}_{B_i^+}$

$$t \in \mathrm{EPath}\left(H_{i-1}^{\mathrm{c}}\right) \iff d_i(t) \in H_i .$$

First assume that $t \in \mathrm{EPath}\left(H_{i-1}^{\mathrm{c}}\right)$. Let $\alpha \in \omega^\omega$ be a branch such that $t(\alpha) \notin H_{i-1}$. Let $\beta = j_i(d_i(t)) \in (B_{i+1})^\omega$. We show that $\beta \models \varphi_i$. Take as G the set containing i-blocks corresponding to the vertices of α. Then the set of addresses of i-blocks of G is obviously narrow and deep (one vertex at each level of the ω-tree). Additionally, the set of $(i-1)$-blocks occurring in bodies of i-blocks in G is exactly the set

$$\{[_{i-1} \cdot (t(\alpha))_n \cdot]_{i-1} \ : n \in \mathbb{N}\} .$$

Language H_{i-1} is monotone, so, by Fact 9.86, since $t(\alpha) \notin H_{i-1}$, the set G satisfies Item 3 in the definition of φ_i.

The other direction is a little more tricky. Assume that $j_i(d_i(t)) \models \varphi_i$. Let G be as in the definition of φ_i. Then the set of addresses of i-blocks of G is narrow and deep. Let $B \subseteq \omega^*$ be the set of nodes corresponding to these addresses and let T be the closure of B under prefixes, i.e.:

$$T = \left\{v \in \omega^* : \exists_{v' \in B}\ v \preccurlyeq v'\right\} .$$

As in Fact 9.85, there exists an infinite branch $\alpha \in \omega^\omega$ of T. Observe that the set

$$\{[_{i-1} \cdot (t(\alpha))_n \cdot]_{i-1} : n \in \mathbb{N}\}$$

is contained in the set of $(i-1)$-blocks in bodies of i-blocks in G. Because of the monotonicity of H_{i-1} and Item 3 in the definition of φ_i, $t(\alpha) \notin H_{i-1}$. ■

Lemma 9.44. *For $i > 0$ the function c_i is a reduction of* IF^i *to* $\mathrm{EPath}\left(H^{\mathrm{c}}_{i-1}\right)$.

Proof. Take $i = 1$. An ω-tree $t \in \omega\mathrm{PTr}^1$ contains an infinite branch if and only if $c_1(t)$ contains a branch labelled by words of the form $(a^*b)^*|_0 a$ if and only if $c_1(t) \in \mathrm{EPath}\left(H^{\mathrm{c}}_0\right)$.

Induction step: $i > 1$. Take a multi-branching tree $t \in \omega\mathrm{PTr}^i$. The following conditions are equivalent:

$$\begin{array}{ll} t \in \mathrm{IF}^i & \\ \exists_{\alpha\in\omega^\omega}\ t\restriction_\alpha \notin \mathrm{IF}^{i-1} & \text{by the definition of } \mathrm{IF}^i \\ \exists_{\alpha\in\omega^\omega}\ c_{i-1}(t\restriction_\alpha) \notin \mathrm{EPath}\left(H^{\mathrm{c}}_{i-2}\right) & \text{by the inductive assumption} \\ \exists_{\alpha\in\omega^\omega}\ r_{i-1}(t\restriction_\alpha) \notin H_{i-1} & \text{by Lemma 43} \\ \exists_{\alpha\in\omega^\omega}\ c_i(t)(\alpha) \notin H_{i-1} & \text{by Lemma 42} \\ \quad c_i(t) \in \mathrm{EPath}\left(H^{\mathrm{c}}_{i-1}\right) & \text{by the definition of } \mathrm{EPath}\,(L). \end{array}$$

■

It concludes the proof of the fact that r_i reduces IF^i to H_i.

9.5 Upper Bounds

To complete the proof of Theorem 9.7 we need to show the following lemma.

Lemma 9.45. *The languages* $\mathrm{L}(\varphi_i)$ *belong to* $\mathbf{\Sigma}^1_i$.

The rest of this section is devoted to proving this lemma. The proof is inductive: we assume inductively that $\mathrm{L}(\varphi_{i-1}) \in \mathbf{\Pi}^1_i$ and show that $\mathrm{L}(\varphi_i) \in \mathbf{\Sigma}^1_i$, so in particular $\mathrm{L}(\varphi_i) \in \mathbf{\Pi}^1_{i+1}$.

Clearly $\mathrm{L}(\varphi_0)$ is a Borel language, so $\mathrm{L}(\varphi_0) \in \mathbf{\Pi}^1_1$.

The following fact expresses that the conditions of deepness and narrowness are in fact Borel (see [HST10, Proposition 2]).

Fact 9.87. *The set of pairs (β, G) such that:*

- *$\beta \in B^\omega_{i+1}$ is an infinite sequence of i-blocks,*
- *$G \subseteq \omega$ be a set containing only whole i-blocks in β,*
- *the set of addresses of i-blocks in G is deep,*
- *the set of addresses of i-blocks in G is narrow.*

is Borel.

Proof. All the conditions except the last one are explicitly Borel.

We say that a set $P \subseteq \omega$ is *well-formed* if $P \subseteq G$ and P contains prefixes of some i-blocks in G. If P is well-formed then by $\max \sharp_b(P)$ let us denote the maximal number of letters b in P among all the i-blocks. By the definition, G is narrow if and only if for every r and well-formed set P such that $\max \sharp_b(P) \leqslant r$, the lengths of sequences a^* in P are bounded.

Note that for each $r \in \mathbb{N}$ there is a maximal well-formed set $P_r \subseteq G$ such that $\max \sharp_b(P_r) \leqslant r$ — we take maximal prefixes of all the i-blocks in G until the $(r+1)$'th letter b in each i-block. Observe that for a given $r \in \mathbb{N}$ the set P_r depends continuously on (β, G). Also if $P \subseteq P'$ are well-formed then the lengths of sequences a^* are bounded in P only if they are bounded in P'. Therefore, G is narrow if and only if

$$\forall_{r\in\mathbb{N}}\ \exists_{n\in\mathbb{N}}\ \text{for every sequence } a^* \text{ in } P_r \text{ the length of } a^* \text{ is at most } n.$$

This definition is clearly Borel. ■

Therefore, an ω-word satisfies φ_i if there exists a set G satisfying Conditions 1 and 2 in the definition of φ_i and such that the bodies of the i-blocks of G form an ω-word satisfying $\neg\varphi_{i-1}$. By the inductive assumption, all these three conditions are $\mathbf{\Sigma}^1_i$ conditions, so $\mathrm{L}(\varphi_i)$ is a projection of a $\mathbf{\Sigma}^1_i$ language and it is itself $\mathbf{\Sigma}^1_i$.

9.5.1 Proof of Theorem 9.7

Now we can combine the previous results to prove Theorem 9.7.

Theorem 9.7. *There exists an alphabet A such that for every $i > 0$ there exists an* MSO+U *formula φ_i such that the language $\mathrm{L}(\varphi_i) \subseteq A^\omega$ of ω-words satisfying φ_i is $\mathbf{\Sigma}^1_i$-complete.*

Proof. Let $A = \{0, 1\}$. Take $i \in \mathbb{N}$ and φ_i. Functions c_i, d_i, j_i are continuous by Lemma 9.40 and the definition of j_i. Moreover, using the definition of H_i and Lemmas 9.43 and 9.44 their composition reduces IF^i to $\mathrm{L}(\varphi_i)$. Thanks to Fact 9.84, the set IF^i is $\mathbf{\Sigma}^1_i$-hard.

Lemma 9.45 shows that $\mathrm{L}(\varphi_i)$ belongs to $\mathbf{\Sigma}^1_i$.

By standard methods we can encode all the alphabets B_i into A using binary coding. This additional coding does not influence the topological complexity of the languages. ■

9.6 Conclusions

This chapter is devoted to a construction of examples of MSO+U-definable languages of ω-words that lie arbitrarily high in the projective hierarchy. Since every MSO+U-definable language of ω-words or infinite trees is somewhere in the projective hierarchy, it closes the question about bounds on topological complexity of MSO+U.

Already these examples show that there is no *simple* model of automata with countably many states that would capture MSO+U on ω-words. Since the argument is topological, it covers wide range of complicated models, e.g. automata with counters, stacks, tapes, etc. Most of the known decidability results for variants of MSO involve some automata equivalent in expressive power. This result can be seen as a witness that decidability of MSO+U on ω-words (if holds at all) requires some essentially new techniques.

As discussed in Chap. 10, the examples constructed in this chapter can be used to prove that in some sense MSO+U logic is undecidable on infinite trees.

This chapter is based on [HS12].

Chapter 10
Undecidability of MSO+U

As explained in Chap. 9, MSO+U logic is an extension of MSO that allows to express quantitative properties of structures. One of the consequences of the big expressive power of MSO+U is that many decision problems about other quantitative formalisms can be reduced to MSO+U. An example is the reduction [CL08] of the non-deterministic index problem to a certain boundedness problem that can be further reduced to MSO+U on infinite trees. Therefore, decidability of MSO+U would be a very desirable result.

In this chapter we show how topological hardness of MSO+U on ω-words from Chap. 9 can be used to study decidability of MSO+U on infinite trees. This methods lead to the following theorem from [BGMS14] stating that under a certain set-theoretic assumption the MSO+U theory of the complete binary tree is undecidable. Intuitively, the assumption that V=L states that all sets in the universe of set theory are *constructible*.

Theorem 10.88 (Bojańczyk Gogacz Michalewski S. [BGMS14]). *Assuming* V=L, *it is undecidable if a given sentence of* MSO+U *is true in the complete binary tree* $(\{\mathrm{L},\mathrm{R}\}^*, \preceq, \leqslant_{\mathrm{lex}})$.

The proof of this theorem is divided into two parts by introducing an intermediate object called *proj*-MSO — a logic evaluated on Polish spaces where every monadic quantifier ranges over sets from an explicitly declared level of the projective hierarchy (i.e. for each n there is a quantifier $\exists_{X\in\boldsymbol{\Sigma}^1_n}$).

The first part of the proof of Theorem 10.88 is expressed by the following theorem (it does not rely on the V=L assumption).

Theorem 10.8. *The proj*-MSO *theory of* $\{\mathrm{L},\mathrm{R}\}^{\leq\omega}$ *with prefix* $\preceq$ *and lexicographic* $\leqslant_{\mathrm{lex}}$ *orders effectively reduces to the* MSO+U *theory of the complete binary tree* $(\{\mathrm{L},\mathrm{R}\}^*, \preceq, \leqslant_{\mathrm{lex}})$.

Already this reduction is a strong indication that MSO+U should not be decidable. This indication is discussed in Sect. 10.3 of this chapter where we give an easy

M. Skrzypczak, *Set Theoretic Methods in Automata Theory*, LNCS 9802
DOI: 10.1007/978-3-662-52947-8_10

argument showing that if MSO+U on the complete binary tree would be decidable in the standard sense then it would have unexpectedly strong consequences regarding set theory (namely, it would imply that analytic determinacy does not hold).

This chapter is focused on the first part of the proof of Theorem 10.88, that is on Theorem 10.8.

The second part of the proof of Theorem 10.88 in [BGMS14] is an adaptation of the techniques of Shelah [She75] (see also [GS82]) who proves that the MSO theory of the real line $(\mathbb{R}, \leqslant)$ is undecidable. On page 410 of the cited paper Shelah observes:

Aside from countable sets, we can use only a set constructible from any well – ordering of the reals. (10.1)

The assumption V=L used in Theorem 10.88 exploits this observation by guaranteeing that there exists such a well-ordering that is projective. By adjusting the reasoning of Shelah, one gets the following proposition.

Proposition 10.25. (Bojańczyk Gogacz Michalewski S. [BGMS14]**).** *Assuming that* V=L*, the proj-*MSO *theory of the Cantor set* $(\{\mathtt{L}, \mathtt{R}\}^\omega, \leqslant_{\text{lex}})$ *is undecidable.*

This result together with the reduction from Theorem 10.8 concludes the proof of Theorem 10.88, see Sect. 10.4.1. A standalone proof of Proposition 10.25 is given in [BGMS14]. Since this proposition is not in the scope of this thesis, we only sketch a proof of it in Sect. 10.4.

The following corollary expresses in what sense Theorem 10.88 implies undecidability of MSO+U. It uses another important feature of the V=L assumption: if ZFC is consistent (i.e. there exists a model of set theory) then there exists a model satisfying V=L.

Corollary 10.2. *If* ZFC *is consistent then there is no algorithm which decides the* MSO+U *theory of the complete binary tree* $\left(\{\mathtt{L}, \mathtt{R}\}^*, \preceq, \leqslant_{\text{lex}}\right)$ *and has a proof of correctness in* ZFC.

Proof. (The following proof is in ZFC) Assume that ZFC is consistent and let M be a model of ZFC. Then Gödel's constructible universe L of M is also a model of ZFC. In Gödel's constructible universe L, the assumption V=L holds. Therefore, if ZFC is consistent then by Theorem 10.88 it has a model where the MSO+U theory of $\{\mathtt{L}, \mathtt{R}\}^*$ is undecidable. ■

The chapter is organised as follows. In Sect. 10.1 we introduce basic notions, in particular proj-MSO. Section 10.2 is devoted to a proof of Theorem 10.8. In Sect. 10.3 we show that already this theorem implies that it is unlikely to prove decidability of MSO+U in ZFC. In Sect. 10.4 we sketch a proof of Proposition 10.25 and show how to entail Theorem 10.88. Finally, in Sect. 10.5 we conclude.

10.1 Basic Notions

We consider the following logical structures: the complete binary tree $\{\mathtt{L}, \mathtt{R}\}^*$, the Cantor set $\{\mathtt{L}, \mathtt{R}\}^\omega$, and the union of the two $\{\mathtt{L}, \mathtt{R}\}^{\leq\omega}$. In the complete binary tree $\{\mathtt{L}, \mathtt{R}\}^*$, the universe consists of finite words over $\{\mathtt{L}, \mathtt{R}\}$, called *nodes*, and there are predicates for the prefix $\preceq$ and lexicographic $\leqslant_{\text{lex}}$ orders. The prefix order corresponds to the ancestor relation. In the Cantor set $\{\mathtt{L}, \mathtt{R}\}^\omega$, the universe consists of ω-words over $\{\mathtt{L}, \mathtt{R}\}$, called *branches*, and there is a predicate for the lexicographic order. Finally, in $\{\mathtt{L}, \mathtt{R}\}^{\leqslant\omega}$, the universe consists of both nodes and branches, and there are predicates for the prefix and lexicographic order. In $\{\mathtt{L}, \mathtt{R}\}^{\leqslant\omega}$, the prefix relation can hold between two nodes, or between a node and a branch. The lexicographic order is a total order on both nodes and branches, e.g. $\mathtt{L} < \mathtt{L}^\omega < \mathtt{LR}$.

10.1.1 Gödel's Constructible Universe

Let us give a short overview of the construction of Gödel's constructible universe [Göd39], following [Jec02, Chap. 13].

Assume that M is a set and $\in$ is a relation on M. We say that a set $X \subseteq M$ is *definable over M* if there exists a formula $\varphi(x, \boldsymbol{a})$ of first-order logic in the language $\{\in\}$ and a tuple of elements $\boldsymbol{a} \in M$ such that

$$X = \{x \in M : (M, \in) \models \varphi(x, \boldsymbol{a})\}.$$

Now let

$$\begin{aligned} L_0 &= \varnothing, \\ L_{\eta+1} &= \{X \subseteq L_\eta : X \text{ is definable over } (L_\eta, \in)\}, \\ L_\eta &= \bigcup_{\eta' < \eta} L_{\eta'} \quad \text{(if } \eta \text{ is a limit ordinal)}, \\ L &= \bigcup_{\eta} L_\eta \qquad \text{(where the sum ranges over all ordinals)}. \end{aligned}$$

Now, let V=L be the axiom stating that: for every set X there exists an ordinal η such that $X \in L_\eta$. Since the above inductive construction can be formalized in ZFC, this axiom can be formalized as a first-order sentence of set theory.

Now, Theorems 13.3, 13.16, and 13.18 in [Jec02] state that:

- L is a model of ZFC,
- L satisfies the axiom V=L (it is not obvious, since the notion of definability in L may a priori be different than in the original model).

Therefore, if ZFC has any model it has a model satisfying V=L. As observed in [Jec02, Theorem 25.26] (see also [Mos80, Sect. 5A]), the following implication holds.

Proposition 10.26. V=L *implies that there exists a well-order* $\leqslant$ *on* $\{\mathrm{L}, \mathrm{R}\}^\omega$ *of length* ω_1 *such that* $\leqslant$ *is a* $\mathbf{\Delta}^1_2$ *relation, i.e.* $\leqslant \in \mathbf{\Delta}^1_2\big(\{\mathrm{L}, \mathrm{R}\}^\omega \times \{\mathrm{L}, \mathrm{R}\}^\omega\big)$.

This concludes the properties of the assumption V=L that are used in Theorem 10.88.

10.1.2 *Projective* MSO

For $n \le \omega$ define the syntax of MSO_n to be the same as the syntax of MSO, except that instead of one pair of set quantifiers $\exists X$ and $\forall X$, there is a pair of quantifiers $\exists_i X$ and $\forall_i X$ for every $i \le n$. To evaluate a sentence of MSO_n on a structure, we need a sequence $\{\mathcal{X}_j\}_{j\le i}$ of families of sets, called the *monadic domains*. The semantics are then the same as for MSO, except that the quantifiers $\exists_j$ and $\forall_j$ are interpreted to range over subsets of the universe that belong to $\mathcal{X}_j$. First-order quantification is as usual, it can quantify over arbitrary elements of the universe. We write $\mathrm{MSO}\big[\mathcal{X}_1, \mathcal{X}_2, \ldots\big]$ for the above logic with the monadic domains being fixed to $\mathcal{X}_1, \mathcal{X}_2, \ldots$. Standard MSO for structures with a universe Ω is the same as $\mathrm{MSO}\big[\mathsf{P}(\Omega)\big]$, i.e. there is one monadic domain for the powerset of the universe. If Ω is equipped with a topology, we define *proj*-MSO on Ω to be

$$\mathrm{MSO}\big[\mathbf{\Sigma}^1_1(\Omega),\ \mathbf{\Sigma}^1_2(\Omega),\ \ldots\big]$$

The expressive power of proj-MSO is incomparable with the expressive power of MSO: although proj-MSO cannot quantify over arbitrary subsets, it can express that a set is in, say, $\mathbf{\Sigma}^1_1$.

Example 10.6. In the structure $\{\mathrm{L}, \mathrm{R}\}^{\leqslant\omega}$, being a node is first-order definable: a node is an element of the universe that is a proper prefix of some other element. Since there are countably many nodes, every set of nodes is Borel, and therefore in $\mathbf{\Sigma}^1_1(\{\mathrm{L}, \mathrm{R}\}^{\leqslant\omega})$. Therefore, in proj-MSO on $\{\mathrm{L}, \mathrm{R}\}^{\leqslant\omega}$ one can quantify over arbitrary sets of nodes. It is easy to see that a subset of $\{\mathrm{L}, \mathrm{R}\}^{\leqslant\omega}$ is in $\mathbf{\Sigma}^1_n(\{\mathrm{L}, \mathrm{R}\}^{\leqslant\omega})$ if and only if it is a union of a set of nodes and a set from $\mathbf{\Sigma}^1_n(\{\mathrm{L}, \mathrm{R}\}^{\omega})$.

Therefore, we obtain the following remark.

Remark 10.12. proj-MSO on $\{\mathrm{L}, \mathrm{R}\}^{\leqslant\omega}$ effectively has the same expressive power as the logic

$$\mathrm{MSO}\big[\mathsf{P}\big(\{\mathrm{L}, \mathrm{R}\}^*\big),\ \mathbf{\Sigma}^1_1\big(\{\mathrm{L}, \mathrm{R}\}^\omega\big),\ \mathbf{\Sigma}^1_2\big(\{\mathrm{L}, \mathrm{R}\}^\omega\big),\ \ldots\big].$$

The following example presents certain properties of sets that can easily be expressed in proj-MSO.

Example 10.7. In proj-MSO on $\{\text{L},\text{R}\}^{\leqslant\omega}$, one can say that a set of branches is countable. This is by using notions of interval, closed set, and perfect. A set of branches is open if and only if for every element, it contains some open interval around that element. A *perfect* is a set of branches which is closed (i.e. its complement is open) and contains no isolated points. The notions of open interval, closed set, and perfect are first-order definable. By [Kec95, Theorem 29.1], a set of branches is countable if and only if it is in $\mathbf{\Sigma}^1_1(\{\text{L},\text{R}\}^\omega)$ and does not contain any perfect subset, which is a property definable in proj-MSO.

10.2 Reduction

In this section we prove the following theorem.

Theorem 10.8. *The proj-*MSO *theory of* $\{\text{L},\text{R}\}^{\leq\omega}$ *with prefix* $\preceq$ *and lexicographic* $\leqslant_{\text{lex}}$ *orders effectively reduces to the* MSO+U *theory of the complete binary tree* $(\{\text{L},\text{R}\}^*, \preceq, \leqslant_{\text{lex}})$.

In Sect. 10.3 we observe that this reduction itself gives an evidence that MSO+U should not be decidable. The crucial ingredient of the proof of Theorem 10.8 is Theorem 9.7 (see Chap. 9, page 159) stating that it is possible to define in MSO+U languages of ω-words that are arbitrarily high in the projective hierarchy. The following lemma shows how these languages can be used in the reduction.

Lemma 10.46. *Suppose that* $L_1, L_2, \ldots \subseteq A^\omega$ *are definable in* MSO+U, *and let*

$$\mathcal{X}_i \stackrel{\text{def}}{=} \{f^{-1}(L_i) : f : \{\text{L},\text{R}\}^\omega \to A^\omega \text{ is a continuous function}\}. \qquad (10.2)$$

Then for every sentence of $\text{MSO}[\mathsf{P}(\{\text{L},\text{R}\}^*), \mathcal{X}_1, \mathcal{X}_2, \ldots]$ *on* $\{\text{L},\text{R}\}^{\leqslant\omega}$, *one can compute an equivalently satisfiable sentence of* MSO+U *on* $\{\text{L},\text{R}\}^*$.

The proof of this lemma is based on the observation that, using quantification over sets of nodes, one can quantify over continuous functions $\{\text{L},\text{R}\}^\omega \to A^\omega$. The construction is similar in the spirit to the one from [Skr13] (such encodings in the case of $\mathbf{\Sigma}^0_2$- and $\mathbf{\Delta}^0_3$-sets date back probably to Büchi [Büc83a]).

Proof. Call a mapping $f\colon \{\text{L},\text{R}\}^* \to A \sqcup \{\epsilon\}$ *proper* if on every infinite path in $\{\text{L},\text{R}\}^*$, the labelling f contains infinitely many letters different than ϵ. If f is proper then define $\hat{f}\colon \{\text{L},\text{R}\}^\omega \to A^\omega$ to be the function that maps a branch to the concatenation of the values under f of the nodes on the branch (such concatenation erases symbols ϵ).

Assume that $L_1, L_2, \ldots \subseteq A^\omega$ is a sequence of MSO+U-definable sets. For $i > 0$ and a proper mapping $f\colon \{\text{L},\text{R}\}^* \to A \sqcup \{\epsilon\}$ define

$$[f]_i \stackrel{\text{def}}{=} \{\alpha \in \{\text{L},\text{R}\}^\omega : \hat{f}(\alpha) \in L_i\},$$

$$\text{reduces}(L_i) \stackrel{\text{def}}{=} \{L \subseteq \{\text{L},\text{R}\}^\omega : L \text{ reduces continously to } L_i\} \quad \text{(see (10.2))}.$$

Proposition 2.6 in [Kec95] implies that

$$\{[f]_i : f \text{ is proper}\} = \text{reduces}(L_i). \tag{10.3}$$

Since a mapping $f\colon \{\text{L}, \text{R}\}^* \to A \sqcup \{\epsilon\}$ can be encoded as a family of disjoint sets $\{X_a \subseteq \{\text{L}, \text{R}\}^*\}_{a\in A}$, we will use quantification over sets of nodes to simulate quantification over continuous functions $g\colon \{\text{L}, \text{R}\}^\omega \to A^\omega$.

The reduction in the statement of the lemma works as follows. First-order quantification over branches is replaced by (monadic second-order) quantification over paths, i.e. subsets of $\{\text{L}, \text{R}\}^*$ that are totally ordered and maximal for that property. For a formula $\exists_i X.\, \varphi$, we replace the quantifier by existential quantification over a family of disjoint subsets $\{X_a\}_{a\in A}$ which encode a continuous function. In the formula φ, we replace a subformula $x \in X$, where x is now encoded as a path, by a formula which says that the image of x, under the function encoded by $\{X_a\}_{a\in A}$, belongs to the language L_i. In order to verify if a given element belongs to the language L_i definable in MSO+U on ω-words, we can use a formula of MSO+U on infinite trees.

More formally, our translation inputs a formula of MSO$\big[$reduces(L_1), reduces$(L_2), \ldots\big]$ and outputs a formula of MSO+U on $\{\text{L}, \text{R}\}^*$. It interprets:

- a branch $x \in \{\text{L}, \text{R}\}^\omega$ by the path $B_x = \{v \prec x\} \subseteq \{\text{L}, \text{R}\}^*$,
- a set $X_i \in \text{reduces}(L_i)$ by a labelling $f^i_X\colon \{\text{L}, \text{R}\}^* \to A \sqcup \{\epsilon\}$ such that $[f^i_X]_i = X_i$,
- a condition $v \prec x$ by $v \in B_x$,
- a condition $x \in X_i$ by checking that the formula defining L_i is true on the labelling f^i_X on the nodes in B_x.

Equation (10.3) says that the quantifications over $X_i \in \text{reduces}(L_i)$ and over proper labellings f^i_X are equivalent. ∎

Proof of Theorem 10.8. Theorem 2.7 from Chap. 9 shows that there is an alphabet A such that for every $i \geq 1$, there is a language $L_i \subseteq A^\omega$ which is definable in MSO+U on ω-words and complete for $\mathbf{\Sigma}^1_i(\{\text{L}, \text{R}\}^\omega)$. Apply Lemma 10.46 to these languages. By their completeness, the classes $\mathcal{X}_1, \mathcal{X}_2, \ldots$ in Lemma 10.46 are exactly the projective hierarchy on $\{\text{L}, \text{R}\}^\omega$, and therefore Theorem 10.8 follows thanks to Remark 10.12. ∎

10.3 Projective Determinacy

In this section we present an example of a non-trivial property that can be expressed in proj-MSO on $\{\text{L}, \text{R}\}^{\leqslant\omega}$. It implies that any algorithm deciding MSO+U on the complete binary tree would have strong set theoretic consequences.

A Gale-Stewart game with winning condition $W \subseteq \{\text{L}, \text{R}\}^\omega$ is the following two-player game. For ω rounds, the players propose directions $d \in \{\text{L}, \text{R}\}$ in an alternating fashion, with the first player proposing a direction in even-numbered rounds, and the second player proposing a directions in odd-numbered rounds. At the end of such a play, an infinite sequence $\alpha = d_0 d_1 \ldots$ is produced, and the first player wins if this sequence belongs to W, otherwise the second player wins. Such a game is called *determined* if either the first or the second player has a winning strategy, see [Kec95, Chap. 20] or [Jec02, Chap. 33] for a broader reference. Martin [Mar75] proved that the games are determined if W is a Borel set (see Theorem 1.2 on page 6).

We show that for every $i > 0$, the statement

"every Gale-Stewart game with a winning condition in $\mathbf{\Sigma}^1_i$ is determined" (10.4)

can be formalised as a sentence φ^i_{det} of proj-MSO on $\{\mathrm{L}, \mathrm{R}\}^{\leqslant\omega}$.

Assume that a formula even(u) (resp. odd(u)) expresses that a given node is at the even (resp. odd) depth in the complete binary tree $\{\mathrm{L}, \mathrm{R}\}^*$. By $s_{\mathrm{L}}(u)$ and $s_{\mathrm{R}}(u)$ we denote the respective successors of u in the tree, i.e. $S_d(u) = ud$.

First, we define that a set of nodes encodes a strategy for the first player in the Gale-Stewart game:

$$\begin{aligned} S_{\mathrm{I}}(\sigma) = \epsilon \in \sigma \wedge \\ &\forall_{u\in\sigma}\ \mathrm{even}(u) \implies (s_{\mathrm{L}}(u) \in \sigma \Leftrightarrow s_{\mathrm{R}}(u) \notin \sigma) \wedge \\ &\forall_{u\in\sigma}\ \mathrm{odd}(u) \implies (s_{\mathrm{L}}(u) \in \sigma \wedge s_{\mathrm{R}}(u) \in \sigma). \end{aligned}$$

The formula $S_{\mathrm{II}}(\sigma)$ defining a strategy for the second player is analogous except that the predicates even and odd are interchanged.

The following formula says that σ is a winning strategy for the first player for a winning condition $W \subseteq \{\mathrm{L}, \mathrm{R}\}^\omega$:

$$\mathrm{win}_{\mathrm{I}}(\sigma, W) = S_{\mathrm{I}}(\sigma) \wedge \forall_{\alpha\in\{\mathrm{L},\mathrm{R}\}^\omega}\ (\forall_{u\prec\alpha}\ u \in \sigma) \Rightarrow \alpha \in W.$$

Similarly we define

$$\mathrm{win}_{\mathrm{II}}(\sigma, W) = S_{\mathrm{II}}(\sigma) \wedge \forall_{\alpha\in\{\mathrm{L},\mathrm{R}\}^\omega}\ (\forall_{u\prec\alpha}\ u \in \sigma) \Rightarrow \alpha \notin W.$$

Finally, Statement (10.4), namely the determinacy of all the Gale-Stewart games with winning conditions in $\mathbf{\Sigma}^1_i$ is expressed by

$$\varphi^1_{\mathrm{det}} \stackrel{\mathrm{def}}{=} \forall_{W\in\mathbf{\Sigma}^1_i}\ \exists_{\sigma\in\mathrm{P}(\{\mathrm{L},\mathrm{R}\}^*)}\ \mathrm{win}_{\mathrm{I}}(\sigma, W) \vee \mathrm{win}_{\mathrm{II}}(\sigma, W).$$

As we show below, the ability to formalise determinacy of Gale-Stewart games with winning conditions in $\mathbf{\Sigma}^1_1$ already indicates that it is unlikely that proj-MSO on $\{\mathrm{L}, \mathrm{R}\}^{\leqslant\omega}$ is decidable.

Indeed, suppose that there is an algorithm P deciding the proj-MSO theory of $\{\mathrm{L}, \mathrm{R}\}^{\leqslant\omega}$ with a correctness proof in ZFC. Note that by Theorem 10.8, this would be the case if there was an algorithm deciding the MSO+U theory of $\{\mathrm{L}, \mathrm{R}\}^*$ with a correctness proof in ZFC. Run the algorithm on φ^1_{det} obtaining an answer, either "yes" or "no". The algorithm together with its proof of correctness and the run on φ^1_{det} form a proof in ZFC resolving Statement (10.4) for $i = 1$. The determinacy of all $\mathbf{\Sigma}^1_1$ games cannot[1] be proved in ZFC, because it does not hold if V=L, see [Jec02, Corollary 25.37 and Sect. 33.9], and therefore P must answer "no" given input φ^1_{det}.

This means that a proof of correctness for P would imply a ZFC proof that Statement (10.4) is false for $i = 1$. Such a possibility is considered very unlikely by set theorists, see [FFMS00] for a discussion of plausible axioms extending the standard set of ZFC axioms.

A similar example regarding the MSO theory of $(\mathbb{R}, \leqslant)$ and the Continuum Hypothesis was provided in [She75].

10.4 Undecidability of proj-MSO on $\{\mathrm{L}, \mathrm{R}\}^\omega$

The undecidability of MSO+U (see Theorem 10.88) follows from the reduction in Theorem 10.8 and Proposition 10.25 below.

Proposition 10.25 (Bojańczyk Gogacz Michalewski S. [BGMS14]). *Assuming that* V=L, *the proj-*MSO *theory of the Cantor set* $(\{\mathrm{L}, \mathrm{R}\}^\omega, \leqslant_{\text{lex}})$ *is undecidable.*

This proposition is not in the scope of the thesis and we do not prove it here in detail. Instead, in this section we show how this result can be obtained by adjusting the reasoning in [She75, Theorem 7.1] by following the suggestion of Shelah, see Quotation (10.1) on page 170 of the thesis.

There are three adjustments needed:

1. Instead of working on the real line $\mathbb{R}$ we use here the Cantor set $\{\mathrm{L}, \mathrm{R}\}^\omega$.
2. We have to repeat the inductive construction of a set Q from [She75 Lemma 7.4] in such a way to guarantee that Q is $\mathbf{\Sigma}^1_n$ for some $n \in \mathbb{N}$.
3. We have to argue that the resulting formula $G(\theta)$ is a proj-MSO formula.

The second adjustment above uses the assumption that V=L to construct a projective set Q. Having done this, it is enough to carefully read the formula $G(\theta)$ of Shelah: it quantifies existentially over sets Q, countable sets D, arbitrary subsets of D, perfects, and intervals. All these quantifiers are projective, see Example 10.7.

10.4.1 Proof of Theorem 10.88

Now we can combine the above results to prove the undecidability result.

Theorem 10.88. *Assuming* V=L, *it is undecidable if a given sentence of* MSO+U *is true in the complete binary tree* $\left(\{\mathrm{L}, \mathrm{R}\}^*, \preceq, \leqslant_{\text{lex}}\right)$.

[1] Except for the case if ZFC is not consistent and it is possible to prove everything in ZFC.

Proof. Assume V=L. In that case the proj-MSO theory of the Cantor set $(\{\mathrm{L}, \mathrm{R}\}^*, \leqslant_{\mathrm{lex}})$ is undecidable by Proposition 10.25. By Remark 10.12 it can be reduced to the proj-MSO theory of $(\{\mathrm{L}, \mathrm{R}\}^{\leqslant\omega}, \preceq, \leqslant_{\mathrm{lex}})$. Theorem 10.8 implies that the latter can be reduced to the MSO+U theory of the complete binary tree. Therefore, this theory is undecidable. ■

10.5 Conclusions

This chapter presents a reduction from a logic called proj-MSO to MSO+U on infinite trees. The reduction involves the topologically hard languages constructed in Chap. 9. As shown in [BGMS14], assuming that V=L, the proj-MSO theory of the Cantor set is undecidable. Therefore, the two results together imply that (assuming V=L) MSO+U logic is undecidable on infinite trees.

As shown in the above chapter, it is possible to express in proj-MSO some deep properties of the universe of set theory. Therefore, any algorithm solving MSO+U on infinite trees would have some remarkable knowledge about this universe. As an example, it is shown that any such algorithm (with its proof of correctness) implies that analytic determinacy is provably false (in ZFC). The latter possibility is considered very unlikely by set theorists. These intuitions are expressed by the following conjecture.

Conjecture 10.8. It is possible to prove in ZFC that the MSO+U theory of the complete binary tree is undecidable.

The undecidability result about proj-MSO makes a strong link between topological complexity and decidability. What is in fact proved in [BGMS14] is that under the assumption that V=L, even a weaker variant of proj-MSO where set quantifiers range over sets up to the sixth level of the projective hierarchy (i.e. $\boldsymbol{\Sigma}^1_6$-sets) is undecidable. On the other hand, if we restrict set quantifiers to $\boldsymbol{\Sigma}^0_2$ then the theory becomes decidable. It somehow justifies the impression that the more complicated sets are allowed, the more undecidable the theory is. It should be related to the following conjecture of Shelah.

Conjecture 10.9 ([She75, Conjecture 7B]). The monadic theory of $(\mathbb{R}, \leqslant)$ where the set quantifiers range over Borel sets is decidable.

As Shelah comments, the above conjecture is motivated by Borel determinacy (that was proved by Martin [Mar75], see Theorem 1.2 on page 6). On the other hand, the assumption that V=L implies that projective determinacy fails. Therefore, one can state the following question.

Question 10.2. Assume that all analytic ($\boldsymbol{\Sigma}^1_1$) games are determined. Does it imply that the monadic second-order theory of $(\mathbb{R}, \leqslant)$ where the set quantifiers range over $\boldsymbol{\Sigma}^1_1$-sets is decidable?

This chapter is based on [BGMS14].

Chapter 11
Separation for ωB- and ωS-regular Languages

In this chapter we study the classes of ωB- and ωS-regular languages, introduced by Bojańczyk and Colcombet in [BC06]. These languages of ω-words are defined as those that can be recognised by a certain model of *counter automata* with asymptotic acceptance condition. Both these classes are strictly contained in the class of MSO+U-definable languages, the advantage of these classes is that they admit effective constructions. A standard example of an ωB-regular language is the following

$$\left\{a^{n_0}ba^{n_1}ba^{n_2}b\ldots :\ \text{the sequence } n_i \text{ is bounded}\right\} \subseteq \{a, b\}^\omega.$$

The main technical contribution of [BC06] states that the complement of an ωB-regular language is effectively ωS-regular and vice versa; and the emptiness problem is decidable for both these classes. Although these languages do not form a Boolean algebra, these properties guarantee some kind of robustness of these two classes.

In this chapter we show that both classes of ωB- and ωS-regular languages admit the separation property with respect to ω-regular languages (see Definition 1.3 on page 24 in Sect. 1.7.5), as expressed by the following theorem.

Theorem 11.9. *If L_1, L_2 are disjoint languages of ω-words both recognised by ωB-(respectively ωS)-automata then there exists an ω-regular language L_{sep} such that*

$$L_1 \subseteq L_{\mathrm{sep}} \quad \text{and} \quad L_2 \subseteq L_{\mathrm{sep}}^{\mathrm{c}}.$$

Additionally, the construction of L_{sep} is effective.

The result is especially interesting since these are two mutually dual classes (see Theorem 11.96) — usually exactly one class from a pair of dual classes has the separation property, see Sect. 1.7.5, page 24.

As a consequence of the separation property we obtain the following corollary.

Corollary 11.3. *If a given language of ω-words L and its complement L^{c} are both ωB-regular (resp. ωS-regular) then L is ω-regular.*

M. Skrzypczak, *Set Theoretic Methods in Automata Theory*, LNCS 9802
DOI: 10.1007/978-3-662-52947-8_11

Proof. Let L be a language of ω-words such that L and L^c are both ωT-regular (for $T \in \{B, S\}$). By Theorem 2.9 there exists an ω-regular language L_{sep} that separates L and L^c. But in that case $L_{sep} = L$ so L is ω-regular. ■

The above corollary was independently known by some researchers in the area (with a proof not involving separation). Nevertheless, to the best of the author's knowledge, it has never been published before [Skr14].

To prove Theorem 2.9 we reduce the separation property of ω-word languages to the case of profinite words. For this purpose we use B- and S-automata introduced in [Col09]. As shown in [Tor12] it is possible to define a language recognised by a B- or S-automaton as a subset of the profinite monoid $\widehat{A^*}$. An intermediate step in our reasoning is proving the separation property for B- and S-regular languages of profinite words.

The chapter is organised as follows. In Sect. 11.1 we introduce basic notions including the profinite monoid $\widehat{A^*}$. Section 11.2 defines the automata models we use. In Sect. 11.3 we prove separation results for languages of profinite words recognised by B- and S-automata. Section 11.4 contains the crucial technical tool, Theorem 11.98, that enables to transfer separation results for languages of profinite words to the case of ω-words. In Sect. 11.5 we use this theorem to show that ωB- and ωS-regular languages have the separation property. Finally, Sect. 11.6 is devoted to conclusions.

11.1 Basic Notions

We work with two models of automata (ωB and ωS) at the same time. Therefore, we introduce a notion ωT to denote one of the models: ωB or ωS. By T we denote the corresponding model of automata on finite words (B or S).

11.1.1 Monoid of Runs

We define here a monoid representing possible *runs* of a non-deterministic automaton. It can be seen as an algebraic formalisation of the structure used by Büchi [Büc62] in his famous complementation lemma. A general introduction to monoids is given in Sect. 1.5.1 (see page 11).

Let $\mathcal{A}$ be a non-deterministic automaton. Define $M_{trans}(\mathcal{A})$ as $\mathsf{P}(Q^{\mathcal{A}} \times Q^{\mathcal{A}})$. Let the neutral element be $\{(q, q) : q \in Q^{\mathcal{A}}\}$ and product:

$$s \cdot s' = \left\{(p, r) : \exists_{q \in Q^{\mathcal{A}}} (p, q) \in s \wedge (q, r) \in s'\right\}.$$

Let $f_{\mathcal{A}} \colon A^* \to M_{trans}(\mathcal{A})$ map a given finite word u to the set of pairs (p, q) such that the automaton $\mathcal{A}$ has a run over u starting in p and ending in q.

It is easy to check that $M_{trans}(\mathcal{A})$ is a finite monoid and $f_{\mathcal{A}}$ is a homomorphism.

11.1.2 Profinite Monoid

In this subsection we introduce the profinite monoid $\widehat{A^*}$. A formal introduction to profinite structures can be found in [Alm03] or [Pin09]. We refer to [Pin09]. A construction of the profinite monoid using purely topological methods is given in [Skr11].

First we provide a construction of the profinite monoid $\widehat{A^*}$. The idea is to enhance the set of all finite words by some *virtual* elements representing sequences of finite words that are more and more similar.

Let $K_0, K_1, \ldots$ be a list of all regular languages of finite words. Let $X = 2^\omega$. Each element $x \in X$ can be seen as a sequence of bits, the bit $x(n)$ indicates whether our *virtual* word belongs to the language K_n.

Define $\mu\colon A^* \to X$ by the following equation:

$$\mu(u)_n = \begin{cases} 1 & \text{if } u \in K_n, \\ 0 & \text{if } u \notin K_n. \end{cases}$$

The function μ defined above is injective. Let $\widehat{A^*} \subseteq X$ be the closure of $\mu(A^*)$ in X with respect to the product topology of X. Therefore, $\widehat{A^*}$ contains $\mu(A^*)$ and the limits of its elements. To simplify the notion we identify $u \in A^*$ with its image $\mu(u) \in \widehat{A^*}$.

Example 11.8 (Proposition 2.5 in [Pin09]). Let $u_n = a^{n!}$ for $n \in \mathbb{N}$. A simple automata-theoretic argument shows that for every regular language K, either almost all words $(u_n)_{n\in\mathbb{N}}$ belong to K or almost all do not belong to K. Therefore, the sequence $(\mu(u_n))_{n\in\mathbb{N}}$ is convergent coordinate-wise in X. The limit of this sequence is an element of $\widehat{A^*} \setminus \mu(A^*)$.

The following fact summarises basic properties of $\widehat{A^*}$.

Fact 11.89 (Proposition 2.1, Proposition 2.4, and Theorem 2.7 in [Pin09]). *$\widehat{A^*}$ is a compact metric space. A^* (formally $\mu(A^*)$) is a countable dense subset of $\widehat{A^*}$. $\widehat{A^*}$ has a structure of a monoid that extends the structure of A^* and the product is continuous.*

It turns out that the operation assigning to every regular language of finite words $K \subseteq A^*$ its topological closure $\overline{K} \subseteq \widehat{A^*}$ has good properties (see Theorem 11.91). Therefore, we introduce the following definition.

Definition 11.22. *A* profinite-regular *language is a subset of $\widehat{A^*}$ of the form $\overline{K}$ for some regular language $K \subseteq A^*$.*

Using this definition, we can denote a generic profinite-regular language as $\overline{K}$ for K ranging over regular languages. Using the definition of μ one can show the following easy fact.

Fact 11.90. *A language of profinite words $M \subseteq \widehat{A^*}$ is profinite-regular if and only if it is of the form*

$$M = \{x \in 2^\omega : x \in \widehat{A^*} \wedge x_n = 1\}, \tag{11.1}$$

for some $n \in \mathbb{N}$. In that case $M = \overline{K_n}$.

The structures of profinite-regular and regular languages are in some sense identical. This is expressed by the following theorem.

Theorem 11.91 (Theorem 2.4 in [Pin09]). *The function $K \mapsto \overline{K} \subseteq \widehat{A^*}$ is an isomorphism of the Boolean algebra of regular languages and the Boolean algebra of profinite-regular languages. Its inverse is $M \mapsto \mu^{-1}(M) \subseteq A^*$ (when identifying A^* with $\mu(A^*)$ we can write $M \mapsto M \cap A^* \subseteq A^*$).*

By the definition of $\widehat{A^*}$ and the fact that regular languages are closed under finite intersection, we obtain the following important fact.

Fact 11.92. *The family of profinite-regular languages is a basis of the topology of $\widehat{A^*}$.*

The topology of $\widehat{A^*}$ is the product topology. Therefore, a sequence of finite words $U = u_0, u_1, \ldots$ is convergent to $u \in \widehat{A^*}$ if and only if $(\mu(u_n))_{n\in\mathbb{N}} \subseteq X$ is convergent coordinate-wise to u. The following fact formulates this condition in a more intuitive way.

Fact 11.93. *A sequence of finite words $U = u_0, u_1, \ldots$ is convergent to $u \in \widehat{A^*}$ if and only if for every profinite-regular language $\overline{K}$ either:*

– *$u \in \overline{K}$ and almost all words u_n belong to K,*
– *$u \notin \overline{K}$ and almost all words u_n do not belong to K.*

The topology of $\widehat{A^*}$ is defined in such a way that it corresponds precisely to profinite-regular languages. The following fact summarises this correspondence.

Fact 11.94 (Proposition 4.2 in [Pin09]). *A language $M \subseteq \widehat{A^*}$ is profinite-regular if and only if it is a closed and open (clopen) subset of $\widehat{A^*}$.*

Proof. First assume that $M = \overline{K}$ is a regular language of profinite words. Equation (11.1) in Fact 11.90 defines a closed and open set.

Now assume that M is a closed and open subset of $\widehat{A^*}$. Recall that profinite-regular languages form a basis for the topology of $\widehat{A^*}$ (Fact 11.92). Since M is open so it is a union of base sets $\bigcup_{j\in J} \overline{K_j}$. Since M is a closed subset of a compact space $\widehat{A^*}$, M is compact. Therefore, only finitely many languages among $\{\overline{K_j}\}_{j\in J}$ form a cover of M. But a finite union of profinite-regular languages is a profinite-regular language. Therefore, M is profinite-regular. ■

11.1.3 Ramsey-Type Arguments

In this section we introduce an extension of Ramsey's theorem (see Sect. 1.5.4, page 13) to the case where colours come from the profinite monoid. To state it formally we use the following definitions.

Definition 11.23. *Assume that* $U = u_0, u_1, \ldots$ *is a sequence of finite words. We say that* $W = w_0, w_1, \ldots$ *is a* grouping *of* U *if there exists an increasing sequence of numbers* $0 = i_0 < i_1 < \ldots$ *such that for every* $n \in \mathbb{N}$ *we have*

$$w_n = u_{i_n} u_{i_n+1} \ldots u_{i_{n+1}-1}.$$

Observe that if $W = w_0, w_1, \ldots$ is a grouping of $U = u_0, u_1, \ldots$ then $u_0 u_1 \cdots = w_0 w_1 \cdots$.

We will use the notion of the f-type of a decomposition $\alpha = u_0 u_1 \ldots$ from Definition 1.1 on page 13. Recall also that $t = (s, e)$ is called a *linked pair* if $s \cdot e = s$ and $e \cdot e = e$. By the definition, if $t = (s, e)$ is an f-type of a decomposition of some ω-word then t is a linked pair.

Note that if U is a decomposition of an ω-word α and U is of f-type $t = (s, e)$ then every grouping of U is also a decomposition of α of f-type t. The notion of grouping introduces a stronger version of convergence.

Definition 11.24. *We say that a sequence of finite words* $U = u_0, u_1, \ldots$ *is* strongly convergent *to a profinite word* u *if every grouping of* U *is convergent to* u.

The following result is an extension of Ramsey's theorem to the case of the profinite monoid.

Theorem 11.95 (Bojańczyk Kopczyński Toruńczyk [BKT12]). *Let* $U = u_0, u_1, \ldots$ *be an infinite sequence of finite words. There exists a grouping* Z *of* U *such that* Z *strongly converges in* $\widehat{A^*}$.

For the sake of completeness we give a proof of this fact below. The theorem holds in general, where instead of $\widehat{A^*}$ is any compact metric monoid. Also, the notion of convergence can be strengthened in the thesis of the theorem: all the groupings of U converge in a *uniform way*. In this chapter we use only the above, simplified form.

Proof. Let K be a regular language and $W = w_0, w_1, \ldots$ be a sequence of finite words. Define a function $\alpha_{K,W} \colon [\mathbb{N}]^2 \to \{0, 1\}$ that takes a pair of numbers $i < j$ and returns 1 if and only if $w_i w_{i+1} \ldots w_{j-1}$ belongs to K. By Theorem 1.1 from page 3, there exists a monochromatic set $S \subseteq \mathbb{N}$ with colour $c \in \{0, 1\}$ such that for every pair $i < j \in S$ we have $\alpha_{K,W}(\{i, j\}) = c$.

Now, take a sequence of finite words U. We will construct a sequence of words z_i using a diagonal construction. Let $K_0, K_1, \ldots$ be an enumeration of all regular languages and let $U^0 = U$. We proceed by induction for $i = 0, 1, \ldots$. Assume that after i'th step a sequence $U^i = u^i_0, u^i_1, \ldots$ is defined. First define z_i as u^i_0. Now, let $S = \{n_0, n_1, \ldots\}$ be an infinite monochromatic set with respect to α_{K_i, U^i}. Define

$$U^{i+1} = \left(u^i_{n_0} u^i_{n_0+1} \dots u^i_{n_1-1}\right), \left(u^i_{n_1} u^i_{n_1+1} \dots u^i_{n_2-1}\right), \left(u^i_{n_2} u^i_{n_2+1} \dots u^i_{n_3-1}\right), \dots$$

Note that U^{i+1} is a suffix of a grouping of U^i. Since S is monochromatic and by the definition of $\alpha_{K,W}$, we know that:
($*$) For every grouping of U^{i+1} either all words in the grouping belong to K_i or all of them do not belong.

We claim that our sequence $Z = z_0, z_1, \dots$ is strongly convergent. Let W be a grouping of Z and let $K = K_i$ be a regular language. Observe that almost all words in W (all except first at most i words) are obtained by grouping words in U^{i+1}. Therefore, by ($*$), either almost all words of W belong to K or almost all of them do not belong to K. Fact 11.93 implies that W is convergent in $\widehat{A^*}$.

Now observe that almost all words in W belong to K_i if and only if almost all the words in Z belong to K_i. Therefore, the limit of W does not depend on the choice of W. It means that Z is strongly convergent in $\widehat{A^*}$. ∎

11.1.4 Notation

In this chapter we deal with three types of languages: of finite words, of profinite words, and of ω-words. To simplify reading of the chapter, we use the following conventions:

- finite and profinite words are denoted by u, w,
- sequences of finite words are denoted by U, W, Z,
- ω-words are denoted by α, β,
- regular languages of finite words are denoted by K,
- profinite-regular languages are, using Theorem 11.91, denoted by $\overline{K}$,
- general languages of profinite words are denoted by M,
- languages of ω-words (both ω-regular and not) are denoted by L.

11.2 Automata

In this section we provide definitions of four kinds of automata: B-, S-, ωB- and ωS-automata. B- and S-automata read finite words while ωB- and ωS-automata read ω-words.

The ωB- and ωS-automata models were introduced in [BC06], we follow the definitions from this work. The B- and S-automata models were defined in [Col09]. For the sake of simplicity, we use only the operations {**nil**, **inc**, **reset**} (without the *check* operation). As noted in Remark 1 in [Col09] (see also [BC06]), this restriction does not influence the expressive power.

The four automata models we study here are part of a more general theory of regular cost functions that is developed mainly by Colcombet [Col09, Col13]. In particular, the theory of B- and S-automata has been extended to finite trees in [CL10].

All four automata models we deal with are built on the basis of a *counter automaton*. The difference is the acceptance condition that we introduce later.

Definition 11.25. A counter automaton *is a tuple* $\mathcal{A} = \langle A^{\mathcal{A}}, Q^{\mathcal{A}}, I^{\mathcal{A}}, \Gamma^{\mathcal{A}}, \delta^{\mathcal{A}} \rangle$, *where:*

- $A^{\mathcal{A}}$ *is an input alphabet,*
- $Q^{\mathcal{A}}$ *is a finite set of states,*
- $I^{\mathcal{A}} \subseteq Q^{\mathcal{A}}$ *is a set of initial states,*
- $\Gamma^{\mathcal{A}}$ *is a finite set of counters,*
- $\delta^{\mathcal{A}} \subseteq Q^{\mathcal{A}} \times A^{\mathcal{A}} \times \{\mathbf{nil}, \mathbf{inc}, \mathbf{reset}\}^{\Gamma^{\mathcal{A}}} \times Q^{\mathcal{A}}$ *is a transition relation.*

All counters store natural numbers and cannot be read during a run. The values of the counters are only used in an acceptance condition.

In the initial configuration all counters equal 0. A transition $(p, a, o, q) \in \delta^{\mathcal{A}}$ (sometimes denoted $p \xrightarrow{a,o} q$) means that if the automaton is in a state p and reads a letter a then it can perform counter operations o and go to the state q. For a counter $c \in \Gamma^{\mathcal{A}}$ a counter operation $o(c)$ can:

$o(c) = \mathbf{nil}$	leave the counter value unchanged,
$o(c) = \mathbf{inc}$	increment the counter value by one,
$o(c) = \mathbf{reset}$	reset the counter value to 0.

A run ρ of the automaton $\mathcal{A}$ over a word (finite or infinite) is a sequence of transitions as for standard non-deterministic automata. Given a run ρ, a counter $c \in \Gamma^{\mathcal{A}}$, and a position r_c of a word where the counter c is reset, we define $\mathrm{val}(c, \rho, r_c)$ as the value stored in the counter c at the moment before the reset r_c in ρ.

To simplify the constructions we allow ϵ-transitions in our automata. The only requirement is that there is no cycle consisting of ϵ-transitions only. ϵ-transitions can be removed using non-determinism of an automaton and by combining a sequence of counter operations into one operation. Such a modification may change the exact values of counters, for instance when we replace **inc**, **reset** by **reset**. However, the limitary properties of the counters are preserved (the values may be disturbed only by a linear factor).

11.2.1 ωB- and ωS-automata

First we deal with automata for ω-words, following the definitions in [BC06]. An ωT-*automaton* (for $\omega\mathrm{T} \in \{\omega\mathrm{B}, \omega\mathrm{S}\}$) is just a counter automaton. A run ρ of an ωT-automaton over an ω-word α is accepting if it starts in an initial state in $I^{\mathcal{A}}$, every counter is reset infinitely many times, and the following condition is satisfied:

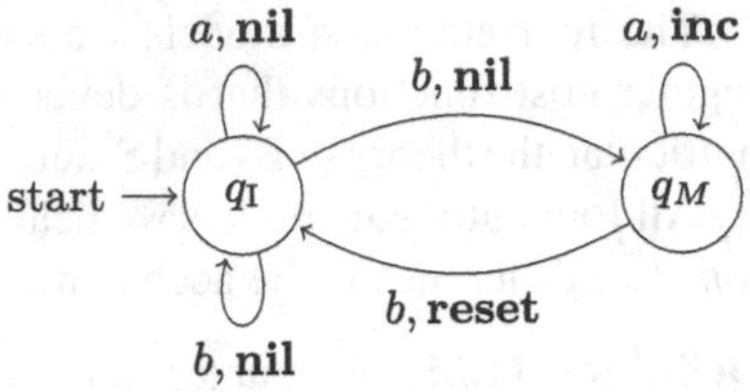

Fig. 11.1 An example of an ωB-automaton $\mathcal{A}_{\omega\text{B}}$.

ωB-automaton the values of all counters are bounded during the run,
ωS-automaton for every counter c the values of c during resets in ρ tend to infinity (i.e. the limit of the values of c is ∞).

An ωT-automaton $\mathcal{A}$ accepts an ω-word if it has an accepting run on it. The set of all ω-words accepted by $\mathcal{A}$ is denoted $L(\mathcal{A})$.

Example 11.9. Consider the ωB-automaton $\mathcal{A}_{\omega\text{B}}$ depicted on Fig. 11.1. $\mathcal{A}_{\omega\text{B}}$ guesses (by moving to the state q_M) to measure the length of some blocks of letters a. It accepts an ω-word α if and only if it is of the form

$$\alpha = a^{n_0} b a^{n_1} b \ldots \quad \text{with} \quad \liminf_{i\to\infty} n_i < \infty.$$

We can also treat $\mathcal{A}_{\omega\text{B}}$ as an ωS-automaton. In that case it accepts an ω-word α if and only if it is of the form

$$\alpha = a^{n_0} b a^{n_1} b \ldots \quad \text{with} \quad \limsup_{i\to\infty} n_i = \infty.$$

It is easy to check that a non-deterministic Büchi automaton can be transformed into an equivalent ωB- (resp. ωS)-automaton. Therefore, all ω-regular languages are both ωB- and ωS-regular.

The following theorem summarizes properties of ωB- and ωS-regular languages.

Theorem 11.96 ([BC06, Theorem 4.1]). *The complement of an ωB-regular language is effectively ωS-regular and vice versa.*

The emptiness problem is decidable for ωB- and ωS-regular languages.

11.2.2 B- and S-automata

In the finite word models the situation is a little more complicated than in the ωB- and ωS-automata models. The automaton not only accepts or rejects a given word but also it assigns a *value* to a word.

Formally, a T-*automaton* (for T $\in$ {B, S}) is a counter automaton that is additionally equipped with a set of final states $F^{\mathcal{A}} \subseteq Q^{\mathcal{A}}$. An accepting run ρ of an automaton over a finite word u is a sequence of transitions starting in some initial state in $I^{\mathcal{A}}$ and ending in some final state in $F^{\mathcal{A}}$.

The following equations define $\mathrm{val}(\mathcal{A}, u)$ — the value assigned to a given finite word by a given automaton. We use the convention that if a set of values is empty then the minimum of this set is ∞ and the maximum is 0. The variable ρ ranges over all accepting runs, c ranges over counters in $\Gamma^{\mathcal{A}}$, while r_c ranges over positions where the counter c is reset in ρ. As noted at the beginning of this section, we do not allow explicit *check* operation, we only care about the values of the counters before resets.

B-automaton $\mathcal{A}_{\mathrm{B}}$

$$\mathrm{val}(\mathcal{A}_{\mathrm{B}}, u) = \min_{\rho} \mathrm{val}(\rho) \quad \text{and} \quad \mathrm{val}(\rho) = \max_{c} \max_{r_c} \ \mathrm{val}(c, \rho, r_c),$$

S-automaton $\mathcal{A}_{\mathrm{S}}$

$$\mathrm{val}(\mathcal{A}_{\mathrm{S}}, u) = \max_{\rho} \mathrm{val}(\rho) \quad \text{and} \quad \mathrm{val}(\rho) = \min_{c} \min_{r_c} \ \mathrm{val}(c, \rho, r_c).$$

The following simple observation is crucial in the subsequent definitions.

Lemma 11.47. *For a given number n, a* B*-automaton $\mathcal{A}_{\mathrm{B}}$, and an* S*-automaton $\mathcal{A}_{\mathrm{S}}$ the following languages of finite words are regular:*

$$L(\mathcal{A}_{\mathrm{B}} \leqslant n) \stackrel{def}{=} \{u : \ \mathrm{val}(\mathcal{A}_{\mathrm{B}}, u) \leqslant n\},$$
$$L(\mathcal{A}_{\mathrm{S}} > n) \stackrel{def}{=} \{u : \ \mathrm{val}(\mathcal{A}_{\mathrm{S}}, u) > n\}.$$

Proof. We can encode a bounded valuation of the counters into a state of a finite automaton. ■

11.2.3 Languages

The above definitions give semantics of a T-automaton in terms of a function

$$\mathrm{val}(\mathcal{A}, .) \colon \left(A^{\mathcal{A}}\right)^* \to \mathbb{N} \sqcup \{\infty\}.$$

As noted in [Tor12], it is possible to define the language recognised by such an automaton as a subset of the profinite monoid $\widehat{A^*}$. We successively define it for B-automata and S-automata. In both cases the construction is justified by Lemma 11.47.

B **case:** Fix a B-automaton $\mathcal{A}_{\mathrm{B}}$ and define

$$L(\mathcal{A}_{\mathrm{B}}) \stackrel{\mathrm{def}}{=} \bigcup_{n \in \mathbb{N}} \overline{L(\mathcal{A}_{\mathrm{B}} \leqslant n)} \subseteq \widehat{A^*}. \tag{11.2}$$

S case: Fix an S-automaton $\mathcal{A}_S$ and define

$$L(\mathcal{A}_S) \stackrel{\text{def}}{=} \bigcap_{n\in\mathbb{N}} \overline{L(\mathcal{A}_S > n)} \subseteq \widehat{A^*}. \tag{11.3}$$

Note that the sequences of languages in the above equations are monotone: increasing in (11.2) and decreasing in (11.3).

There exists another, equivalent way of defining languages recognised by these automata [Tor12]. One can observe that the function $\mathrm{val}(\mathcal{A}, .)$ assigning to every finite word its value has a unique continuous extension on $\widehat{A^*}$. The languages recognised by B- and S-automata can be defined as $\mathrm{val}(\mathcal{A}, .)^{-1}(\mathbb{N})$ and $\mathrm{val}(\mathcal{A}, .)^{-1}(\{\infty\})$ respectively. In this chapter we only refer to the definitions (11.2) and (11.3).

Example 11.10. Consider the S-automaton $\mathcal{A}_S$ depicted in Fig. 11.2. The automaton measures the number of letters a in a given word. Then it guesses that the word is finished and moves to the accepting state. For every finite word u the value $\mathrm{val}(\mathcal{A}_S, u)$ equals the number of letters a in u.

The language $L(\mathcal{A}_S)$ does not contain any finite word. It contains a profinite word u if for every n the word u belongs to the profinite-regular language defined by the formula "the word contains more than n letters a" (i.e. $u \in \overline{L(\mathcal{A}_S > n)}$). In particular, the limit of the sequence $(a^{n!})_{n\in\mathbb{N}}$ from Example 11.8 belongs to $L(\mathcal{A}_S)$.

Lemma 11.48. *Every* B*-regular language is an open subset of* $\widehat{A^*}$ *and dually every* S*-regular language is closed.*

Proof. By Eqs. (11.2) and (11.3), a B-regular language is a sum of profinite-regular languages and an S-regular language is an intersection of profinite-regular languages. By Fact 11.94, profinite-regular languages are closed and open, therefore their sum is open and the intersection is closed. ■

The converse of Lemma 11.48 is false as there are uncountably many open subsets of $\widehat{A^*}$ — there are some open subsets of $\widehat{A^*}$ that are not B-regular.

We finish the definitions of automata models by recalling the following theorem.

Theorem 11.97 (Fact 2.6 and Corollary 3.4 in [BC06], Theorem 8 and paragraph *Closure properties* in [Tor12]). *Let* T $\in$ {B, S, ωB, ωS}*. The class of* T*-regular languages is effectively closed under union and intersection. The emptiness problem for* T*-regular languages is decidable.*

Therefore, it is decidable whether given two T*-regular languages are disjoint.*

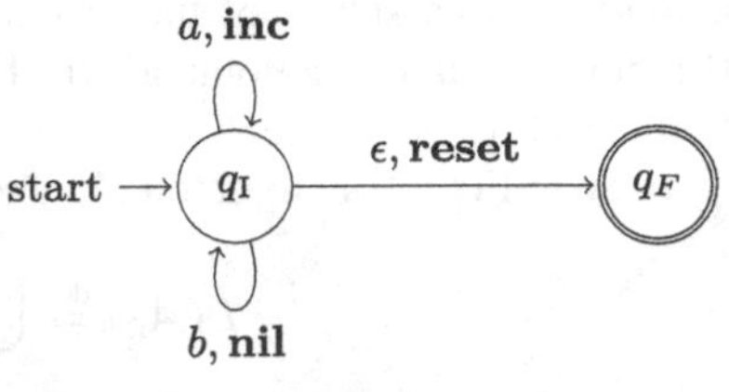

Fig. 11.2 An example of an S-automaton $\mathcal{A}_S$.

11.3 Separation for Profinite Languages

In this section we show the following proposition.

Proposition 11.27. *Let* T $\in$ {B, S}. *Assume that languages of profinite words* $M_1, M_2 \subseteq \widehat{A^*}$ *are recognised by* T*-automata and* $M_1 \cap M_2 = \emptyset$. *Then there exists a profinite-regular language* $K_{\mathrm{sep}} \subseteq \widehat{A^*}$ *such that*

$$M_1 \subseteq \overline{K_{\mathrm{sep}}} \text{ and } M_2 \subseteq \overline{K_{\mathrm{sep}}}^{\mathrm{c}}.$$

Additionally, the language $\overline{K_{\mathrm{sep}}}$ *can be computed effectively basing on* M_1 *and* M_2.

The proof of the proposition consists of two parts, one for each of the two cases of T $\in$ {B, S}: Lemma 11.49 and Proposition 11.28.

First we prove the case when T = S. The presented proof uses a general topological fact: the separation property of closed (i.e. $\mathbf{\Pi}^0_1$) sets in a zero-dimensional Polish space (see Sect. 1.6, page 14 for a definition of these spaces).

Lemma 11.49. *A pair of disjoint* S*-regular languages of profinite words can be separated by a profinite-regular language.*

Proof. Take two S-regular languages $M_1, M_2 \subseteq \widehat{A^*}$.

Observe that $\widehat{A^*}$ is a closed subset of a zero-dimensional Polish space 2^ω, therefore $\widehat{A^*}$ is also zero-dimensional Polish space. Therefore, the $\mathbf{\Pi}^0_1$-separation property holds for $\widehat{A^*}$ (see [Kec95 Theorem 22.16]). By Lemma 11.48 every S-regular language is $\mathbf{\Pi}^0_1$ in $\widehat{A^*}$, therefore M_1, M_2 can be separated in $\widehat{A^*}$ by a set M_{sep} that is closed and open in $\widehat{A^*}$. By Fact 11.94, the language M_{sep} is profinite-regular. ∎

Instead of using the $\mathbf{\Pi}^0_1$-separation property, one can provide the following straightforward argument that uses the compactness of $\widehat{A^*}$. We know that M_1 is a closed subset of a compact space $\widehat{A^*}$ so M_1 is compact itself. Assume that M_2 is recognised by an S-automaton $\mathcal{A}_S$. By (11.3) we obtain

$$M_2 = \bigcap_{n\in\mathbb{N}} \overline{L(\mathcal{A}_S > n)} \subseteq \widehat{A^*}.$$

For $n \in \mathbb{N}$ define $N_n \stackrel{\mathrm{def}}{=} \overline{L(\mathcal{A}_S > n)}^{\mathrm{c}}$ — the complement of the profinite-regular language $\overline{L(\mathcal{A}_S > n)}$. Clearly $M_1 \subseteq \bigcup_n N_n$ because M_1 and M_2 are disjoint. Fact 11.94 and Lemma 11.47 imply that the sets N_n are open subsets of $\widehat{A^*}$. Therefore, the family $(N_n)_{n\in\mathbb{N}}$ is an open cover of M_1. Since M_1 is compact, there is $n_0 \in \mathbb{N}$ such that

$$M_1 \subseteq N_0 \cup N_1 \cup \ldots \cup N_{n_0} = N_{n_0}.$$

Therefore, N_{n_0} is a profinite-regular language that separates M_1 and M_2.

Remark 11.13. The language N_{n_0} can be computed effectively.

Proof. It is enough to observe that n_0 can be taken as the minimal n such that M_1 does not intersect the profinite-regular language $\overline{L(\mathcal{A}_S > n)}$. Such n exists by the above argument. ■

Now we proceed with the separation property for B-regular languages. By Lemma 11.48 we know that B-regular languages are open sets in $\widehat{A^*}$. An easy exercise shows that in general open sets do not have the separation property. Thus, to show the following proposition we need an argument that is a bit more involved than in the case of S-regular languages.

Proposition 11.28. *A pair of disjoint* B*-regular languages of profinite words can be separated by a profinite-regular language.*

We obtain the above proposition by applying the following observation.

Lemma 11.50. *For every* B*-regular language* $M_B \subseteq \widehat{A^*}$ *there exists a profinite-regular language* $\overline{K_R} \subseteq \widehat{A^*}$ *such that*

$$M_B \subseteq \overline{K_R} \quad \text{and} \quad M_B \cap A^* \;=\; \overline{K_R} \cap A^*.$$

Moreover, the language $\overline{K_R}$ *can be computed effectively.*

Proof. Take a B-automaton $\mathcal{A}_B$ recognising M_B. Define a new automaton $\mathcal{A}_R$ by removing from $\mathcal{A}_B$ all the counters and all the counter operations. What remains are transitions, initial states, and final states. Put $\overline{K_R} = \overline{L(\mathcal{A}_R)} \subseteq \widehat{A^*}$. Of course $M_B \subseteq \overline{L(\mathcal{A}_R)}$ by the definition of M_B. Clearly $\overline{L(\mathcal{A}_R)} \cap A^* = L(\mathcal{A}_R)$ by Theorem 11.91. What remains to show is that $L(\mathcal{A}_R) \subseteq M_B$.

Take a finite word $u \in L(\mathcal{A}_R)$. Observe that $\mathcal{A}_B$ has an accepting run on u because $u \in L(\mathcal{A}_R)$. So $\mathrm{val}(\mathcal{A}_B, u) \leqslant |u|$ because $\mathcal{A}_B$ cannot do more increments than the number of positions of the word. Therefore $u \in M_B$. ■

Poof of Proposition 11.28. Take two disjoint B-regular languages $M_1, M_2 \subseteq \widehat{A^*}$. Define $\overline{K_{\mathrm{sep}}}$ to be the language $\overline{K_R}$ from Lemma 11.50 for M_1. Thus we know that $M_1 \subseteq \overline{K_R}$. We only need to show that $M_2 \cap \overline{K_R} = \emptyset$. Assume the contrary, that $M_I \stackrel{\text{def}}{=} M_2 \cap \overline{K_R} \neq \emptyset$. Since B-regular languages are open sets in $\widehat{A^*}$, M_I is an open set. Since A^* is dense in $\widehat{A^*}$ so M_I contains a finite word $u \in A^*$. But by the definition of $\overline{K_R}$ in that case $u \in M_1$. So $u \in M_1 \cap M_2$ — a contradiction to the disjointness of M_1, M_2. ■

Remark 11.14. Both separation results for B- and S-regular languages are effective: there is an algorithm that inputs two counter automata, verifies that the intersection of the languages is empty, and outputs an automaton recognising a separating language.

Proof. By Theorem 11.97 it is decidable if two B-(resp. S)-regular languages are disjoint. As observed in Remark 11.13 and Lemma 11.50, both constructions can be performed effectively. ■

This concludes the proof of Proposition 11.27 in both cases T = B and T = S.

11.4 Reduction

This section contains a proof of our crucial technical tool — Theorem 11.98. It is inspired by the *reduction theorem* from [Tor12], however, the statements of these theorems are incomparable.

Intuitively, ωB- and ωS-automata are composed of two *independent* parts, we can call them the ω-regular part and the asymptotic part. The ω-regular part corresponds to states and transitions of the automaton, while the asymptotic part represents quantitative conditions that can be measured by counters. In this section we show how to formally state this division. It can be seen as an extension of the technique presented in [BC06].

Recall from Sect. 11.1.1 (see page 178) that $M_{\text{trans}}(\mathcal{A})$ is the monoid of state transformations of a non-deterministic automaton $\mathcal{A}$. The canonical homomorphism from finite words to $M_{\text{trans}}(\mathcal{A})$ is denoted $f_{\mathcal{A}}$.

Theorem 11.98. *Let* T $\in$ {B, S}. *Fix an* ωT*-automaton* $\mathcal{A}$ *and a linked pair* $t = (s, e)$ *in the trace monoid* $M_{\text{trans}}(\mathcal{A})$. *There exists a* T*-regular language of profinite words* $M_t \subseteq \widehat{A^*}$ *with the following property:*

If α *is an* ω*-word and* $U = u_0, u_1, \ldots$ *is a decomposition of* α *of type* t *then the following conditions are equivalent:*

1. $\alpha \in L(\mathcal{A})$,
2. *there exists a grouping* W *of* U *that strongly converges to a profinite word* $w \in M_t$,
3. *there exists a grouping* W *of* U *that converges to a profinite word* $w \in M_t$.

Additionally, one can ensure that $M_t \subseteq \overline{f_{\mathcal{A}}^{-1}(e)}$. *The construction of a* T*-automaton recognising* M_t *is effective given* $\mathcal{A}$ *and* t.

The rest of this section is devoted to showing the above theorem. We fix for the whole proof an ωT-automaton $\mathcal{A} = \langle A, Q, I, \Gamma, \delta\rangle$ and a type $t = (s, e)$ in $M_{\text{trans}}(\mathcal{A})$.

Intuitively, the requirement for a decomposition U to be of the type t corresponds to the ω-regular part of $\mathcal{A}$ while the convergence of U to an element of M_t takes care of the asymptotic part of $\mathcal{A}$.

Let us put $K_e = f_{\mathcal{A}}^{-1}(e)$ and assume that $\mathcal{B}_e = \langle A, Q_e, \{q_{I,e}\}, \delta_e, F_e\rangle$ is a deterministic finite automaton recognising the regular language K_e. We will ensure that $M_t \subseteq \overline{K_e}$.

First we show how to construct a language M_t, later we prove its properties. The definition of M_t depends on whether T = B or T = S. The first case is a bit simpler.

Case T = B The language M_t is obtained as the union of finitely many B-regular languages indexed by states $q \in Q$:

$$M_t = \bigcup_{q \in Q} L(\mathcal{A}_q),$$

for B-automata $\mathcal{A}_q$ that we describe below. Intuitively, an automaton $\mathcal{A}_q$ measures *loops* in $\mathcal{A}$ starting and ending in q.

If for no $q_0 \in I$ we have $(q_0, q) \in s$ or if $(q, q) \notin e$ then $L(\mathcal{A}_q) = \varnothing$. Assume otherwise. First we give an informal definition of $\mathcal{A}_q$:

- it is obtained from $\mathcal{A}$ by interpreting it as a finite word B-automaton,
- it has initial and final state set to q,
- it checks that all the counters are reset in a given word,
- it checks that a given word belongs to K_e,
- it resets all the counters at the end of the word.

Now we give a precise definition of $\mathcal{A}_q = \langle A, Q_q, I_q, \Gamma_q, \delta_q, F_q \rangle$. Let:

- $Q_q = \{\star\} \sqcup Q \times Q_e \times \{\bot, \top\}^\Gamma$,
- $I_q = \{(q, q_{I,e}, (\bot, \bot, \ldots, \bot))\}$,
- $\Gamma_q = \Gamma$,
- $F_q = \{\star\}$,

and let δ_q contain the following transitions:

- $(p, r, b) \xrightarrow{a,o} (p', r', b')$ if $p \xrightarrow{a,o} p' \in \delta$, $r \xrightarrow{a} r' \in \delta_e$ and for every $c \in \Gamma$ we have $b'(c) = b(c) \vee (o(c) = \mathbf{reset})$,
- $(q, r, (\top, \top, \ldots, \top)) \xrightarrow{\epsilon,o} \star$ for $o = (\mathbf{reset}, \mathbf{reset}, \ldots, \mathbf{reset})$ if $r \in F_e$.

The state $\star$ is the only final state used to perform the reset at the end of a word. During a run, the automaton $\mathcal{A}_q$ simulates $\mathcal{A}$ and $\mathcal{B}_e$ in parallel, using Q and Q_e. Additionally, a vector in $\{\bot, \top\}^\Gamma$ denotes for every counter whether it was already reset in a word or not.

Case T = S In that case the language M_t is obtained as the union of finitely many S-regular languages indexed by pairs $(q, \tau) \in Q \times \{\leftarrow, \rightarrow\}^\Gamma$:

$$M_t = \bigcup_{(q,\tau)} L(\mathcal{A}_{q,\tau}).$$

Intuitively, an automaton $\mathcal{A}_{q,\tau}$ recognises loops $q \rightarrow^* q$ as before. Additionally, the vector τ denotes whether a given counter $c \in \Gamma$ obtains bigger values before the first reset ($\tau(c) =\rightarrow$) or after the last reset ($\tau(c) =\leftarrow$) on a given finite word. The following definition formalises this property. A similar technique of assigning a *reset type* to a finite run can be found in [BC06].

Definition 11.26. *Let ρ be a run of some counter automaton $\mathcal{A}$ over an ω-word α. Let $k \in \mathbb{N}$ be a position in α and let $c \in \Gamma$ be a counter of $\mathcal{A}$. Let:*

- *V_{L} be the number of increments of c between the last reset before k and k,*
- *V_{R} be the number of increments of c between k and the first reset after k.*

If there is no reset of c at some side of k then the respective value is 0. Define the end-type *of c on ρ in k (denoted as* $\mathrm{Etp}(c, \rho, k)$*) by the following equation:*

$$\mathrm{Etp}(c, \rho, k) = \begin{cases} \rightarrow & \text{if } V_{\mathrm{L}} < V_{\mathrm{R}}, \\ \leftarrow & \text{if } V_{\mathrm{L}} \geqslant V_{\mathrm{R}}. \end{cases}$$

As before if for no $q_0 \in I$, we have $(q_0, q) \in s$ or if $(q, q) \notin e$ then $L(\mathcal{A}_{q,\tau}) = \varnothing$. Assume otherwise. We start with an informal definition of $\mathcal{A}_{q,\tau}$:

- it is obtained from $\mathcal{A}$ by interpreting it as a finite word S-automaton,
- it has initial and final state set to q,
- it checks that all the counters are reset in a given word,
- it checks that a given word belongs to K_e,
- for every counter $c \in \Gamma$:
  - if $\tau(c) = \leftarrow$ then $A_{q,\tau}$ skips the first reset of c and all the previous increments of c but resets c at the end of a given word,
  - if $\tau(c) = \rightarrow$ then $A_{q,\tau}$ acts on c exactly as $\mathcal{A}$ (with no additional reset at the end of the word).

Formally, let $\mathcal{A}_{q,\tau} = \langle A, Q_{q,\tau}, I_{q,\tau}, \Gamma_{q,\tau}, \delta_{q,\tau}, F_{q,\tau} \rangle$ such that

- $Q_{q,\tau} = \{\star\} \sqcup Q \times Q_e \times \{\bot, \top\}^{\Gamma}$,
- $I_{q,\tau} = \{(q, q_{I,e}, (\bot, \bot, \ldots, \bot))\}$,
- $\Gamma_{q,\tau} = \Gamma$,
- $F_{q,\tau} = \{\star\}$,

and $\delta_{q,\tau}$ contains the following transitions:

- $(p, r, b) \xrightarrow{a,o'} (p', r', b')$ if $p \xrightarrow{a,o} p' \in \delta$, $r \xrightarrow{a} r' \in \delta_e$, and for every $c \in \Gamma$ we have:
  - $b'(c) = b(c) \vee (o(c) = \textbf{reset})$,
  - if $b(c) = \bot$ and $\tau(c) = \leftarrow$ then $o'(c) = \textbf{nil}$, otherwise $o'(c) = o(c)$,
- $(q, r, (\top, \top, \ldots, \top)) \xrightarrow{\epsilon,o} \star$ if $r \in F_e$ and for every $c \in \Gamma$ we have $o(c) = \textbf{reset}$ if $\tau(c) = \leftarrow$ and $o(c) = \textbf{nil}$ otherwise.

Now we proceed with the proof that the above constructions give us the desired language M_t. First note that in both cases the constructed automata explicitly verify that a given word belongs to K_e. Therefore, $M_t \subseteq \overline{K_e}$.

We start by taking an ω-word α and its decomposition $U = u_0, u_1, \ldots$ of the type t.

11.4.1 Implication (1) ⇒ (2)

We need to prove that if $\alpha \in L(\mathcal{A})$ and u is a decomposition of α of type t then there exists a grouping W of U that strongly converges to a profinite word $w \in M_t$.

Assume that there exists an accepting run ρ of $\mathcal{A}$ over α. We want to construct a grouping $W = w_0, w_1, \ldots$ of U such that:

S.1 for $n > 0$ we have $w_n \in K_e$,
S.2 all counters in Γ are reset by ρ in every word w_n,

S.3 the state that occurs in the run ρ at the end-points of all the words w_n is some fixed state $q \in Q$,
S.4 there exists a vector $\tau \in \{\leftarrow, \rightarrow\}^\Gamma$ such that for every counter c and every position k between successive words w_n, w_{n+1} in α we have $\mathrm{Etp}(c, \rho, k) = \tau(c)$,
S.5 the sequence of words W is strongly convergent to some profinite word w.

The grouping Z is obtained in steps. Observe that all the above properties are preserved when taking a grouping of a sequence. Condition S.1 is already satisfied by the sequence U. First, we group words of U in such a way to satisfy Condition S.2 using the fact that the run ρ is accepting. Then we further group the sequence to satisfy Conditions S.3 and S.4 — some state and value of Etp must appear in infinitely many end-points. Finally, we apply Theorem 11.95 to group the sequence into a strongly convergent one.

Now, it suffices to show that $w \in M_t$. First, observe that ρ is a witness that there is a path from I to q and from q to q in $\mathcal{A}$.

We consider two cases:

Case T = B Since ρ is accepting, there exists a constant l such that the values of all counters during ρ are bounded by l. We show that for every $n > 0$ we have $w_n \in L(\mathcal{A}_q \leqslant l)$. It implies that $w \in \overline{L(\mathcal{A}_q \leqslant l)}$ and therefore $w \in L(\mathcal{A}_q) \subseteq M_t$.
Observe that ρ induces a run ρ_n of $\mathcal{A}_q$ on w_n. By Conditions S.1, S.2, and S.3 we know that ρ_n is an accepting run of $\mathcal{A}_q$ — it starts in the only initial state and ends in $\star$. Since $\mathcal{A}_q$ simulates all the resets of $\mathcal{A}$, we know that $\mathrm{val}(\rho_n) \leqslant l$ and therefore $\mathrm{val}(\mathcal{A}_q, w_n) \leqslant l$.

Case T = S We show that for every $l \in \mathbb{N}$ the sequence W from some point on satisfies $\mathrm{val}(\mathcal{A}_{q,\tau}, w_n) > \frac{l}{2}$. It implies that for every l we have $w \in \overline{L(\mathcal{A}_{q,\tau} > l)}$ and therefore $w \in L(\mathcal{A}_{q,\tau})$.
Since ρ is accepting, for every constant l, from some point on, all the counters are reset with a value greater than l. Assume that the last reset with the value at most l occurs before the word w_N. We show that for $n \geqslant N$ we have $\mathrm{val}(\mathcal{A}_{q,\tau}, w_n) > \frac{l}{2}$. Let ρ'_n be the sequence of transitions of ρ on w_n. Observe that ρ'_n induces a run ρ_n of $\mathcal{A}_{q,\tau}$ on w_n. As before, ρ_n is accepting by Conditions S.1, S.3, and S.2. Take a counter $c \in \Gamma$ and a reset of this counter r_c in ρ_n. Consider the following cases, recalling Definition 11.26:

- r_c corresponds to the first reset of c in the run ρ'_n. Since $\mathcal{A}_{q,\tau}$ did not skip r_c, $\tau(c) = \rightarrow$. Therefore, c has more increments after the beginning of w_n than before it in ρ. Therefore $\mathrm{val}(c, \rho_n, r_c) > \frac{l}{2}$.
- r_c corresponds to a reset of c in the run ρ'_n but not the first one. In that case $\mathrm{val}(c, \rho_n, r_c) = \mathrm{val}(c, \rho'_n, r_c) > l$.
- r_c is the additional reset performed by $\mathcal{A}_{q,\tau}$ at the end of the word w_n. In that case $\tau(c) = \leftarrow$ so c has greater or equal number of increments before the end of the word w_n than after it in ρ. Therefore $\mathrm{val}(c, \rho_n, r_c) > \frac{l}{2}$.

In all three cases $\mathrm{val}(c, \rho_n, r_c) > \frac{l}{2}$. So we have shown that

$$\mathrm{val}(\mathcal{A}_{q,\tau}) \geqslant \mathrm{val}(\rho_n) > \frac{l}{2}.$$

This concludes the proof of the implication (1) ⇒ (2).

11.4.2 Implication (2) ⇒ (3)

This implication is trivial since strong convergence entails convergence.

11.4.3 Implication (3) ⇒ (1)

Now we want to prove that if U is a decomposition of α of type t and there exists a grouping W of U that converges to a profinite word $w \in M_t$ then $\alpha \in L(\mathcal{A})$.

Let W be a grouping of U such that W converges to a limit $w \in M_t$.

We consider two cases:

Case T = B Since $w \in M_t$, there exists a state $q \in Q$ such that $w \in L(\mathcal{A}_q)$. Therefore, $w \in \overline{L(\mathcal{A}_q \leqslant l)}$ for some l. Since $\overline{L(\mathcal{A}_q \leqslant l)}$ is an open set and w is a limit of W, almost all elements of W belong to $L(\mathcal{A}_q \leqslant l)$. Assume that for $n \geqslant N$ we have $w_n \in L(\mathcal{A}_q \leqslant l)$. Let ρ_n be a run that witnesses this fact. By the construction of $\mathcal{A}_q$, the run ρ_n induces a run ρ'_n of $\mathcal{A}$ on w_n. Also, since ρ_n is accepting, ρ'_n resets all the counters at least once.

By the assumption about t, there exists a run ρ'_0 of $\mathcal{A}$ on w_0 that starts in some state in I and ends in q, and a sequence of runs ρ'_n on w_n for $0 < n < N$ that lead from q to q. Therefore, we can construct an infinite run ρ of $\mathcal{A}$ on α being the concatenation of the runs ρ'_n on the words w_n for $n \in \mathbb{N}$. We show that if r_c is a reset of a counter c in ρ that appears after the word w_N then $\mathrm{val}(c, \rho, r_c) \leqslant 2 \cdot l$. Since there are only finitely many resets of counters before the word w_N, this bound suffices to show that the run ρ is accepting.

Observe that the increments in ρ correspond to the increments in the runs ρ_n. Also, ρ performs all the resets that appear in runs ρ_n except the resets at the end of the words. There can be at most one such skipped reset in a row because every counter is reset in every run ρ'_n. Therefore, $\mathrm{val}(c, \rho, r_c) \leqslant 2 \cdot l$.

Case T = S Let q, τ be parameters such that $w \in L(\mathcal{A}_{q,\tau})$. Therefore, for every $l \in \mathbb{N}$ we have $w \in \overline{L(\mathcal{A}_{q,\tau} > l)}$. As W is convergent to w and languages $\overline{L(\mathcal{A}_{q,\tau} > l)}$ are open, it means that

$$\forall_l \, \exists_N \, \forall_{n \geqslant N} \; \mathrm{val}(\mathcal{A}_{q,\tau}, w_n) > l. \tag{11.4}$$

As above we construct a run ρ over α that first leads on w_0 from some state of I to q and later consists of a concatenation of runs over words w_n. Let ρ'_0 be any run of $\mathcal{A}$ that leads from I to q on w_0. For $n > 0$ we pick a run ρ_n in such a way that it is accepting and[1]

$$\text{val}(\rho_n) = \text{val}(\mathcal{A}_{q,\tau}, w_n).$$

Observe that by (11.4), we obtain

$$\lim_{n\to\infty} \text{val}(\mathcal{A}_{q,\tau}, w_n) = \lim_{n\to\infty} \text{val}(\rho_n) = \infty. \tag{11.5}$$

For $n > 0$ by ρ'_n be denote the run of $\mathcal{A}$ on w_n induced by ρ_n. Similarly as in the previous case, runs ρ'_n for $n \in \mathbb{N}$ can be combined into a run ρ of $\mathcal{A}$ on α. By the construction of $\mathcal{A}_{q,\tau}$, ρ resets every counter infinitely often.
Let r_c be a position in α where a counter $c \in \Gamma$ is reset during ρ. Assume that r_c is contained in a word w_n and $n > 1$ — we do not care about first two words. Consider two cases:

($\tau(c) = \rightarrow$) In that case ρ performs the same increments and resets of c as the runs ρ_n. Therefore, $\text{val}(c, \rho, r_c) \geqslant \text{val}(\rho_n)$.

($\tau(c) = \leftarrow$) If r_c is not the first reset of c in ρ'_n then the value of c before r_c in ρ is the same as in ρ_n. Assume that r_c is the first reset of c in ρ'_n. Note that ρ_{n-1} performs an additional reset of c at the end of w_{n-1}. This reset does not appear in ρ so $\text{val}(c, \rho, r_c) \geqslant \text{val}(\rho_{n-1})$.

In all the cases

$$\text{val}(c, \rho, r_c) \geqslant \min\left(\text{val}(\rho_{n-1}), \text{val}(\rho_n)\right),$$

so the values of c before successive resets tend to infinity by (11.5). It means that ρ is an accepting run and $\alpha \in L(\mathcal{A})$.

This concludes the proof of (3) ⇒ (1) and of Theorem 11.98.

11.5 Separation for ω-languages

In this section we show the main result of the chapter. The technique is to lift the separation results for T-regular languages of profinite words into the ω-word case.

Theorem 11.9. *If L_1, L_2 are disjoint languages of ω-words both recognised by ωB-(respectively ωS)-automata then there exists an ω-regular language L_{sep} such that*

$$L_1 \subseteq L_{\text{sep}} \quad \text{and} \quad L_2 \subseteq L_{\text{sep}}^{\text{c}}.$$

[1] Since there are only finitely many runs of an automaton on a finite word, there always exists a run realising the value $\text{val}(\mathcal{A}_{q,\tau}, w_n)$, no matter whether the value is finite or not.

Additionally, the construction of L_{sep} is effective.

The rest of the section is devoted to showing this theorem.

Let $i \in \{1, 2\}$ and M^i_{trans} denote the monoid of transitions for an ωT-automaton $\mathcal{A}_i$ recognising L_i. Let $f_i = f_{\mathcal{A}_i}$ be the canonical homomorphisms from A^* to M^i_{trans}. Define Tp^i as the set of types $t_i = (s_i, e_i)$ in the monoid of transitions M^i_{trans}.

For every type $t_i = (s_i, e_i) \in \mathrm{Tp}^i$ define $M^i_{t_i} \subseteq \widehat{A^*}$ as the T-regular language of profinite words given by Theorem 11.98 for $\mathcal{A} = \mathcal{A}_i$ and $t = t_i$. By the statement of the theorem we know that $M^i_{t_i} \subseteq \overline{f_i^{-1}(e_i)}$.

Definition 11.27. *For a pair of types $t_1 = (s_1, e_1) \in \mathrm{Tp}^1$, $t_2 = (s_2, e_2) \in \mathrm{Tp}^2$, we say that t_1, t_2 are* coherent *if there exist finite words $u_s, u_e \in A^*$ such that: $f_i(u_s) = s_i$ and $f_i(u_e) = e_i$ for $i = 1, 2$.*

An important application of Theorem 11.98 is the following lemma.

Lemma 11.51. *If a pair of types $t_1 \in \mathrm{Tp}^1$, $t_2 \in \mathrm{Tp}^2$ is coherent then the languages $M^1_{t_1}$, $M^2_{t_2}$ are disjoint.*

Proof. Take coherent types $t_1 = (s_1, e_1)$ and $t_2 = (s_2, e_2)$.

Assume that there exists a profinite word $u \in M^1_{t_1} \cap M^2_{t_2}$. Since $u \in \overline{f_i^{-1}(e_i)}$ for $i = 1, 2$, there exists a sequence $U = u_1, u_2, \ldots$ of finite words converging to u such that $f_1(u_n) = e_1$ and $f_2(u_n) = e_2$ for all $n > 0$. Moreover, by coherency of t_1, t_2 there exists a finite word u_0 such that $f_1(u_0) = e_1$ and $f_2(u_0) = e_2$. Let $\alpha = u_0 u_1 u_2 \ldots$ We show that $\alpha \in L_1 \cap L_2$ — a contradiction.

Take $i \in \{1, 2\}$. Observe that $\alpha = u_0 u_1 \ldots$ is a decomposition of α of f_i-type t_i. Additionally observe that the sequence U converges to u and u belongs to $M^i_{t_i}$. So, by Theorem 11.98 we have $\alpha \in L_i$. ■

Take a pair of coherent types t_1, t_2. Since the languages $M^1_{t_1}$, $M^2_{t_2}$ are disjoint, we can use Proposition 11.27 to find a separating profinite-regular language $\overline{R_{t_1,t_2}} \subseteq \widehat{A^*}$ such that

$$M^1_{t_1} \subseteq \overline{R_{t_1,t_2}} \quad \text{and} \quad M^2_{t_2} \subseteq \overline{R_{t_1,t_2}}^{\,c}.$$

Now we can introduce the ω-regular language L_{sep} separating L_1 and L_2.

Definition 11.28. *Consider a coherent pair of types (t_1, t_2). Let S_{t_1,t_2} be defined as follows: S_{t_1,t_2} is the language of ω-words α such that there exists a decomposition $\alpha = u_0 u_1 \ldots$ of types t_1, t_2 with respect to f_1, f_2, such that every grouping of $(u_n)_{n\in\mathbb{N}}$ from some point on belongs to the regular language R_{t_1,t_2}.*

Note that the above definition can be expressed in MSO *so S_{t_1,t_2} is an ω-regular language.*

Let L_{sep} be the ω-regular language defined as

$$L_{\mathrm{sep}} = \bigcup_{(t_1,t_2)} S_{t_1,t_2},$$

where the sum ranges over pairs of coherent types.

Clearly L_{sep} is an ω-regular language. What remains is to show the following lemma.

Lemma 11.52. *The language L_{sep} separates L_1 and L_2.*

Proof. First observe that $L_1 \subseteq L_{\text{sep}}$. Take $\alpha \in L_1$. We want to construct a decomposition $U = u_0, u_1, \ldots$ of α such that:

- the f^i-type of U is t_i for $i = 1, 2$ and some pair of coherent types (t_1, t_2) in $\text{Tp}^1 \times \text{Tp}^2$,
- the sequence U is strongly convergent to some profinite word $u \in \widehat{A^*}$.

The sequence U is obtained in steps. First we use Theorem 1.1 to find a decomposition of α with respect to both monoids M^1_{trans}, M^2_{trans} at the same time. Such decomposition satisfies the first bullet above. Then, using Theorem 11.95, we can group our sequence into U in such a way that U is strongly convergent.

By Theorem 11.98, there exists a grouping W of U that converges to a profinite word $w \in M^1_{t_1} \subseteq \overline{R_{t_1,t_2}}$. But since U is strongly convergent, $w = u$. Therefore, by the strong convergence of U, every grouping of U converges to $u \in \overline{R_{t_1,t_2}}$. So every grouping of α from some point on belongs to R_{t_1,t_2} as in the definition of L_{sep}. Therefore, $\alpha \in L_{\text{sep}}$.

Now we show that $L_2 \cap L_{\text{sep}} = \emptyset$. Assume otherwise, that there exists an ω-word $\alpha \in L_2 \cap L_{\text{sep}}$. Since $\alpha \in L_{\text{sep}}$, there exists a coherent pair of types (t_1, t_2) such that $\alpha \in S_{t_1,t_2}$. Therefore, α can be decomposed as $\alpha = u_0 u_1 \ldots$ of types t_1, t_2 respectively. Let $U = u_0, u_1, \ldots$ Because $\alpha \in L_2$ so by Theorem 11.98 there exists a grouping W of U with a limit $w \in M^2_{t_2}$. But by the definition of S_{t_1,t_2} almost all words in W belong to R_{t_1,t_2} so $w \in \overline{R_{t_1,t_2}}$. Since $\overline{R_{t_1,t_2}} \cap M^2_{t_2} = \emptyset$, we have the required contradiction. ■

This concludes the proof of Theorem 2.9.

11.6 Conclusions

The main result of this chapter states that both ωB- and ωS-regular languages have the separation property with respect to ω-regular languages. Therefore, it gives some understanding how these quantitative models extend ω-regular languages. In particular, from the results of the chapter it follows that if a given language is both ωB- and ωS-regular then it is ω-regular.

The crucial technical part of the proof is Sect. 11.4 (a variant of the *reduction theorem* from [Tor12]) that enables to reduce separation of ωB- and ωS-regular languages of ω-words to the separation of B- and S-regular languages of profinite words. The reduction depends highly on compactness arguments and an appropriate Ramsey's theorem. The presented proof explicitly distinguishes between two *orthogonal* parts of ωB- and ωS-regular languages: ω-regular part and asymptotic part.

After the reduction, the study of the separation property in the profinite monoid is relatively easy. In the case of S-languages of profinite words the separation result follows directly from general topological argument (separation property of $\mathbf{\Pi}^0_1$-sets). In the case of B-languages a simple automata theoretic construction is given.

As proved by Bojańczyk and Colcombet [BC06], the classes of ωB- and ωS-regular languages are dual: a language is ωB-regular if and only if its complement is ωS-regular. A usual pattern in descriptive set theory is that from a pair of dual classes, exactly one has the separation property and the other does not have. What is somehow surprising in the case of ωB- and ωS-regular languages both classes have the separation property. It may be a witness that these classes are in some sense meager — they do not contain enough languages to reveal an inseparable pair of sets.

The area of quantitative extensions of regular languages is still developing (see e.g. [BC06, Boj11, Col13, BT09, BT12]). A number of formalisms was proposed but it is still not clear which of them is the most robust. The results of this chapter may help to better understand how these formalisms are related and in what directions they extend ω-regular languages.

This chapter is based on [Skr14].

Chapter 12
Conclusions

The results of the thesis involve a number of methods of descriptive set theory. One of the most common examples is topological hardness: sometimes it is enough to find topologically hard language to obtain some negative results of non-definability. An instance of this approach are Chaps. 9 and 10 where negative results about decidability of MSO+U are given. First, in Chap. 9 examples of MSO+U-definable languages lying arbitrarily high in the projective hierarchy are given. In consequence there can be no *simple* automata model capturing MSO+U on ω-words. The topological hardness of MSO+U is later used in Chap. 10 to prove that the MSO+U theory of the complete binary tree cannot be decidable in the standard sense. Also, topological hardness is used in Chap. 5 to prove that index bounds computed by the proposed algorithm (see Sect. 5.4.1, page 79) are tight. Additionally, the dichotomy proved in Chap. 6 involves topological hardness: a regular language of thin trees is either WMSO-definable or $\boldsymbol{\Pi}^1_1$-hard.

Another important notion that is used in various contexts are ranks. Chapter 4 introduces a new rank based on a given Büchi automaton. It is shown that this rank corresponds in a very precise sense to the descriptive complexity of the language. The whole idea to study such a rank is based on one of the fundamental results of descriptive set theory — the boundedness theorem. Ranks also appear in the study of thin trees in Chap. 6: they turn out to be the combinatorial core of the characterisation of languages that are WMSO-definable among all trees. Also in this chapter, the related construction of derivatives is used to give tight upper bounds on the topological complexity of regular languages of thin trees.

The study of the class of bi-unambiguous languages yielded a new conjecture of non-uniformizability (Conjecture 2.1). While this notion seems to be well understood for sets studied in descriptive set theory, there is only few results about uniformizability in the class of MSO-definable languages of infinite trees. Some new negative results of this kind are given in Chap. 8. Also, consequences of the newly proposed conjecture regarding the class of bi-unambiguous languages are listed in Chap. 7. Hopefully, Conjecture 2.1 will be proved at some point extending our understanding of ability to uniformize certain relations in MSO.

M. Skrzypczak, *Set Theoretic Methods in Automata Theory*, LNCS 9802
DOI: 10.1007/978-3-662-52947-8_12

A distinct descriptive set theoretic notion that is used in the thesis is the separation property. First, it is used in Chap. 3 to construct certain automata of small index. A similar construction appears also in Chap. 7. On the other hand, Chap. 11 provides a new separation result about certain quantitative extensions of ω-regular languages.

One of the most fundamental notions in topology is compactness. A combinatorial counterpart of it is König's lemma. In various cases it is possible to use a compactness argument instead of pumping. One of the examples is the reduction from languages of ω-words to profinite words from Chap. 11. Also, the main idea behind the languages constructed in Chap. 9 is based on an appropriate application of König's lemma. Convergence in a compact space turns out to be useful when proving equivalence between various measures of complexity of trees in Chap. 4.

Regardless of the fact that the involved topological methods are not *effective*, in most of the cases the final statements of the presented results are very concrete: they consist mainly of new decision procedures and computable constructions. Even the negative results have consequences expressible in the language of theoretical computer science, for instance Chap. 10 uses topological methods to prove non-existence of a certain algorithm.

Hopefully, the interplay between topological and automata theoretic methods presented in the thesis will motivate some further development of such techniques.

References

[ADMN08] Arnold, A., Duparc, J., Murlak, F., Niwiński, D.: On the topological complexity of tree languages. In: Logic and Automata, pp. 9–28 (2008)

[AL13] Afshari, B., Leigh, G.E.: On closure ordinals for the modal mu-calculus. In: CSL, pp. 30–44 (2013)

[Alm03] Almeida, J.: Profinite semigroups and applications. In: Structural Theory of Automata, Semigroups, and Universal Algebra, pp. 7–18 (2003)

[AMN12] Arnold, A., Michalewski, H., Niwiński, D.: On the separation question for tree languages. In: STACS, pp. 396–407 (2012)

[AN07] Arnold, A., Niwiński, D.: Continuous separation of game languages. Fundamenta Informaticae **81**(1–3), 19–28 (2007)

[Arn99] Arnold, A.: The mu-calculus alternation-depth hierarchy is strict on binary trees. ITA **33**(4/5), 329–340 (1999)

[AS05] Arnold, A., Santocanale, L.: Ambiguous classes in μ-calculi hierarchies. TCS **333**(1–2), 265–296 (2005)

[Ban83] Banaschewski, B.: The birkhoff theorem for varieties of finite algebras. Algebra Universalis **17**(1), 360–368 (1983)

[BC06] Bojańczyk, M., Colcombet, T.: Bounds in ω-regularity. In: LICS, pp. 285–296 (2006)

[BGMS14] Bojańczyk, M., Gogacz, T., Michalewski, H., Skrzypczak, M.: On the decidability of MSO+U on infinite trees. In: Esparza, J., Fraigniaud, P., Husfeldt, T., Koutsoupias, E. (eds.) ICALP 2014, Part II. LNCS, vol. 8573, pp. 50–61. Springer, Heidelberg (2014)

[BI09] Bojańczyk, M., Idziaszek, T.: Algebra for infinite forests with an application to the temporal logic EF. In: Bravetti, M., Zavattaro, G. (eds.) CONCUR 2009. LNCS, vol. 5710, pp. 131–145. Springer, Heidelberg (2009)

[Bil11] Bilkowski, M.: Strongly unambiguous regular languages of infinite trees. Talk at Young Researchers Forum during MFCS 2011 (2011)

[BIS13] Bojańczyk, M., Idziaszek, T., Skrzypczak, M.: Regular languages of thin trees. In: STACS 2013. LIPIcs, vol. 20, pp. 562–573 (2013)

[BKKS13] Boker, U., Kuperberg, D., Kupferman, O., Skrzypczak, M.: Nondeterminism in the presence of a diverse or unknown future. In: Fomin, F.V., Freivalds, R., Kwiatkowska, M., Peleg, D. (eds.) ICALP 2013, Part II. LNCS, vol. 7966, pp. 89–100. Springer, Heidelberg (2013)

[BKR11] Bárány, V., Kaiser, Ł., Rabinovich, A.: Expressing cardinality quantifiers in monadic second-order logic over chains. J. Symb. Log. **76**(2), 603–619 (2011)

[BKT12] Bojańczyk, M., Kopczyński, E., Toruńczyk, S.: Ramsey's theorem for colors from a metric space. Semigroup Forum **85**, 182–184 (2012)

M. Skrzypczak, *Set Theoretic Methods in Automata Theory*, LNCS 9802
DOI: 10.1007/978-3-662-52947-8

[BL69] Büchi, J.R., Landweber, L.H.: Solving sequential conditions by finite-state strategies. Trans. Am. Math. Soc. **138**, 295–311 (1969)

[Blu11] Blumensath, A.: Recognisability for algebras of infinite trees. Theor. Comput. Sci. **412**(29), 3463–3486 (2011)

[Blu13] Blumensath, A.: An algebraic proof of Rabin's tree theorem. Theor. Comput. Sci. **478**, 1–21 (2013)

[BNR+10] Bojańczyk, M., Niwiński, D., Rabinovich, A., Radziwończyk-Syta, A., Skrzypczak, M.: On the Borel complexity of MSO definable sets of branches. Fundamenta Informaticae **98**(4), 337–349 (2010)

[Boj04] Bojańczyk, M.: A bounding quantifier. In: Marcinkowski, J., Tarlecki, A. (eds.) CSL 2004. LNCS, vol. 3210, pp. 41–55. Springer, Heidelberg (2004)

[Boj10a] Bojańczyk, M.: Algebra for trees. A draft version of a chapter that will appear in the AutomathA handbook (2010)

[Boj10b] Bojańczyk, M.: Beyond ω-regular languages. In: STACS, pp. 11–16 (2010)

[Boj11] Bojańczyk, M.: Weak MSO with the unbounding quantifier. Theory Comput. Syst. **48**(3), 554–576 (2011)

[BP12] Bojańczyk, M., Place, T.: Regular languages of infinite trees that are boolean combinations of open sets. In: Czumaj, A., Mehlhorn, K., Pitts, A., Wattenhofer, R. (eds.) ICALP 2012, Part II. LNCS, vol. 7392, pp. 104–115. Springer, Heidelberg (2012)

[Bra98] Bradfield, J.: Simplifying the modal mu-calculus alternation hierarchy. In: STACS, pp. 39–49 (1998)

[BS81] Burris, S., Sankappanavar, H.P.: A Course in Universal Algebra. Graduate Texts in Mathematics, vol. 78. Springer, New York (1981)

[BS13] Bilkowski, M., Skrzypczak, M.: Unambiguity and uniformization problems on infinite trees. In: CSL. LIPIcs, vol. 23, pp. 81–100 (2013)

[BT09] Bojańczyk, M., Toruńczyk, S.: Deterministic automata and extensions of weak MSO. In: FSTTCS, pp. 73–84 (2009)

[BT12] Bojańczyk, M., Toruńczyk, S.: Weak MSO+U over infinite trees. In: STACS, pp. 648–660 (2012)

[Büc62] Büchi, J.R.: On a decision method in restricted second-order arithmetic. In: Proceedings of 1960 International Congress for Logic, Methodology and Philosophy of Science, pp. 1–11 (1962)

[Büc83a] Büchi, J.R.: State-strategies for games in $f_{\sigma\delta} \bigcap g_{\delta\sigma}$. J. Symbolic Logic **48**(4), 1171–1198 (1983)

[Büc83b] Büchi, J.R.: State-strategies for games in F-sigma-delta intersected G-delta-sigma. J. Symb. Log. **48**(4), 1171–1198 (1983)

[BW08] Bojańczyk, M., Walukiewicz, I.: Forest algebras. In: Logic and Automata, pp. 107–132 (2008)

[CDFM09] Cabessa, J., Duparc, J., Facchini, A., Murlak, F.: The Wadge hierarchy of max-regular languages. In: FSTTCS, pp. 121–132 (2009)

[CKLV13] Colcombet, T., Kuperberg, D., Löding, C., Vanden Boom, M.: Deciding the weak definability of Büchi definable tree languages. In: CSL, pp. 215–230 (2013)

[CL07] Carayol, A., Löding, C.: MSO on the infinite binary tree: choice and order. In: Duparc, J., Henzinger, T.A. (eds.) CSL 2007. LNCS, vol. 4646, pp. 161–176. Springer, Heidelberg (2007)

[CL08] Colcombet, T., Löding, C.: The non-deterministic Mostowski hierarchy and distance-parity automata. In: Aceto, L., Damgård, I., Goldberg, L.A., Halldórsson, M.M., Ingólfsdóttir, A., Walukiewicz, I. (eds.) ICALP 2008, Part II. LNCS, vol. 5126, pp. 398–409. Springer, Heidelberg (2008)

[CL10] Colcombet, T., Löding, C.: Regular cost functions over finite trees. In: LICS, pp. 70–79 (2010)

[CLNW10] Carayol, A., Löding, C., Niwiński, D., Walukiewicz, I.: Choice functions and well-orderings over the infinite binary tree. Cent. Europ. J. Math. **8**, 662–682 (2010)

[Col09] Colcombet, T.: The theory of stabilisation monoids and regular cost functions. In: Albers, S., Marchetti-Spaccamela, A., Matias, Y., Nikoletseas, S., Thomas, W. (eds.) ICALP 2009, Part II. LNCS, vol. 5556, pp. 139–150. Springer, Heidelberg (2009)

[Col13] Colcombet, T.: Fonctions régulières de coût. Habilitation thesis, Université Paris Diderot-Paris 7 (2013)

[CPP07] Carton, O., Perrin, D., Pin, J.-É.: Automata and semigroups recognizing infinite words. In: Logic and Automata, History and Perspectives, pp. 133–167 (2007)

[Cza10] Czarnecki, M.: Analiza formuł modalnego rachunku mu pod wzgl edem szybkości osi agania punktów stałych. Master's thesis, University of Warsaw (2010)

[DFM11] Duparc, J., Facchini, A., Murlak, F.: Definable operations on weakly recognizable sets of trees. In: FSTTCS, pp. 363–374 (2011)

[DFR01] Duparc, J., Finkel, O., Ressayre, J.-P.: Computer science and the fine structure of borel sets. Theoret. Comput. Sci. **257**(1–2), 85–105 (2001)

[DFR13] Duparc, J., Finkel, O., Ressayre, J.-P.: The wadge hierarchy of Petri Nets ω-languages. In: Artemov, S., Nerode, A. (eds.) LFCS 2013. LNCS, vol. 7734, pp. 179–193. Springer, Heidelberg (2013)

[EJ91] Emerson, A., Jutla, C.: Tree automata, mu-calculus and determinacy. In: FOCS 1991, pp. 368–377 (1991)

[FFMS00] Feferman, S., Friedman, H.M., Maddy, P., Steel, J.R.: Does mathematics need new axioms? Bull. Symbolic Logic **6**(4), 401–446 (2000)

[Fin06] Finkel, O.: Borel ranks and Wadge degrees of context free omega-languages. Math. Struct. Comput. Sci. **16**(5), 813–840 (2006)

[FMS13] Facchini, A., Murlak, F., Skrzypczak, M.: Rabin-Mostowski index problem: a step beyond deterministic automata. In: LICS, pp. 499–508 (2013)

[FS09] Finkel, O., Simonnet, P.: On recognizable tree languages beyond the Borel hierarchy. Fundam. Inform. **95**(2–3), 287–303 (2009)

[FS14] Finkel, O., Skrzypczak, M.: On the topological complexity of w-languages of non-deterministic Petri nets. Inf. Process. Lett. **114**(5), 229–233 (2014)

[GH82] Gurevich, Y., Harrington, L.: Trees, automata, and games. In: STOC, pp. 60–65 (1982)

[GMMS14] Gogacz, T., Michalewski, H., Mio, M., Skrzypczak, M.: Measure properties of game tree languages. In: Csuhaj-Varjú, E., Dietzfelbinger, M., Ésik, Z. (eds.) MFCS 2014, Part I. LNCS, vol. 8634, pp. 303–314. Springer, Heidelberg (2014)

[Göd39] Gödel, K.: Consistency-proof for the generalized continuum-hypothesis. Proc. Natl. Acad. Sci. USA **25**(4), 220–224 (1939)

[Gre51] Green, J.A.: On the structure of semigroups. Ann. Math. **54**(1), 163–172 (1951)

[GS82] Gurevich, Y., Shelah, S.: Monadic theory of order and topology in ZFC. Ann. Math. Logic **23**(2–3), 179–198 (1982)

[GS83] Gurevich, Y., Shelah, S.: Rabin's uniformization problem. J. Symb. Log. **48**(4), 1105–1119 (1983)

[HMN09] Hummel, S., Michalewski, H., Niwiński, D.: On the Borel inseparability of game tree languages. In: STACS, pp. 565–575 (2009)

[HP06] Henzinger, T.A., Piterman, N.: Solving games without determinization. In: Ésik, Z. (ed.) CSL 2006. LNCS, vol. 4207, pp. 395–410. Springer, Heidelberg (2006)

[HS12] Hummel, S., Skrzypczak, M.: The topological complexity of MSO+U and related automata models. Fundamenta Informaticae **119**(1), 87–111 (2012)

[HST10] Hummel, S., Skrzypczak, M., Toruńczyk, S.: On the topological complexity of MSO+U and related automata models. In: Hliněný, P., Kučera, A. (eds.) MFCS 2010. LNCS, vol. 6281, pp. 429–440. Springer, Heidelberg (2010)

[Hum12] Hummel, S.: Unambiguous tree languages are topologically harder than deterministic ones. In: GandALF, pp. 247–260 (2012)

[Idz12] Idziaszek, T.: Algebraic methods in the theory of infinite trees. Ph.D. thesis, University of Warsaw (2012) (unpublished)

[Imm99] Immerman, N.: Descriptive Complexity. Graduate Texts in Computer Science. Springer, New York (1999)

[Jec02] Jech, T.: Set Theory. Springer, New York (2002)

[JPZ08] Jurdziński, M., Paterson, M., Zwick, U.: A deterministic subexponential algorithm for solving parity games. SIAM J. Comput. **38**(4), 1519–1532 (2008)

[JW96] Janin, D., Walukiewicz, I.: On the expressive completeness of the propositional mu-calculus with respect to monadic second order logic. In: Sassone, V., Montanari, U. (eds.) CONCUR 1996. LNCS, vol. 1119, pp. 263–277. Springer, Heidelberg (1996)

[Kec95] Kechris, A.: Classical Descriptive Set Theory. Springer, New York (1995)

[KSV96] Kupferman, O., Safra, S., Vardi, M.Y.: Relating word and tree automata. In: LICS, pp. 322–332. IEEE Computer Society (1996)

[KV99] Kupferman, O., Vardi, M.Y.: The weakness of self-complementation. In: Meinel, C., Tison, S. (eds.) STACS 1999. LNCS, vol. 1563, p. 455. Springer, Heidelberg (1999)

[KV11] Kuperberg, D., Boom, M.V.: Quasi-weak cost automata: a new variant of weakness. In: FSTTCS. LIPIcs, vol. 13, pp. 66–77 (2011)

[LS98] Lifsches, S., Shelah, S.: Uniformization and skolem functions in the class of trees. J. Symb. Log. **63**(1), 103–127 (1998)

[Mar75] Martin, D.A.: Borel determinacy. Ann. Math. **102**(2), 363–371 (1975)

[McN66] McNaughton, R.: Testing and generating infinite sequences by a finite automaton. Inf. Control **9**(5), 521–530 (1966)

[MN12] Michalewski, H., Niwiński, D.: On topological completeness of regular tree languages. In: Constable, R.L., Silva, A. (eds.) Logic and Program Semantics, Kozen Festschrift. LNCS, vol. 7230, pp. 165–179. Springer, Heidelberg (2012)

[Mos80] Moschovakis, Y.N.: Descriptive Set Theory. Studies in Logic and Fundations of Mathematics. North-Holland Publishing Company, Amsterdam (1980)

[Mos84] Mostowski, A.W.: Regular expressions for infinite trees and a standard form of automata. In: Symposium on Computation Theory, pp. 157–168 (1984)

[Mos91] Mostowski, A.W.: Games with forbidden positions. Technical report, University of Gdańsk (1991)

[MP71] McNaughton, R., Papert, S.: Counter-free automata. M.I.T. Press Research Monographs. M.I.T. Press, Cambridge (1971)

[MS95] Muller, D.E., Schupp, P.E.: Simulating alternating tree automata by nondeterministic automata: new results and new proofs of the theorems of Rabin, McNaughton and Safra. Theor. Comput. Sci. **141**(1&2), 69–107 (1995)

[MS14] Michalewski, H., Skrzypczak, M.: Unambiguous Büchi is weak. arXiv:1401.4025 (2014)

[Mur08] Murlak, F.: The Wadge hierarchy of deterministic tree languages. Logical Meth. Comput. Sci. **4**(4), 1–44 (2008)

[Niw86] Niwiński, D.: On fixed-point clones. In: Kott, L. (ed.) ICALP 1986. LNCS, vol. 226, pp. 464–473. Springer, Heidelberg (1986)

[Niw97] Niwiński, D.: Fixed point characterization of infinite behavior of finite-state systems. Theor. Comput. Sci. **189**(1–2), 1–69 (1997)

[NW96] Niwiński, D., Walukiewicz, I.: Ambiguity problem for automata on infinite trees (1996) (unpublished)

[NW98] Niwiński, D., Walukiewicz, I.: Relating hierarchies of word and tree automata. In: STACS, pp. 320–331 (1998)

[NW03] Niwiński, D., Walukiewicz, I.: A gap property of deterministic tree languages. Theor. Comput. Sci. **1**(303), 215–231 (2003)

[NW05] Niwiński, D., Walukiewicz, I.: Deciding nondeterministic hierarchy of deterministic tree automata. Electr. Notes Theor. Comput. Sci. **123**, 195–208 (2005)

[Pin09] Pin, J.-É.: Profinite methods in automata theory. In: STACS, pp. 31–50 (2009)

[PP04] Perrin, D., Pin, J.É.: Infinite Words: Automata, Semigroups, Logic and Games. Elsevier, Amsterdam (2004)

[PZ14] Place, T., Zeitoun, M.: Going higher in the first-order quantifier alternation hierarchy on words. In: Esparza, J., Fraigniaud, P., Husfeldt, T., Koutsoupias, E. (eds.) ICALP 2014, Part II. LNCS, vol. 8573, pp. 342–353. Springer, Heidelberg (2014)

[Rab69] Rabin, M.O.: Decidability of second-order theories and automata on infinite trees. Trans. Am. Math. Soc. **141**, 1–35 (1969)

[Rab70] Rabin, M.O.: Weakly definable relations and special automata. In: Proceedings of the Symposium on Mathematical Logic and Foundations of Set Theory, pp. 1–23. North-Holland (1970)

[Rab07] Rabinovich, A.: On decidability of monadic logic of order over the naturals extended by monadic predicates. Inf. Comput. **205**(6), 870–889 (2007)

[RR12] Rabinovich, A., Rubin, S.: Interpretations in trees with countably many branches. In: LICS, pp. 551–560. IEEE (2012)

[RS59] Rabin, M.S., Scott, D.: Finite automata and their decision problems. IBM J. Res. Dev. **3**(2), 114–125 (1959)

[Sch65] Schützenberger, M.P.: On finite monoids having only trivial subgroups. Inf. Control **8**(2), 190–194 (1965)

[She75] Shelah, S.: The monadic theory of order. Ann. Math. **102**(3), 379–419 (1975)

[Sie75] Siefkes, D.: The recursive sets in certain monadic second order fragments of arithmetic. Arch. Math. Logik **17**(1–2), 71–80 (1975)

[Sim75] Simon, I.: Piecewise testable events. In: Brakhage, H. (ed.) Automata Theory and Formal Languages. LNCS, vol. 33, pp. 214–222. Springer, Heidelberg (1975)

[Skr11] Skrzypczak, M.: Equational theories of profinite structures. CoRR, abs/1111.0476 (2011)

[Skr13] Skrzypczak, M.: Topological extension of parity automata. Inf. Comput. **228**, 16–27 (2013)

[Skr14] Skrzypczak, M.: Separation property for wB- and wS-regular languages. Logical Meth. Comput. Sci. **10**(1), 1–20 (2014)

[Sku93] Skurczyński, J.: The Borel hierarchy is infinite in the class of regular sets of trees. Theoret. Comput. Sci. **112**(2), 413–418 (1993)

[Sri98] Srivastava, S.M.: A Course on Borel Sets. Graduate Texts in Mathematics, vol. 180. Springer, New York (1998)

[Tho79] Thomas, W.: Star-free regular sets of omega-sequences. Inf. Control **42**(2), 148–156 (1979)

[Tho80] Thomas, W.: Relationen endlicher valenz über der ordnung der natürlichen zahlen. Habilitationsschrift, Universitat Freiburg, April 1980

[Tho96] Thomas, W.: Languages, automata, and logic. In: Handbook of Formal Languages, pp. 389–455. Springer (1996)

[TL93] Thomas, W., Lescow, H.: Logical specifications of infinite computations. In: REX School/Symposium, pp. 583–621 (1993)

[Tor12] Toruńczyk, S.: Languages of profinite words and the limitedness problem. In: Czumaj, A., Mehlhorn, K., Pitts, A., Wattenhofer, R. (eds.) ICALP 2012, Part II. LNCS, vol. 7392, pp. 377–389. Springer, Heidelberg (2012)

[Tra62] Trakhtenbrot, B.A.: Finite automata and the monadic predicate calculus. Siberian Math. J. **3**(1), 103–131 (1962)

[Van11] Vanden Boom, M.: Weak cost monadic logic over infinite trees. In: Murlak, F., Sankowski, P. (eds.) MFCS 2011. LNCS, vol. 6907, pp. 580–591. Springer, Heidelberg (2011)

[Wad83] Wadge, W.: Reducibility and determinateness in the Baire space. Ph.D. thesis, University of California, Berkeley (1983)

[Wil93] Wilke, T.: An algebraic theory for regular languages of finite and infinite words. Int. J. Alg. Comput. **3**, 447–489 (1993)

[Wil98] Wilke, T.: Classifying discrete temporal properties. Habilitationsschrift, Universitat Kiel, April 1998

Printed in the United States
By Bookmasters

Printed in the United States
By Bookmasters